REAGAN RISING

Also by Craig Shirley

*LAST ACT: THE FINAL YEARS AND EMERGING LEGACY OF
RONALD REAGAN*

*DECEMBER 1941: 31 DAYS THAT CHANGED AMERICA AND
SAVED THE WORLD*

*RENDEZVOUS WITH DESTINY: RONALD REAGAN AND THE
CAMPAIGN THAT CHANGED AMERICA*

*REAGAN'S REVOLUTION: THE UNTOLD STORY OF THE
CAMPAIGN THAT STARTED IT ALL*

REAGAN RISING

THE DECISIVE YEARS, 1976–1980

CRAIG SHIRLEY

BROADSIDE BOOKS
An Imprint of HarperCollinsPublishers

HarperCollins books may be purchased for educational, business, or sales promotional use. For information, please email the Special Markets Department at SPsales@harpercollins.com.

Broadside Books™ and the Broadside logo are trademarks of HarperCollins Publishers.

FIRST EDITION

Frontispiece: Courtesy of Dennis Cook/AP/Rex/Shutterstock

Designed by William Ruoto

Library of Congress Cataloging-in-Publication Data has been applied for.

ISBN 978-0-06-245655-7

17 18 19 20 21 LSC 10 9 8 7 6 5 4 3 2 1

For Zorine, my wife, my love, my best friend, my rock, the center of all things for me.

And the best editor and confidante ever.

Contents

Foreword

by Jon Meacham

I t began, as is so often the case, with an ending. At the 1976 Republican National Convention in Kansas City, Ronald Reagan—former sportscaster, movie actor, union president, and governor of California—had come up just short of defeating the incumbent president, Gerald R. Ford, to win their party's nomination. On the day he was heading home from Missouri to California, Reagan said a few words to gathered staff and supporters. "We lost, but the cause—the cause goes on," he said, quoting an old ballad: "'I'll lay me down and bleed awhile; although I am wounded, I am not slain. I shall rise and fight again.'"

Truer words have rarely been spoken in American politics. To some, at sixty-five (a greatly advanced age in the context of the times, the year of standard retirement), Reagan seemed to have missed his chance at history. To others (including to himself), the eternally optimistic "Dutch" Reagan, a son of the Midwest who became an emblem of Sunbelt possibility and prosperity, was only getting started. He believed in himself and in his ideas and in the reality of destiny, and he would not give up.

In the following pages, Craig Shirley continues his own monumental project of chronicling what shall most likely come to be known as the Age of Reagan. In accounts of the 1976 and 1980 campaigns, as well as of the former president's long good-bye in the twilight of Alzheimer's, Shirley, a historian and practitioner of politics, brings a deep love of detail and a romantic's passion for narrative to the task of capturing the how and the why of Reagan. The present project is to explore what Winston Churchill, in his own case, thought of as the wilderness years, the excruciating interval in which the principal was without power but was driven forward, through the thickets, by the enduring conviction that he was *right* after all.

As usually told, the Reagan saga skips from the near-miss in Kansas City to the triumph of 1980, when Reagan fought off a surprisingly strong George H. W. Bush in the primaries, won the nomination in Detroit, nearly offered the vice presidency to the defeated Ford, and then buried Jimmy Carter in a landslide after a reassuring debate performance in Cleveland a week before the voting.

The virtue of Shirley's *Reagan Rising* lies in the author's insistence that the popular version of, well, Reagan's rise is incomplete without an understanding of the hero's four years in California between the 1976 and 1980 campaigns. Here is a portrait of an America that was in many ways suffering a kind of nervous breakdown—high inflation, punishing interest rates, fear of Soviet advances, and the debilitating Iranian hostage crisis. Here, too, is a portrait of a man biding his time, learning, watching, and preparing for his moment—a moment that, truth be told, not many people in real time thought would ultimately come.

The great New York columnist Murray Kempton once remarked that "the better story is in the loser's locker room." Shirley shares that view, and his unique viewpoint as a conservative in the arena gives his histories a palpable immediacy and authenticity. You will find the ensuing account of Reagan, in his short tenure as a loser, compelling, thorough, and convincing.

Back in Kansas City, on the final night of the convention, Gerald Ford, in a characteristic act of grace, invited Nancy and Ronald Reagan to come to the podium after the president's own acceptance speech. The last word of the evening, then, belonged not to Ford but to Reagan.

"If I could just take a moment," Reagan said on that distant summer night, "I had an assignment the other day. Someone asked me to write a letter for a time capsule that is going to be opened in Los Angeles a hundred years from now." Reagan thought it would be a simple matter, he said, but it proved more difficult than he had anticipated. "Then as I tried to write—let your own minds turn to that task," Reagan said. "You are going to write for people a hundred years from now, who know all about us. We know nothing about them. We don't know what kind of a world they will be living in."

We now know what kind of a world President Reagan left us. Craig Shirley's fine book tells us much about the moment that summoned him to greatness, and the man who answered that call. It's a story we all need to know.

Preface

And for conservatives, it was their Camelot.

As historian Doug Brinkley once told me, the world of Ronald Reagan scholarship is just beginning to open up. Though over a thousand books have been written about the Gipper, there remain unreported or underreported aspects to his monumental life and times. Many of these books, written by people with an agenda, are quite bad. Others, written by objective historians, are quite good. History is already sorting out the factual from the farcical.

My first book, *Reagan's Revolution*, was on his failed but vitally important and historic 1976 campaign, and *Rendezvous with Destiny* was primarily about his successful 1980 campaign for president. When researching and writing *Rendezvous* between 2006 and 2009, it dawned on me that I'd given only a cursory glance to these four wilderness years for the Gipper. What happened in those four years to prepare him in a way that he was not prepared for 1976? How did he gain so much momentum between failing and winning? Surely the reason for his popularity was more than just taking on issues, taking on Jimmy Carter, and being present and accounted for. What else was there? These questions and others were always at the back of my mind when writing about the 1980 election.

This is why I decided to focus this book on the all-important details of Reagan's growth and political maturation and seasoning in the years of late 1976 to late 1979, as he emerged from his losing campaign to wrest the Republican nomination away from Gerald Ford and moved to yet another grab at the brass ring in 1980. Unlike most men, who arrive at a settled worldview in their thirties and forties, Reagan was dynamic, not static, and his position and views kept becoming more refined, intellectual, and nuanced, even into his fifties and sixties. In 1975, in an interview with *Reason* magazine, he said that he believed "the very heart and soul of conservatism is libertarianism," and then launched into a detailed explanation for his belief.[1]

But that's not all. Coming to power in London was Margaret Thatcher, and in Rome was Pope John Paul II. Also Mikhail Gorbachev was rising in several short years in the Soviet Union, though, in the late 1970s, the gangster Leonid Brezhnev still ruled with an iron fist. All these people and events would have a profound effect on Reagan, America, and the world.

So much happened in America, the world, and to him that little can be overlooked as he mulled over his third and final run for the GOP nomination. These truly were decisive years as he honed and sharpened his philosophy, which gave intellectual underpinnings to his pro-freedom, anti-Communist message. Working against him was the fact that he was a two-time loser and he would be sixty-nine years old in 1980, and many Americans, sometimes a majority, thought he was too old to handle the burdens of being president.

The late 1970s saw the reemergence of the Committee on the Present Danger, which first highlighted the Communist threat in the early 1950s, led this time around by Dick Allen and Jeane Kirkpatrick and Ken Adelman, which put important sunlight on the Soviet threat. Dr. Kirkpatrick had written a vitally important article, "Dictators and Double Standards," which Reagan read avidly and which piqued his interest in how actually to defeat the Soviet Union. The article became a "must read" inside the conservative movement and established Kirkpatrick as a future star.

Reagan was also motivated by the emergence of the fight over the Panama Canal Treaties, which helped the conservative movement get back on its feet, with Reagan leading the charge against the treaties. Though losing narrowly to Ford, Reagan won politically, and Jimmy Carter lost politically. By the time the treaties passed, they had become hugely unpopular.

The era also saw the emergence of supply-side economics, so important to Reagan's populist, economic message of more self-reliance and less dependence. He met with Margaret Thatcher for the first time; both were out of power, but it led to the historic alliance that later defeated Soviet communism. It also saw the rise of Pope John Paul II, and the release and emergence of Alexander Solzhenitsyn as a moral and cultural force in America and the world.

The late 1970s may have been fallow as far as national and international political leaders, but in those fields were planted the seeds of the leaders of the 1980s. They quite literally changed the world, proving Thomas Paine right when he wrote, "We have it in our power to begin the world over again," which may explain why Reagan liked quoting the intellectual writer of the American Revolution so frequently.

It was in this time that Reagan became fully entrenched as the leader of American conservatism, what with his speeches, his daily radio commentaries, and his twice-a-week column. But he wanted more than just to be the leader of the American conservative movement.

He wanted to be president of the United States, and Carl Jung's synchronicity—that events are interrelated, connected—lent him a hand. And for conservatives, it was their Camelot.

REAGAN RISING

Introduction

"The cause goes on."

The rise of Ronald Wilson Reagan began with a fall.

Between August 16 and August 19, 1976, in Kansas City, Missouri, the Republican National Convention set the stage for the party to come to its final death blows and its rebirth. Those four days would change the eventually reborn Republican Party; ironically, though, it would change not because of the winner, but because of the loser.

When Ronald Reagan officially announced his run on November 20, 1975, he had stated that he believed his "candidacy will be healthy for the nation and the party . . . because I have become increasingly concerned about the course of events in the United States and the world."[1] This was to be his second run; the first, in 1968. Reagan also struck a decidedly populist tone, attacking the bigness of corrupt Washington, including big government, big labor, big business, and big lobbyists who used the nation's capital as a "buddy system" to enrich themselves at the expense of the "taxpayer" who funded this corrupt and crooked oligarchy. He also went hard after President Gerald Ford on all manner of things, including détente, Henry Kissinger, and the Panama Canal Treaties. Reagan and Ford really did not like each other. Gerald Ford once said, "Reagan and I both played football. I played for Michigan and he played for Warner Brothers." Without missing a beat, Reagan retorted, "Well, at least when I played football, I played football with my helmet on."[2]

Even before his announcement in November, Reagan had been highly critical of the ineptitude of the Ford camp: there was a trend in the United States, he said, that was "contrary to all the principles that I grew up believing in," and that the "free world is crying out for strong American leadership."[3] The blame was put primarily on President Ford and his Administration and policies. Ford was not elected president; instead, then–House Minority Leader Ford was appointed as vice president by Richard Nixon in late 1973, after the resignation of Spiro Agnew that October; upon the resignation of President Nixon amid the Watergate scandal ten months later, Vice President Ford took over the reins (and the reign) of the Nixon Administration.

Kansas City was contested, chaotic, and controversial. It was hot— and not just because of the weather. Split between the former governor of California and the incumbent president, the convention did not have a clear victor. A total of 1,130 delegates was needed to win the nomination. Depending on whom you asked, either Reagan, Ford, or neither had enough going in. To make matters more complicated, each publication had a different number of committed delegates: *Time* had Ford nine delegates below the number to secure the nomination, whereas *Human Events* had him three below; the Associated Press had Ford with a relatively staggering twenty-six below.[4]

The rhetoric was strong, the rivals were battling it out, and God only knew what was going to happen next. This brought to mind other, recent elections. Recalling the 1952 Republican convention between Dwight D. Eisenhower and Robert A. Taft, the *New York Times* stated that "there ensued a battle royal for the nomination—a fierce struggle between two closely matched contenders." It would be no different in '76.[5] *Newsweek*, likewise, said that this was the "tightest two-way race for a major-party nomination in modern political history."[6] This wasn't lost on Ford or Reagan, at all. They knew what was at stake and how close the vote was. In fact, the pressure was on for Ford as well. Perhaps it was due to his uncertainty, insecurities—or the opposite, vanity—but rumors had swirled in the month before that Ford would break with tradition and arrive early to the convention:[7] a small but significant step to ensure control of a potentially out-of-control vote.

Reagan, in the hope of picking up the uncommitted delegates necessary to win the nomination, had announced his running mate *before* the vote—another break from tradition this run. There he chose Senator Richard Schweiker of Pennsylvania, a moderate Republican. In this way, Reagan hoped to woo the more centrist and liberal GOPers. It was a "wild card," as the *Economist* noted, and it was held up to some ridicule: Senator Edward Brooke of Massachusetts joked "that Mr. Reagan had picked Mr. Schweiker because he could not get Senator [Walter] Mondale, already pre-empted by Mr. Jimmy Carter for the Democratic ticket."[8] During the convention, on Tuesday August 17, the Reagan camp called a vote, in what is now the notorious rule 16-C, to force Gerald Ford to announce his running mate. It failed to pass, 1,180 to 1,069.[9] Actually, both ploys worked, at least partially. The choice of Schweiker kept Reagan's campaign alive for the three weeks leading up to Kansas City, and the vote on 16-C kept things murky for a time. This was the goal all along of Reagan's campaign manager, John Sears. As a result, no one would know who would be the nominee until the actual vote. Even with his close shave at the hands of Reagan, President Gerald Ford refused to consider putting him on the ticket, even if it was the only way to unify the party. And his campaign staff grumbled and leaked against the Gipper to the media.

———

By Wednesday evening, a vote was called to decide the 1976 Republican nominee. It was close, too close, to call, until West Virginia gave 20 votes to Ford, securing his nomination with more than the 1,130 needed. The final count was 1,187 for Ford, 1,070 for Reagan. Ford was the narrow nominee, and the contested convention of 1976 was over. It wasn't just that Reagan lost—he lost to a political culture of insiders, lost to a party system that didn't think much of him, and lost to a man he often disliked, and many thought rightfully so. After all, just weeks earlier, Ford had run commercials calling Reagan a warmonger.

Upon securing the nomination, Ford gestured Ronald Reagan to the stage, to give a speech—no plans, no teleprompters, no script. And what Reagan gave defined him. The speech was one of the future, one of hope for those who would look at 1976 as a turning point in history. "We carry the message they're waiting for," he said to the convention. "There is no substitute for victory."[10] At this point, it was evident that Reagan was not going away. Not long after that speech, a youthful supporter from Florida believed that "from that day forward, I think American politics changed."[11] Columnist Jack Germond noted that Reagan, right there and then, was to become the "heir apparent" of the GOP.[12]

Ford's invitation to Reagan did not salve the personal deep wounds that had been cut into Reagan by Ford and his operatives over the previous two years. The GOP establishment had targeted Reagan for derision and ridicule—and ground zero for this innuendo campaign was the Ford White House, often led by Ford himself, playing into the theme of Reagan as a warmonger. Reagan was incensed, but it was part of a larger pattern emanating from the Ford White House and the Ford campaign.

The morning after the convention, a somber Reagan, accompanied by a tearful Nancy, spoke for forty-five minutes to his campaign workers. It was a bittersweet moment for all two hundred of them. Surely, some must have thought, this was the end of the line. Reagan was, after all, sixty-five years old . . .

But Reagan smashed those fears, and instead offered a promise to them all: "the cause, the cause goes on." He continued, interrupted by applause and a cracking in his voice, "Sure it's just one more battle in a long war and it's going to go on as long as we all live. Nancy and I, we aren't going to go back and sit in a rocking chair on the front porch and say, 'Well, that's all for us.'"[13] Nancy looked to her husband with obvious admiration and pride; but she had to turn away from the audience as she was overcome with emotion.

"The cause goes on," the Gipper vowed.[14]

All by Himself

*"There is no Soviet domination of Eastern Europe, and
there never will be under a Ford Administration."*

Ronald Reagan's elegant departure from the 1976 campaign reminded some of Democratic nominee Adlai Stevenson's graceful
exit in 1952, after his defeat at the hands of Dwight Eisenhower. "Recalling a Lincoln story, Stevenson said he felt like the boy who had stubbed
his toe in the dark; he was too old to cry but it hurt too much to laugh."[1]

With the nomination in hand, President Gerald Ford discovered he
was running thirty points behind Jimmy Carter in the polls. In fact, the
Republicans and the president could count on precious little support
anywhere in America in 1976.

The GOP had controlled the White House for sixteen out of the
previous twenty-four years. However, the situation was horrible at the
state level. Thirty-seven of the nation's governors were Democratic, and
the GOP controlled the legislatures in only four states—Idaho, Vermont, Kansas, and North Dakota. Both houses of Congress had been
firmly in the hands of the Democrats since 1954. In 1976 there were
states in Dixie that had almost no elected Republican official. Moreover,
the "Grand Old Party" was outnumbered better than two to one in the
House, had fewer than forty senators, and in a remarkable testament to
its fecklessness, only one state in the country, Kansas, had Republican
control of both the governorship and the legislature after the Watergate-

wipeout elections of 1974. The other forty-nine states had near or total Democratic control.

Newspapers openly speculated on whether the GOP "survives at all," and even reliable GOP operatives talked frankly about the party's imminent demise. "The Republican Party may have outlived its usefulness," declared John Deardourff, a consultant to moderate GOP candidates.[2] The Republican Party was so sullied by Watergate, corruption, Vietnam, and its own schizophrenic message that many thought that, at the very least, it should change its name.

⸻

Gerald Ford, even as the long shot against Jimmy Carter, was now holding center stage, not Reagan. Reagan was a man born for the arena, and the thought of simply fading away on someone else's terms cut him deeply. He admitted as much in his autobiography, *An American Life*, when he wrote, with not a little understatement, "It was a big disappointment because I hate to lose."[3] He rather liked the idea of dramatically riding off into the sunset after victory, but it was going to be at a moment of his own choosing. Reagan had an actor's innate sense of timing. He knew how and when to enter (or exit) a political stage.

Back in 1970, when Vice President Spiro Agnew had been performing as Richard Nixon's pit bull, savaging one Nixon critic after another who had angered the White House, Reagan told aides, "The trouble with Spiro is that he doesn't know when to stop. A showman should always know when it's time to leave the stage."[4] Reagan always knew when the curtain had fallen.

Conservatives, furious for years over real and imagined slights at the hands of moderates Tom Dewey, Dwight Eisenhower, Richard Nixon, Ford, and the "Eastern Elites," were in open revolt against the remaining forces of the GOP. A handful went so far as to hold their own convention in Chicago, after Kansas City, under the banner of the American Independent Party. Their plans backfired, however, when a group of racists took over the convention and the conservatives stormed out.[5] Conservatism in America had gained respectability among many as an intellectual force, though it had still not driven out the last dregs of bigots, who'd comprised a small faction twenty years earlier—though, truth be told, racists were still abundant in the Democratic Party of 1976.

What was particularly distressing for conservatives at the time was a Gallup poll showing that though Republicans were about as popular as "ring around the collar," an apparent calamity for men everywhere if Madison Avenue was to be believed, almost 50 percent of Americans called themselves "conservative," while less than a quarter called themselves "liberal." The *Wall Street Journal* lamented that "the WASP small businessman of the GOP has not found a way to make common cause with the Catholic blue-collar worker despite the latter's increasingly conservative political perceptions."[6] Carter had an ongoing problem stitching together the urban northern liberals with the southern traditionalists.

Carter also had a problem with Catholics, who were traditional Democratic voters, in part because of his "fuzzy" abortion position. His campaign went so far as to set up an "ethnic desk" at his headquarters in Atlanta, which was staffed by Terry Sunday, formerly with the National Conference of Catholic Bishops.[7]

Some Catholics, frustrated with their inability to pin down the slippery Carter on cultural hot-button issues, endorsed Ford for president, though the National Coalition of American Nuns supported the Georgian. Carter was also having trouble with Catholics in his own backyard, which was underscored when the wife of the Democratic governor of Louisiana, Edwin Edwards, endorsed Ford over Carter. Elaine Edwards's endorsement was a lonely one, though, as the GOP struggled for relevance. Many in the conservative chattering classes had settled on the analogy between the current GOP and its predecessors, the Whigs, who had come down with a bad case of confusion in the 1850s and died as a result. Amid these problems, Ron and Nancy headed to the Ranch for the balance of September, for some much-needed rest and relaxation.

Despite the healing powers of ranch time, Reagan and some of his supporters were slow to join the Ford bandwagon. Often, staff and friends would go see the Reagans at Rancho del Cielo on a Sunday, only to find themselves drafted for chores. Nancy Reynolds found herself painting one afternoon when her nine-year-old son, Mike, fell asleep in the back of Reagan's pickup truck and Reagan drove off, not knowing the boy was in there. Reagan "jumped" when he stopped and discovered the still-slumbering boy.[8] The Ranch was Reagan's port in a storm, and he had little communication with the outside world while there, save a beat-up old television that barely picked up broadcasts from Los Angeles. By day, he would work and ride; by evening, he'd write, read, and go for long strolls with Nancy.

In early September, the *New York Times* succumbed to the anti-Reagan spin coming from the Ford campaign. They saw Reagan as a sore loser against Gerald Ford, the real '76 GOP nominee.[9] In fact, neither Reagan nor his staff, and especially Mrs. Reagan, would forget or forgive for a very long time. In the case of some staffers, they would never, ever forgive Ford. The paper had it correct the day before, when it was reported that Reagan "has grown impatient with the White House" over the newest round of backbiting against him by Ford operatives.[10]

Reagan and his team had learned a timeless lesson the hard way. They had discovered "that presidential campaigns are a lot like clashes on a battlefield."[11] Like his staff, "Reagan admitted he had been 'naïve,' had lost a lot of his innocence in Kansas City as he saw with his own eyes how politicians used their muscle to block him from taking over the Republican party, which they had controlled for so many years as their personal fiefdoms."[12]

While some Reagan staffers were casting about for jobs, Charlie Black, Lyn Nofziger, and Paul Russo hooked up with Ford's running mate, Kansas senator Robert J. Dole, whose mission was clear from day one: "To try to keep followers of Ronald Reagan . . . from abandoning the party."[13] Dole's campaign approach, charitably freestyle as he continually altered the plans he received from the Ford headquarters, was "giddy, repetitive, relaxed about absurdity, high-spirited but quite uncertain whither it is heading, in November or even a day or two ahead of time."[14] Nofziger compared Dole's campaign style to a "hungry Doberman pinscher."[15]

===

Governor Carter kicked off the fall campaign in September not in Detroit's Cadillac Square, as had been the tradition for many Democratic nominees the past decades, but in his home state of Georgia. He wisely chose Warm Springs, where Franklin D. Roosevelt availed himself of the hot baths as he battled polio for many years and where he died early in 1945. Unlike his political heroes, the urbane and urban FDR and John F. Kennedy, Carter sounded a southern populist theme with agrarian overtones, stressing his "farm boy" roots and down-home perspective. Both FDR and JFK were Harvard alumni who had gone to elite New England prep schools and grown up in great wealth. To be sure, Carter had grown up comfortably as a child, and lived even more so as a successful businessman, but he was truly a populist, suspicious of the Washington culture, unlike the majority of his fellow Democrats.

Ford stuck mostly to the White House, "acting 'Presidential.'"[16] He was seen by many, in contrast to the populist Carter, as an elitist. The perceived contrast between the two men was misleading, at least in terms of their relative wealth. Carter had successfully grown the family peanut business and, by 1975, had a taxable annual income of $122,189. Yet he had paid only $17,500 in taxes due to income averaging and an investment tax credit. He and Mrs. Carter also enjoyed, besides the equity in the peanut business, extensive landholdings and a large stock portfolio, which included Coca-Cola.[17] In contrast, President Ford earned an annual income of $200,000, but his assets were meager due to his years in Congress and the cost of raising four children.[18]

The government's private investigation in 1976 of the Fords' personal finances was splashed all over the front pages of the major newspapers. The investigation began in July, a month before the convention, when FBI director Clarence Kelley, a Nixon appointee, brought Attorney General Edward Levi unconfirmed information "that political contributions from certain named unions had been transmitted to political committees in Kent County, Mich., with the understanding they would be passed on to Mr. Ford for his personal use."[19] Levi passed the buck and asked the Watergate Special Prosecution Force, still in business more than two years after Nixon's resignation, to take on the investigation. Charles F. Ruff, the fourth and final Watergate special prosecutor in charge of that office at the time, "said he took the case at the request of Atty. Gen. Edward Levi and, with the help of FBI agents, investigated the unions' campaign records and Ford's personal finances."[20]

The investigation quickly became the theater of the absurd, however, with IRS agents wondering aloud to the *Wall Street Journal* why Mrs. Ford's checks to her hairdresser, Mrs. Madelyn Bourbeau of the Fairlington Beauty Salon in Alexandria, Virginia, were always for round amounts.[21] "Dozens of reporters including *Washington Post* stars Bob Woodward and Carl Bernstein are on the case, but nobody has turned up much."[22] The nonscandal scandal was all breathlessly reported.

The final report, issued by Ruff on October 14 after a painstaking FBI investigation, found absolutely nothing and exonerated Ford.[23] The political damage done by five months of sensationalized headlines surrounding the probe, however, did little to enhance his chances in November.

———

Gerald Ford's pollster, Bob Teeter, darkly warned in 1976 that America was not a two-party system as much as it was a "1½ party system." A deeper cut came from former governor Tom McCall of Oregon "when he said with gallows wit, 'I thought the party was already six feet under. You should speak more respectfully of the dead.'"[24]

In fact, the GOP wasn't dead, but it was on life support, as one moderate after another thought they'd rather switch than fight. Rita Hauser, a Manhattan attorney and former representative to the United Nations

Commission on Human Rights, mused publicly about the state of the internecine warfare in the GOP when she said, "We are viewed by the right wing as if we were lepers."[25] The prevailing view, according to Ford's strategist F. Clifton White, was that Republicans weren't a barrel of laughs and that they took things too seriously.[26] The party's sullied image had the added burden of being seen as the "eat-your-spinach" gang, who just said "no" to Democratic initiatives.

Republicans were in especially bad shape in Carter's Deep South. From the time of the Civil War and Reconstruction, the South had been solidly Democratic and had become even more so as the party of Lincoln freed the slaves, denounced lynching and the Ku Klux Klan, and supported universal suffrage. Unable to expand into the South, and with the hangover of Watergate, the Republicans in 1976 were facing the prospect of political extermination. The *Wall Street Journal* noted that one bright spot for the GOP was ironically in Carter's home state of Georgia, where a young college professor, Newton Gingrich, was challenging Representative John Flynt.[27]

Bad blood between Reagan and Ford continued as Reagan's Californians told the Ford campaign they would help only if Paul Haerle were banished for his earlier betrayal, having jumped from Reagan to Ford the year before. He'd been Reagan's appointments secretary before Reagan appointed him the Golden State's GOP chairman. It was tit for tat in New York, where the Ford forces excluded Reagan's Brooklyn chairman, the combative but much esteemed George Clark, from the Empire State's efforts for the president. A lot of finger pointing ensued over who was to blame for Reagan's not helping or endorsing Ford. "However, some of Mr. Reagan's aides have felt that the president's political advisers were unduly slow in attempting to recruit the Californian for the campaign this fall," the *New York Times* stated.[28]

Another individual harboring bad feelings was Vice President Nelson Rockefeller, who'd been unceremoniously dumped from the ticket the year before by Ford, when the Reagan challenge was looming and the thought of offering up a sacrifice of "Rocky" himself might appease the conservative natives. It didn't. In late October, Rockefeller went on *The Mike Douglas Show*, where he was supposed to stump for Ford, but spent his appearance dumping on Ford's former campaign manager, Howard "Bo" Callaway, who had publicly embarrassed him over whether he should be dropped from the ticket with Ford.

Still, Rockefeller gamely campaigned with Bob Dole in Binghamton, New York, a hotbed of liberal activism by virtue of the presence of the State University of New York at Binghamton, ironically part of the state's higher education system that Rockefeller considered one of his crowning achievements as governor. Hecklers and students greeted the duo rudely, with loud chants and catcalls. Rocky, cackling, responded to the crowd with his middle finger.[29] A photo of Rockefeller's gesture was sent out on all the wires, and Dole wisecracked that he had trouble with his right hand.

In a debate later in the campaign with Walter Mondale, Dole displayed his wit when he told the American audience that he sought the vice presidency because it was "indoor work and no heavy lifting." He also tweaked his opponent by charging that Mondale was so thoroughly in the grip of organized labor "that AFL-CIO President George Meany was probably his makeup man."[30] But Mondale gave as good as he got, charging Dole with being a "hatchet man," and then he leveled Dole with the Kansan's charge that all the wars of the twentieth century had been "Democrat Wars." Mondale threw a right hook, and Dole leaned into it as Mondale, ironically, lectured the severely wounded war hero about fighting the Nazis, which Dole had done and Mondale had not.

———

Reagan and Ford had only one joint appearance in the fall, and it accomplished little to heal the rift, but Reagan did put in several appearances with Dole, including stops in Denver and Connecticut. Uncomfortable and taut with Ford, Reagan was relaxed with Dole, whose wit and ribald humor he often enjoyed.

Their appearance in Denver had been prearranged, but the stop in Connecticut had been entirely by chance, as Reagan had gone to Yale to visit his youngest son, Ron ("Skipper"), and at the last minute joined Dole in Hartford. Again, "Mr. Reagan did not mention President Ford in his remarks," but he did, yet again, praise the GOP platform.[31] One week earlier, Reagan and Dole also had a joint appearance in California, but it did not go the way the Ford campaign would have liked. Lyn Nofziger, who was traveling with Dole full-time, could not get Reagan pinned down for several days on any type of public event with Dole, despite Nofziger's longtime relationship with Reagan.

For whatever reason, Reagan let Bob Dole cool his heels before finally agreeing to meet and then only at his home in Pacific Palisades. There, they met in private for about forty-five minutes, after which they emerged, along with Mrs. Reagan, for the benefit of the cameras. Posing, they "beamed somewhat awkwardly at one another for the coveted unity pictures."[32] Dole departed before he could be asked any embarrassing questions by the assembled media about the ill-at-ease meeting. That left Reagan to face the reporters. He was asked about Ford's insular, stay-in-the-White-House-and-act-presidential strategy, and he joked, "Well, he's sure got the best-televised Rose Garden in America." Reagan was also put on the spot when the media queried him about the amount of campaigning he was doing for the Ford-Dole ticket. He was forced to defend himself, reminding them of the time he'd spent on the road, the commercials he'd taped, and the time he would be spending, albeit with local candidates and not specifically for the president. Reagan did a bit of a soft-shoe, saying, "Well, you go for the local candidate, but you campaign for the whole ticket and the party. That's understood."[33]

But it wasn't all milk and honey for the Democrats, either. Ominously for Carter, the *New York Times* noted just prior to his election that the candidate "did not reach very deeply into the Democratic Party mainstream for . . . talent," sticking primarily to his fellow Georgians and close friends.[34]

Carter's outsider, go-it-alone strategy had served him well in the Democratic primaries, and in fact, his unconventional campaign was responsible for winning the nomination over a talented field of career politicians. Yet he and his people failed to understand that while it was one thing to win a nomination, it was quite another to win a general election and then govern. General elections are not the place to carry out grudges, especially inside one's own party, and certainly not against the entrenched interests in Washington. Carter and his people were planting unnecessary seeds of discord in Washington, from which he would eventually reap a bitter harvest.

The Washington buddy system, which Reagan had campaigned against, was also a target of Carter's. Carter was genuinely repulsed by the system, especially when he learned from his press secretary, Jody Powell, that an investigative columnist was working on a story that alleged that the Georgian had once had a mistress.[35] The rumor, peddled by Republican operatives, went nowhere, but since Carter had given a controversial interview to *Playboy* magazine just a few weeks earlier, in which he confessed to having "lusted in his heart" for other women, it served to fuel the despicable rumor. He went even further in the *Playboy* interview, which was scheduled for the November issue but leaked at the end of September, telling Americans one controversial thing after another. He said that he considered extramarital sex, sodomy, and homosexuality to be sins, but he assured a shocked nation that he would not be in favor of "breaking down people's doors to see if they were fornicating." He let loose on both Richard Nixon and Lyndon Johnson, saying they had engaged in "lying, cheating and distorting the truth."[36] The bizarre incident did nothing to dispel the notion that Carter was an oddity, despite apologizing to Lady Bird Johnson.

Carter expounded on the media. He told the magazine, "The national news media have absolutely no interest in issues at all . . . There's nobody in the back of this plane who would ask an issue question unless he thought he could trick me into some crazy statement," which is precisely what *Playboy* did.[37] Carter also managed to insult almost every interest group in his own party.

Comedians had a field day, and Bob Hope got in on the fun, as he told a GOP crowd in Beverly Hills, "Carter's slogan is 'The White House or bust!'—and after the *Playboy* interview you kind of wonder which he wants most."[38]

Reagan took note of Carter's comments about adultery in the *Playboy* interview, saying not only that Carter had lost momentum, but also that it appeared to him that "Jimmy Carter can beat Jimmy Carter."[39]

Carter didn't stop with *Playboy*. For an article published in the *New York Times Magazine*, he sat down for an interview with liberal novelist Norman Mailer. He told the writer, "I don't care if people say 'fuck.'" The *New York Times Magazine* censored it.[40] Rather than counteracting the *Playboy* imbroglio and Carter's zealous religiosity problems, the Mailer story only served to stimulate the rumor in Washington circles, fueled by GOP forces, that Carter was unstable.

Carter had a progressive record of integration while governor of Georgia, but the race issue surfaced in an embarrassing fashion for him during the fall campaign. He had refused to quit the Plains Baptist Church where he belonged even though church deacons voted to keep in place a ban, established in the mid-1960s, that prevented blacks and "civil rights agitators" from becoming members. To his credit, Carter protested his church's actions in 1965, and Plains citizens responded by boycotting his business for a time.

Carter responded to media inquiries about his church by saying it would be better to fix things working from the inside rather than the outside. Blacks had been loyal Democratic voters since the New Deal, and even more so to Carter. He had grown up in a Jim Crow culture, and his father was at best "spotty" when it came to blacks, but the southern culture was something no northerner could really understand.

Yet, campaigning in the South, Carter deceitfully praised two old segregationist senators from Mississippi, John Stennis and Jim Eastland, ascribing "courage" to them while calling them "statesmen" on the issue of segregation. When reporters asked the Dixiecrat Stennis about his newfound status as a champion of civil rights, he said, "I'm against it, always have been and always will be."[41] Carter's comments were just plain tomfoolery, and once again raised what his campaign manager, Hamilton Jordan, referred to as "the weirdo problem."[42] In addition, Carter was showing his inexperience as a national candidate, as he attacked George Wallace, Ted Kennedy, and Hubert Humphrey on the stump, all Democrats and all leaders of different but important factions within their party.

While Carter's statements were getting him into trouble, the president was not without his own racial problems. Ford's agriculture secretary, Earl Butz, was an extremely popular Cabinet member with the farm community, but his mouth was too big and too dangerous. During the fall campaign, Butz made a derisive reference to "the coloreds," for which he had to apologize, and was later forced to resign over an extraordinarily bad racial "joke," which *Time* magazine delicately called a "scatological indiscretion."[43] When John Dean of Watergate fame broke the story in an article he'd written for *Rolling Stone*, the proverbial dung hit the fan for Butz, and protesters showed up at the White House with Kick Butz signs.[44] Bipartisanship on racial jokes ruled the roost in 1976, though, as columnist William Safire wrote at the time: "Two of Mr. Carter's top staffers, who are said to have told racial jokes to reporters in wee-hour drinking sessions, live in fear that their own off-color, off-the-record thigh-slappers will be revealed by practitioners of John Dean journalism."[45]

Carter's verbal indiscretions and lack of sophistication about big-league politics began to take their toll. Reporters were boring down on the Georgian—as were the comedians in the hot new, hip show *Saturday Night Live.* Carter was not altogether "ready for prime time." What he and his people, all neophytes to the hardball and highball aspects of Washington politics, did not understand or attempt to cultivate was the all-important Georgetown cocktail crowd that held so much sway over what concerned politicians and the media. The political establishmentarians, fresh from the kills administered to Nixon and Spiro Agnew, were full of themselves and thought at the early stages of the Democratic primaries that the populist Carter was "cute." Yet when the long-shot peanut farmer morphed into the Jolly Green Giant, they were appalled that this hillbilly and his gang of yahoos might actually take over the party and win the nomination and the White House.

Rural populism had sprung up in Carter's South and Reagan's Midwest in the 1890s, and was mostly focused on the power of moneyed eastern interests, especially large banks, railroads, and manufacturers. Reagan, from the time he was a tot, had grown up poor, and though Carter had a more comfortable upbringing, they were both products of their cultures. Big government was also the focus of populists' ire, which Carter acknowledged when he announced his candidacy in 1974.[46] Though some racists were involved in the populist movement in its earlier days, both Carter and Reagan abhorred racism, and their political philosophies originated from many different values.

Washingtonians also did not understand how close Carter was to his wife, Rosalynn. They underestimated this steel magnolia, and for the oft-divorced sophisticates who made up the Washington intelligentsia, such personal closeness between married couples was deemed peculiar. As one member of the intra-Beltway set replied when asked if he was married, "Yes, but I'm not a fanatic about it."[47] A cultural rift was developing between the uncomplicated Georgians and the unctuous Georgetowners, who would bedevil Carter all through the next four years.

Carter's midsummer lead over Ford of 30 percent had melted under the hot glare of the national media pressing him harder and harder each day on his "fuzzy" issue positions. His odd comments and contradictory statements, and his shifting positions on abortion—along with his overly aggressive attacks on Ford by citing the "Nixon-Ford Administration"— were a feeble attempt at guilt by association with Nixon, which even his aides knew was backfiring against "Nice guy Jerry Ford."[48] Voters may have lacked complete confidence in Ford's ability to lead the nation, but they also knew he was an honest man who had tried to lift the country out of the Watergate morass.

Campaigns, however, are not just about raw numbers. They are about a thousand other things, including ideology, geography, luck, and, most important, the character of the candidate himself. John Sears, Reagan's former campaign manager, was invited by the *New York Times* to give his advice to Ford on the fall debates. In typical fashion, Sears did not pull his punches, and after reviewing Ford's assets, he wrote, "[Y]ou are perceived as slow-witted, unimaginative, dull, slow to make a decision, a poor public speaker."[49] The advice went downhill from there. Ford never did engender much fear or respect in his party. No one in his right mind would have dared written such things about Richard Nixon or Lyndon Johnson—and lived to continue in elective politics, that is. However, the fact that a political operative could write so frankly about the leader of his own party was not simply about Ford's weak grip on the party; it was also a testament to the rise of the political consulting class. They, along with the national media, were becoming the real political bosses, not the elected officials they put up with and about whom they made jokes behind their backs. And the real political bosses were certainly not the state and local chairmen.

Still, Sears had some advice that was not only unique but also time-less for politicians when he wrote, "Don't live and die by your polls. Polling cannot tell you what to do; it can only tell you when you are wrong. Pollsters are not politicians; don't use them as a substitute."[50] Sears did not offer any sartorial advice to Ford, but the President Ford Committee had already discovered that voters did not see the president as a leader, and one solution was to firebomb Ford's closet of awful, loud double-knit suits and jackets and have him appear only in dark three-piece suits.

———

In the predebate sessions between Jim Baker and Jody Powell, one thing was vitally important to the Carter people: to have the lecterns as far apart as possible, so as to minimize the six-foot Ford's advantage over the five-nine Carter. Ford had a debate "supporter" in his friend, Michigan senator Robert Griffin, who told reporters, "People don't expect much from Ford and that will be a real advantage."[51] In one way, Ford lived down to those expectations by making a costly gaffe in the second debate, when he bizarrely proclaimed, "There is no Soviet domination of Eastern Europe, and there never will be under a Ford Administration."[52] That must have come as news to the Poles, Czechs, Hungarians, and millions of others behind the Iron Curtain who woke up every morning with Soviet tanks staring at them. Warsaw contained five Soviet divisions alone. Ford then bollixed up things even more when he refused to apologize for his foolish statement for five precious days. He'd been gaining a point a day on Carter, but after his verbal gaffe, his advance faltered until he was finally talked into grudging contrition, which he delivered by phone to the head of the Polish American Congress, Aloysius Mazewski of Chicago.

Ford had actually performed well in this second debate, his liberation of the Warsaw Pact countries notwithstanding, and overnight polls showed that the American people thought the president had won—that is, until the media seized the issue. They began batting Ford around like a surly housecat with a terrified mouse, torturing and pummeling him over his blunder. His momentum was gone, and one week later, everybody in America, possibly including Ford's own family, thought it was Carter who'd won the debate.

The embarrassing gaffe also revived the "dumb" jokes and comments about Ford. "Any dumbbell would know that's not right," explained Helen Obal, seventy, a naturalized Pole, to the *Wall Street Journal*.[53]

In the wake of Ford's faux pas, Walter Mondale, who had one of the most antidefense voting records in the Senate, nonetheless adroitly cast himself as more anti-Communist than Ford when he marched in an ethnic parade in Buffalo and reminded the sympathetic crowd that "it is well known that there are 31 Soviet divisions" in Eastern Europe. He was also greeted with favorable signs such as "Ford Does Not Know What Freedom Is—Jimmy Does." Betty Ford and Bob Dole also marched in the parade, where they were both awkwardly forced to defend Ford's debate comments.[54]

Ford closed fast on Carter, and all knew the election would be a cliff-hanger. Mondale knew how much it had tightened. Late in the campaign, on the last leg of a flight, a nervous young advanceman, Michael McShane, asked Mondale how close the election was. Mondale simply looked at his watch and shook his head.[55] In the end, Ford lost by the narrowest of margins to Carter. In a way, what with Nixon, Watergate, the economy, and his own stumbling Administration, and the fact that he had not been elected to the office in the first place, it was a small miracle Ford had done as well as he had in the election of 1976.

Perhaps Ford might have done better if there had been more Trekkies in America at the time. He'd named the first prototype Space Shuttle *Enterprise* in a bow to pressure from the show's maniacal fans. He would later live long and prosper as an ex-president.

Yet Americans with a vision for the future—save for one sixty-five-year-old Californian who was also trying to figure out his own future—were in short supply at the end of 1976.

While We Were Marching Through Georgetown

"The Republican Party is ill. I think we have some grave problems."

An old adage in politics states that the winner is most often the person who makes the next-to-last mistake, and this held true in 1976. Gerald Ford made the last mistake, and consequently lost the chance to win the presidency in his own right. Jimmy Carter was, as they said down South, "in tall cotton." The slender peanut farmer and his Georgians were invading Washington, the capital of the North, and Union general William T. Sherman was spinning in his grave.

When "The Great Invader" sliced across Georgia in 1864 and burned all in his path while his Union troops pillaged and plundered everything they could from the civilian populace, the proud Georgians were enraged and swore revenge. Adding insult to injury, the Northern troops composed and sang a song, "Marching Through Georgia." The ballad was as hated by the locals as were the troops who sang it. Now the tables were turned, and the South had risen again, just as it had vowed.

Carter was the self-proclaimed "outsider" who was going to clean up the "bloated mess" in Washington that Gerald Ford and his party represented. Carter's anti-Washington rhetoric was appealing to many, including conservatives, who increasingly saw the nation's capital as the problem and not the solution. The crass and arrogant ways of the city by the reeking Potomac River were grating on the American people, as was the growing concentration of power and corruption. "The people are feeling as strongly about the federal government now as they did about the King of England at the time of the Revolution," said Albert Quie, the Republican governor of Minnesota.[1] Quie had been a member of Congress for nearly twenty years but fled as he was filled with revulsion over what he saw there.

———————————

The weekend before the November election, Ronald Reagan had hit the trail again for the GOP, but mentioned Ford's name only in passing. The *New York Times* ran a story by Wallace Turner, saying that "former Gov. Ronald Reagan has succeeded in running out the election campaign without being drawn into direct, full support of President Ford."[2] The story detailed how Reagan was campaigning for conservative Republican candidates and speaking warmly about the platform, but that he rarely mentioned Ford and took only weak shots at Carter. Once again, the fingerprints of the Ford operation were all over the story.

Ford had previously campaigned in California and invited Reagan to join him, but Reagan sent a telegram in which he took a pass on the last-minute invitation, choosing instead to spend his day at his ranch in Santa Barbara. The private telegram to Ford from Reagan was released to the media by the Ford campaign. A Ford operative whined to the paper that Reagan had said he'd help and then accused Reagan of reneging: "Then he said, 'I'm too busy to go everywhere for everybody.'"[3] Reagan figured that two could play at this game of political "chicken." If the Ford operatives wanted to play hardball, that was fine with him. For over two years he'd been the target of personal attacks, the brunt of jokes, and generally dismissed as a "lightweight" by Ford's men.

Ironically, Reagan earned more respect from many Democrats around the country than he had from some members of his own party, starting in the Oval Office. Late in October, in spite of his own mixed feelings, he sent President Ford a telegram promising to do even more campaigning for Ford in at least seven western states. He explained as much and that "it is part of my determination to persuade every American I can reach to join me in voting for you on November 2."[4] Still, he turned down a request from Jim Baker via Holmes Tuttle, of the famous "Kitchen Cabinet," to put in last-minute appearances for Ford in Texas, Florida, and Mississippi. Reagan's schedule was jam-packed, and he simply could not be everywhere for Ford.[5] Reagan was doing just enough to keep up appearances, but not more than enough, as he and his aides were still steamed at Ford and his primary campaign.

Reagan and Mrs. Reagan quietly voted on Election Day near their home in Pacific Palisades. A reporter for UPI asked Reagan if he was "disappointed or bitter that he did not have the opportunity to vote for himself. Mr. Reagan replied, 'No, I'm at peace with the world.'" He also made no bones about the next election, not denying that he would run again. He then told the reporter that the results of the day would have no effect on him. "Regardless of the outcome, I intend to go out on a campaign to use the Republican platform to reach a new majority out there for the GOP."[6]

In retrospect, Reagan's comments provided the eulogy for the President Ford Committee, as it had failed to achieve a "new majority," or any majority for that matter.

The day after the squeaker, Ford conceded the election to the improbable winner, James Earl Carter, former one-term governor from Georgia. Ford had lost his voice in the last several days of the campaign, so his wife, Betty, read his gracious statement. Yet many social conservatives, perturbed at Mrs. Ford's outspoken support for the Equal Rights Amendment and abortion, thought she'd already spoken too much for the president.

———————

Reagan wasted little time moving ahead. He immediately knew that if he ruled himself out, his appeal on the stump would dry up. Exasperated, he later wrote in his memoirs that he and Nancy simply could not believe his aspirations "wouldn't—couldn't—end in Kansas City. After committing ten years of our lives to what we believed in . . ."[7]

Of more immediacy to Reagan than the GOP nomination four years hence was the fact that he had to eat. "Mr. Reagan's stake in the next four years goes beyond political considerations. He has now returned to developing his career as a newspaper and radio pundit and traveling the after-dinner circuit as a highly paid speaker . . . His income from these activities is expected to gross considerably more than $500,000 next year."[8] His old booster William Loeb of the *Manchester Union-Leader* wasted no time himself, as he editorialized on Reagan's future to "form and direct a new political alignment of conservatives from both parties."[9]

Even with the closeness of the outcome—Carter received 297 electoral votes, Ford, 240, and Reagan, 1—no one screamed about conspiracies to steal the election for Carter, though he won Ohio's 25 electoral votes by fewer than 7,500 votes out of over 4 million cast. The margin in Hawaii was even less for Carter (just over 2,000 votes), and these two states, if they'd gone for Ford, possibly could have spelled defeat for the peanut farmer from Plains, Georgia. It was also the closest election, in terms of the popular vote, since 1960, despite Carter's desire for a mandate. For a time, some in the Ford campaign debated a recount, given the history of dirty politics in labor-dominated Ohio, but the president quashed the notion before it got very far.

Ford was intent on leaving with the same dignity and style with which he'd entered the White House. His chief of staff, Richard B. Cheney, handled the transition of power adroitly. Cheney wanted to ensure that Ford's legacy would not simply be that he came into office in turbulent times that shook the nation to its core, but also that he brought his personal decency to the office.

The election of 1976 finished closer than anyone fully realized at the time. Had Hawaii and Ohio gone to Ford, he would have had 269 electoral votes, just one shy of the needed 270. Carter would have had 268 because of one "faithless elector" from Washington State, Mike Padden, who voted for Reagan, despite the fact that the Evergreen State had gone for Ford. There was no legal mechanism to compel Padden to vote for Ford, and thus the outcome of the national election would have been tossed to the House of Representatives.[10]

No one will ever know how that body would have come down. However, because Ford had been a member of it for twenty-five years, with friends on both sides of the aisle, and because Carter had such a tenuous relationship with his own party, no one can say for certain that Ford would not have won, with a little help from his friends. Out of the more than 18,000 Electoral College votes cast in the history of the Republic from 1788 to 1976, only a handful had dared to go against the will of their state.

━━━━━━━

Carter had come a long way from "Jimmy Who?" to "Leader of the Free World." His election reaffirmed the quaint and uniquely American notion that the presidency was not a birthright for the wealthy or privileged and that "anyone can grow up to be president." Shortly into his Administration, however, he would begin to walk into the buzz saw of entrenched interests in Washington, starting with his attempt to reorganize the national government just as he'd done with the state government in Georgia. Oddly, just after the election, Carter told reporters that he was "beginning to feel more and more like a Washington insider."[11] It was anything but the case.

Carter met with outgoing secretary of state Henry Kissinger for a briefing on U.S. foreign policy later that month. Worrisome for conservatives, Carter embraced Kissinger and his accommodationist stance toward the Soviets. He also pronounced Kissinger "my good friend . . . There is no incompatibility among us."[12] At this, jawbones fell agape all over the city-state. During the campaign, the Soviets had become angry with Ford because they believed he'd bowed to the pressure from Reagan over U.S. foreign policy and had departed from the Nixon/Kissinger policy of détente.

The Soviets arrogantly believed that they had a say in American elections and, indeed, had long been rumored to have tried to meddle in the 1960 and 1972 campaigns, aiding both John Kennedy over Richard Nixon and then Richard Nixon over George McGovern.[13] This arrogant mentality of supposed influence from the Soviets continued into the 1980s.[14]

———

Ford's loss did not immediately set off a scramble among other would-be suitors of the White House inside the GOP. Partly, this was due to the party's present state. Though the contest between Carter and Ford was the closest since 1916 in terms of electoral votes, the down-ticket contests went heavily Democratic. In the *New York Times*, Warren Weaver wrote that the GOP "is perhaps closer to extinction than ever before in its 122-year history. Most of its national leaders are either defeated, discredited or too old for any claim on future political influence."[15] One of the few successful GOP candidates in 1976, incoming governor "Big Jim" Thompson of Illinois, told the *New York Times* several weeks earlier, "The Republican Party is ill. I think we have some grave problems."[16]

How badly off was the party, with its squabbling political combatants from the moderate and conservative wings? Bad off enough that Reagan said he could "accept a new name for the party."[17]

———

Toward the end of the '76 campaign, a Ford staffer, speaking on condition of anonymity, told *Time* that if Ford won, "there would be no revolutionary changes, no wrenching of government."[18] That statement summarized better than anything the stark differences between Ford and Reagan. Senator Jesse Helms of North Carolina, a leading conservative, argued for transforming the GOP to a "broad-based conservative party which will bring all of the workers and producers into one camp and leave the special interests and the self-seekers to the liberals." Helms championed a complete house cleaning and echoed Reagan on changing the name of the sordid and disgraced party.[19] Party moderates were appalled that rural conservatives such as Helms were not only attempting to come through the front door of their club but were threatening to break it down.

Meanwhile, the Federal Election Commission released a report on New Year's Day 1977 that said that the entire presidential campaign, financed completely for the first time by the U.S. taxpayer, had cost John and Joan Q. Public $72 million. Back in the summer of 1976, Ford's and Carter's campaigns, once they secured their respective parties' nominations each, had received a check for $21.8 million from the FEC.[20]

Yet the candidate who received the most in the primary period was neither Ford nor Carter, but Reagan, who got a bit more than $5 million,[21] demonstrating his awesome fund-raising appeal among small donors—awesome because federal matching funds were available only to candidates who demonstrated such prowess.

The soon-to-be chairman of the Republican National Committee, Tennessee's outgoing U.S. senator William Brock, would eventually take note of Reagan's abilities and apply them to the party. Brock had been elected to the U.S. Senate in 1970, but lost his bid for reelection six years later.

Another Republican Tennessean, Howard Baker, was ascending in Washington, as he unexpectedly won the election for Senate minority leader over his rival, the more moderate senator Robert Griffin of Michigan. Profiled in the *New York Times*, the affable Baker, whose father-in-law had been the legendary GOP senator Everett "Ev" Dirksen of Illinois, made it clear he would seek counsel from all in the party, including his friend Ronald Reagan. Of the desperate situation that confronted the Republican Party, Baker told the paper "that the extensive Republican losses in the 1974 and 1976 elections 'wiped the slate clean' and that the party 'is looking for a fresh start.'"[22]

Within several days, the members of the Electoral College delivered to Congress their certification of Carter's election with 297 electoral votes, making it official. The fact that this happened so calmly, so matter-of-factly, was a tribute to the wisdom of America's Founding Fathers.

Ford was reluctant just to fade away. The outgoing president proposed to the incoming RNC chairman the creation of a "shadow government" to ensure a GOP response to new initiatives coming from the Carter Administration, but the committee, without comment, tabled the proposal. Bill Brock not only had an independent streak, but he also wanted the Republicans to move on from Ford and Reagan. They were, as far as Brock was concerned, part of the past that the party needed to leave behind. The committee also approved the platform shaped in Kansas City by Reagan partisans. This was not for Reagan's benefit, but simply because the members of the committee had become more conservative over the years and thus agreed with the platform.

Ford's White House staff was scattering to the four winds. For two and a half years, the senior staff had had government limos and drivers at their disposal, and were often picked up each morning at their homes and deposited there again in the evening. For some, their time in the sun had extended back to 1968, when they helped Richard Nixon capture the White House. When they wanted to play tennis, Ford's senior staffers didn't have to wait at one of the public courts in the area but instead had access to the private White House court.

They enjoyed mess privileges in the White House; they had their own pewter nameplates, which immaculate navy stewards placed on the table when they were dining, so their importance could not be mistaken. The food they ate at the mess was world class, as was the food they were served on *Air Force One,* to which they would motorcade, running red lights and having traffic held for them, before driving onto the tarmac right next to the plane at Andrews Air Force Base. These staffers were the unlucky ones. The lucky ones got to "chopper" out to the airfield with the president on *Marine One.*

Once on the plane, there was none of this "fasten your seat belts and put your tray tables in an upright and locked position" nonsense. Staffers stood and chatted, smoked, ate sandwiches, drank coffee—all during takeoffs and landings. In the afternoon and evening, alcohol replaced coffee as the preferred beverage. Again, pleasant navy stewards were on hand for every whim, right down to a hot towel after a meal.

Two sets of White House passes established a caste system in the complex. One pass allowed for access only to the EOB, the Executive Office Building, a gray monstrosity next to the White House that was built during the Lincoln Administration and in which nearly the entire executive branch was housed. The second pass was to the West Wing of the White House itself. Both passes, worn on a chain around the neck, were great icebreakers at the swinging singles bars in Georgetown.

White House staffers all worked hard: six, sometimes seven days a week. In each day, they would often put in fourteen, sixteen, even eighteen hours, nourished by a steady stream of caffeine, nicotine, adrenaline, the stretches of time punctuated with plentiful alcohol, gratuitous sex, and pharmacological enhancements. However, power was the ultimate drug of choice for these aspiring and talented, if excessively insecure, young aides.

It was over all too soon for Ford's staff. The roller-coaster ride they'd been on for the past several years was now ending, and they were faced with the prospect of going back out into "the real world," as these politicos derisively referred to everybody outside their bubble. Cheney's wife, Lynne, told him he shouldn't make any career decisions until he got a good tan. They went off to the Caribbean with Don Rumsfeld and his wife, to unwind after the madness of the past several years.[23]

An eight-year veteran of the Nixon and Ford White House, Agnes Waldron had a good grip on reality when she soberly told her coworkers, "Nobody stays in the White House forever."[24] Had any of the jubilant incoming Carterites been told of Waldron's sagacious advice, none of them would have paid her any mind, but they should have. For one thing, there was no real theme to Jimmy Carter's new Administration, except that it was a break with the "Imperial Presidency" of Richard Nixon. Carter, at his peril, had made making government "down home" a central theme of his candidacy. Yet he unwittingly painted himself into a populist corner, as each time he acted like a president, even for his own safety, the national media complained. Vice President–elect Walter

Mondale saw the coming problems between Carter and the Washington establishment. Carter "was very suspicious of interest groups, of pressure. He knew of the Democratic Party and all its groups . . . he considered a lot of it an albatross around his neck."[25]

Carter's populist image was no pose. He truly was revolted by the cozy relationships in Washington, especially between elected officials and lobbyists.[26] He was the first Deep Southerner elected directly to the presidency since Zachary Taylor (of, at various times, Virginia, Kentucky, and Louisiana) in 1848. Nevertheless, as far as snooty Washington was concerned, no matter how enlightened or charming a southerner might be, or how well he wrote novels, he was still a funny-talking hick who ate grits and strange animal parts.

The bureaucracy of the executive branch had grown exponentially over the years. The White House staff, as an example, had grown from thirty-seven in Herbert Hoover's day to more than five hundred when Carter came to town. Of course, this paled in comparison to the presidential commissions and boards, which grew from a little over a thousand in the second Eisenhower Administration to more than five thousand by the time Carter arrived.

The native and soon-to-be-native Washingtonians were entrenched in the culture of the nation's capital, a culture of jaded courtiers that encompassed the media, Congress, lobbyists, hangers-on, factotums, hod carriers, consorts, talking heads, drunk writers, drunk TV anchors, drunk editorialists, drunk columnists, gofers, minions, lackeys, self-promoters, coat holders, shysters, rumormongers and court jesters—all of whom enjoyed gossip (addictively), fine dining (relatively), cocktails (indulgently), adultery (furtively), and power and politics (excessively).

These "Beautiful People," as heap big charter member Ben Bradlee, editor of the *Washington Post*, referred to them (though the phrase had actually been coined by another charter member, columnist Hugh Sidey of *Time*), were intertwined, inbred, crossbred, and interdependent. They attended the same book receptions and wrote squibs recommending one another's books; their children went to the same private schools; they played tennis at the same clubs; they all shopped at the "Social Safeway"; they complained about the worsening traffic; and they all obsessed about the local football team, the often sad-sack Redskins. To be included in the Beautiful People club, one had to be, above all else, a cynic about everything and everybody who was not a part of their club, and cynical about fellow members, provided their backs were turned. They were especially cynical about "outsiders" such as Carter, that hick from Georgia. Ninety-nine percent of the Beautiful People were liberal, but one conservative was acceptable at a dinner party, so that the other nine liberal attendees could gang up on him, reminding him that conservatives "just don't understand."

Ground zero for the Beautiful People of Washington was Georgetown, the high-end neighborhood of historic town houses and chic bars, restaurants and shops, located in the northwest quadrant of Washington, DC, along the reeking Potomac River waterfront. Blacks were the overwhelming majority in "The District," yet it was rare to see many in this overwhelmingly white and liberal section of Washington, unless it was to drive a limousine or cab, tend someone's garden, or serve cocktails at a Georgetown reception.

Within Georgetown's rarified orbit were the affluent, leafy suburbs of Chevy Chase and Potomac, in Maryland. They might take a flier on McLean, Virginia, occasionally, but that was in the "south," which was hayseed country as far as these sophisticates were concerned. The "hunt country" around Charlottesville or Middleburg was acceptable for a Saturday afternoon ride or a quaint weekend antiquing, but little more. The farther south one went, the more one came across people who talked with an odd accent. A Virginia congressman once quipped that the Fourteenth Street Bridge, which spanned the Potomac River between the District and Virginia, was "the longest bridge in the world." Though it was less than a mile in length, the cultural chasm between the city-state and the gateway to the South was that extensive.[27] The word *tony* was often applied by writers to describe the homes and society of the Beautiful People.

If Carter was going to come from the South, the Beautiful People lamented, why couldn't he have come from a state that had fine wine or great and tortured writers or chic resorts? What did Georgia have besides peaches and sweat? Margaret Mitchell, Georgia-born and -bred author of *Gone with the Wind*, didn't count among the Beautiful People because she hadn't suffered like, say, Tennessee Williams or William Faulkner, as all good southern writers should. Mystified at the culture of Washington, Carter would soon ask himself, "What fresh hell is this?"

The feelings were returned in spades by Carter's Georgians, one who tenderly referred to "the goddam Georgetown cocktail-party liberal a-holes."[28] At his inaugural, Carter's young chief of staff, Hamilton Jordan, foolishly and unnecessarily butted heads with the speaker of the House, Thomas P. O'Neill Jr., over seating. O'Neill not only was one of the most powerful men in Washington, but also had a long memory; after the faux pas, he referred to Jordan as "Hannibal Jerkin."[29] It was the first of many mistakes made by the bourbon drinkers against the wine sippers.

To his deep misfortune, the earnest Georgian would eventually learn the hard way that, in Washington, the first rule of the bureaucracy is to protect the bureaucracy.

Into the Wilderness

"Endangered Species"

The "New Majority" envisioned by Richard Nixon was coming true, but instead for the Democratic Party under Jimmy Carter.[1] After the wipeout of down-ticket races in 1974 and 1976, by early 1977 the GOP had lost complete and utter control in the Old South. "Jimmy-crats" had drowned Nixon's Southern Strategy in a bowl of hominy grits.

U.S. News & World Report devoted a cover story to the nearly extinct GOP, interviewing five prominent Republicans, including Ronald Reagan. The others, among them Howard Baker and John Connally, counseled that the GOP should seek a "moderate" course. Reagan was the only one of the five to argue for bold conservatism.[2] His attitude toward government was bluntly summed up in one simple decree: "Get out of our hair."[3]

The Reaganites believed in their heart of hearts that their man would have run better against Carter and possibly beaten him. They didn't know it, but a never-released CBS poll buttressed that view. It showed that indeed Reagan would have run closer to Carter (though he, too, would have lost) than Ford, and said that Reagan would also have received more Democrats but lost some moderate Republicans.[4] One thing was for sure: Reagan would have done better in the debates than Ford, since he knew that Poland was under the Soviets' thumb.

Reagan, like many conservatives, heard Carter's anti-Washington message and was impressed. He even went so far as to write a column under the heading "Let's Give Carter a Chance."[5] Carter, though disdainful of Reagan's ideology, appreciated his communication skills. He told the *Washington Post* that, had Reagan been nominated rather than Ford, he probably would have ducked debating him. Reagan "was a superb user of the television medium [and] I would have been at a disadvantage then," a remark that also revealed his feelings about Ford as a debater.[6]

Carter's party in 1977 certainly had its own problems, but one of them was not irrelevance. Just the opposite. If possible, it had too many factions—and thus the potential for warfare, unless all sides were adequately and constantly sated. The running joke in Washington was that when Republicans assembled a firing squad, they created a circle. Headlines such as the *Washington Post*'s "The 'Profound Inadequacy' of the GOP," the *National Observer*'s "The GOP: Dying for Real?," and the *New Republic*'s "Feeble Old Party" all told the sad tale.[7]

Though Jimmy Carter's Christian name was James Earl Carter Jr., the American people had come to know him by his down-home, folksy name, and he was thusly affirmed as the nation's thirty-ninth president of the United States on January 20, 1977, the first president to be sworn in using his nickname. Not using his given name caused a bit of unconstitutional indigestion for several constitutional scholars, but the Carterites had checked and affirmed that, indeed, Carter could legally take the oath of office in such a manner.

He gave an unremarkable inaugural address. However, his gracious gesture toward outgoing president Gerald Ford, citing what Ford had done to heal the country after the nightmare of Watergate, began a tradition for all succeeding inaugurals whereby the incoming president finds something nice to say about the man headed for the comfort of boardrooms and exclusive golf courses. Yet, like another populist farmer, Thomas Jefferson, Carter walked rather than rode from the U.S. Capitol to the White House after being sworn in.

In his farewell address to Congress, President Ford said, "I look forward to the status of a private citizen with gladness and gratitude."[8] As one columnist wryly observed, "Now our old kings must go to Palm Springs and play endless rounds of golf with Bob Hope."[9] He would have time to improve on his game, as one friend said (understandably on background) that Ford drove the ball about two hundred yards and putted about the same distance. Nonetheless, in his speech to Congress, he teasingly said, "This report will be my last—maybe."[10]

As a former congressman and president, Ford would not only receive an annual pension of $100,000, but he would also have the opportunity to make even more, serving on corporate boards, lecturing, and receiving a huge advance for his memoirs, *A Time to Heal*. He would also get, as befit a former president, staff, an office allowance, and franking (free postage) privileges for the rest of his life, along with top-secret daily briefings from the CIA on the state of the world.

One former president's prescription for how America should deal with former chief executives was to "chloroform" them and then "cremate" them . . . honorably. That was William Howard Taft. Prior to Taft, a newspaper editor suggested facetiously that all former presidents be branded as public nuisances and hauled before a firing squad. Tongue in cheek, Grover Cleveland said, "An ex-president has already suffered enough."[11]

America in 1977 had two ex-presidents, Ford and Richard Nixon, but Ford ordered his staff to comb invitation lists and public events to make sure Nixon would not be there when he was present, so that no one could snap an embarrassing photo of the two together. They spoke two or three times a year by phone, but that was the extent of their contact, even though for several years they lived relatively close to each other in Southern California.

———————

Democrats were grousing that Carter was not doing more on patronage and other Democratic staples, but these complaints were ignored at the White House. The Democrats had made patronage into a high art form, almost a religion, but Carter was not kneeling in the High Church of Scratch My Back. Meanwhile, GOP operatives were scrounging for work so they could eat.

Rodney Smith was one of those GOP operatives, though he was among the lucky few in that he actually had a job. He was a pleasant, low-key, but effective and creative fund-raiser when he moved over to the National Republican Senatorial Committee (NRSC) from the Republican National Committee in January 1977.[12] Smith had been a young staffer at Nixon's Committee to Re-Elect the President (CREEP) in the summer of 1972. It had been routine for him to see hundreds of thousands of dollars in cash piled on desks at any given time at the committee headquarters, but this was not unusual in politics at the time. Smith's youth and short tenure allowed him to escape the scrutiny of the Watergate special prosecutor. He then went back to the RNC until 1977, when, at the behest of the senatorial committee's new chairman, Bob Packwood of Oregon, he became finance director.

There was a problem, however, one both real and symbolic. The offices of the Republican Senatorial Campaign Committee were in a former men's room on the fourth floor of the Russell Senate Office Building. Though the urinals had been removed, the walls were still lined with the marble found in most antiquated men's facilities at the time. It also reeked of urine. That the committee responsible for helping the GOP gain Senate seats was headquartered in a men's toilet invited just too many jokes and metaphors for the state of the Republican Party in early 1977.[13]

Money was nearly nonexistent, and without the levers of power, there was little the GOP could offer businessmen to gain their financial contributions. It was iron lung time for the Republican Party. The party was dying.

"It sometimes happens in a people among whom various opinions prevail that the balance of parties is lost and one of them obtains an irresistible preponderance, overpowers all obstacles, annihilates its opponents, and appropriates all the resources of society to its own use.

The vanquished despair of success, hide their heads, and are silent."[14] This was Alexis de Tocqueville's description of the disappearance of the aristocratic Federalist Party in the third decade of nineteenth-century America, and a prophetic prediction of the demise of the party that arose to replace it in 1833, the Whig Party, which itself dissolved barely two decades later. It seemed applicable to the GOP in the twentieth century.

In the 1970s, a debate was emerging inside the GOP, focusing on whether it should continue to be the tool of big business or go off in a populist direction. Leading the anti–big business charge was Reagan, supported by Senator Paul Laxalt, and former Nixon speechwriter and syndicated columnist Pat Buchanan. Ironically, big business had been giving more heavily to the Democratic Party than the Republicans for years for one simple reason: Democrats controlled the Congress, and Capitol Hill controlled the purse strings, along with the legislation that could either help or harm business operations.

Other nascent debates were about economics. Richard Nixon, who had no core philosophical beliefs, had told reporters when he was president, "We are all Keynesians now."[15] John Maynard Keynes, the patron saint of liberal economists, had written a book in 1926 entitled *The End of Laissez-Faire*, which argued for a "command economy" managed and regulated by government, rather than the free market. Only three years before the Great Depression, the book could not have been more perfectly timed for those who blamed the unrestricted free market for the worst economic catastrophe ever to hit the United States.

Nixon took the United States off the gold standard, issued wage and price controls, and behaved as if he had invented Keynesian economics. Now, however, intellectuals of the right were openly praising the radical economic theories of Dr. Arthur Laffer and Jude Wanniski and their protégé, Congressman Jack Kemp. Crucial to their eventual ascendancy, these economic "supply-siders," as they were called, were given a bully pulpit by the conservative editor of the *Wall Street Journal*, Robert Bartley.

The supply-siders' new economics, often touted to the public in the widely influential op-ed pages of the paper, was in fact old Adam Smith economics. They argued that, to help Smith's "invisible hand" of self-interest perform its magic, policymakers must get rid of excessive regulations, tear down trade barriers, get government out of the business of managing the economy, and most important, slash taxes to the bone.

Not only would the boom-and-bust cycle of the American economy disappear, the thinking went, but also the result would be low unemployment, low inflation, and ever-increasing revenues for the federal treasury. They pushed their supply-side prescriptions, heretical at the time, with evangelical fervor. Reagan, predisposed to optimistic pro-growth ideas, happily jumped on the bandwagon.

Their argument of cutting taxes to increase revenue seemed counterintuitive, but no one could deny that the Keynesian economy of the 1970s had ground to a halt. Battle lines were slowly forming in the GOP between the old, sober, green-eyeshade-wearing budget balancers of the aged party and the new, young, tax-cutting life-is-great entrepreneurial let's-go-to-the-future supply-siders. Reagan had begun his politics with the old-timers, but he now saw that there was little in the way of hope to offer the American people. In late 1976 he changed his emphasis. The Gipper became a supply-sider and one of its marquee acolytes.

Being dour was simply out of character for Reagan. It was a critical decision in his formulation of projecting a philosophy of hope and opportunity. What better way to inspire people than to tell them they should keep more of their own money (to spend, to save, to do what they want with it), to reduce their dependence on government and to nourish their spirits through greater self-reliance? After multiple recessions since the end of World War II, it was time to try something else.

Coming on the heels of the election disasters of 1974 and 1976 for the Republicans, any healthy debate was a sign of a pulse, and this was a good thing. Still, 1978 was shaping up to be problematic as well. Of the

GOP's meager thirty-eight senators, seventeen would be up for reelection; while, for the Democrats, only sixteen of their sixty-two senators were facing reelection campaigns.[16] Legendary football coach Paul Bryant of Alabama used to refer to it as "schedule luck." The GOP wasn't getting any schedule luck, and it was deeply divided.

Comfortable in his own skin, Ronald Reagan celebrated the New Year quietly with Nancy, but before long, he was back on the hustings.

It was different, though.

Everywhere he went, everywhere he spoke, cops, flight attendants, housewives, doormen, farmers, cleaning women, waitresses, executives, Americans from all walks of life implored him to run for president . . . just one more time. "Governor, you've just got to do it!" Reagan heard this repeatedly. Both Peter Hannaford and Mike Deaver, Reagan's political consultants, took note of the dramatic uptick in grassroots support. Incoming mail to Reagan was also up markedly, with hundreds of invitations to speak, and many more letters from just plain Americans flooded the offices of Deaver and Hannaford Inc., for Helene von Damm, Reagan's secretary, to pore over.[17]

Now his outlook was more national, more international, and more comprehensive, and audiences loved hearing his views. They clamored for more, and Reagan and his team saw the opportunity that accompanied his continually growing popularity. "After the '76 campaign, once the dust had settled, it turned out we had about a million and a half dollars left in the bank," Hannaford explained years later in an oral history interview at the University of Virginia's Miller Center: "This is a huge contrast, because in mid-campaign we were poor as church mice."[18]

Reagan tapped his longtime aide Lyn Nofziger, who had proposed the operation, to run the newly formed Citizens for the Republic (CFTR). It would start out with a budget of $1.2 million (slightly less than Hannaford recalled, but still a hefty sum at the time), which was left over from the Citizens for Reagan committee from the '76 campaign, though about a third would sooner or later be returned to the FEC.[19] Nofziger later chortled that his nemesis, Reagan's former campaign manager John Sears, was unaware that the campaign had any money left over. The aching irony was not lost on Reaganites that, had Sears better managed the money at Citizens for Reagan, this surplus could have been used to win the nomination by more vociferously contesting the primaries of Ohio and New Jersey, where Ford won the vast bulk of delegates, albeit with narrow statewide majorities. It was just another item on a long laundry list of complaints filed by the conservatives against the brilliant but quirky and ideologically suspect Sears. Reagan was the only candidate in 1976 other than the thrifty senator from Washington, Henry Jackson, to end with a surplus.

Nofziger was to be paid $45,000 per year, and by 1977, his reputation in politics was already legendary. At a hotel in Washington, Nofziger and Paul Laxalt bumped into Henry Kissinger, who acidly said, "I remember you. You're the guy who chopped me up in the 1976 elections."[20] Reagan would serve as chairman of CFTR, and his old friend Senator Laxalt of Nevada, who had been the chairman of the campaign, served as head of the steering committee. Ed Meese became general counsel as well as representing Deaver and Hannaford.[21] CFTR began in better financial shape than would most other GOP and conservative committees, and everyone involved took note of Reagan's fund-raising prowess. A reception was held for the launch of CFTR, and Governor and Mrs. Reagan both attended and were in good spirits. The steering committee consisted of plenty of old friends, including Marty Anderson, Ernie Angelo, Dave Keene, Ed Meese, Paul Laxalt, and Senator Richard Schweiker, and a host of political operatives, business executives, and scholars.[22] "Citizens for the Republic would be a volunteer campaign training organization which would put out educational informational materials, too," Reagan adviser Hannaford said later at the Miller Center.[23]

One of Reagan's first forays in 1977 was, ironically, to speak at the annual GOP dinner in Rhode Island. Rhode Island had a New England liberal tradition, and he had failed to garner any of its delegates in 1976. During the speech, he took the opportunity to slap around Carter, who in one of his first acts as president pardoned the Vietnam War draft dodgers. Reagan also publicly defended Nixon, saying that the forth-coming memoirs by the disgraced president would help bring about a more "balanced view" of the thirty-seventh chief executive.[24]

Yet the stench of Watergate still hung over the GOP, as a business associate of Spiro Agnew's, Walter Dilbeck, was sentenced to prison for fil-ing false tax claims. Dilbeck had contributed $140,000 to Reagan's 1968 campaign for the GOP nomination but had then turned around and given $200,000 to Senator Hubert Humphrey's campaign for the 1972 Demo-cratic nomination.[25] It was scandal after scandal for the Republicans.

Carter's Administration started with an embarrassing hiccup when he was forced to withdraw his nomination of Theodore C. "Ted" Sorensen as director of the CIA. Conservative senators from both sides of the aisle objected when it became known that Sorensen had been a conscientious objector during the Korean War, had been less than honest with the Senate Intelligence Committee on how much President Kennedy knew about CIA attempts to assassinate Fidel Castro, and had taken classified documents from the White House after JFK's assassination on Novem-ber 22, 1963.[26] Sorensen bitterly told reporters, "[S]ome say that Carter helped put the noose around my neck but that's not true. He helped kick the chair out from underneath."[27] The president later, in February 1977, tapped retired admiral Stansfield Turner, a former classmate of his, as the next CIA head.[28] Carter was also faced with shoring up morale at the FBI, as it had declined precipitously since the death of J. Edgar Hoover in 1972 and one embarrassing revelation followed another in the firestorm that ensued.

Reagan jumped into the spat with both feet, telling Carter he should keep George H. W. Bush as head of the CIA, saying the president "could do worse" than reappointing Bush. Reagan was appalled that Carter would try to foist a Kennedy loyalist and onetime pacifist like Sorensen onto the national intelligence community.[29]

The problem was the new president was also politically tone-deaf. As House Speaker Tip O'Neill later wrote in his memoirs, *Man of the House*, Carter and company "didn't know much about Washington, but that didn't prevent them from being arrogant . . . [D]uring the Carter years, congressional Democrats often had the feeling that the White House was actually working against us."[30]

===

Part of Carter's attack on the existing order included cutting defense spending even more than Ford and Nixon had. This was in the face of an aggressive Soviet military buildup, according to a report issued by the Pentagon and the CIA. Called the National Intelligence Estimate (NIE), the report was designed to challenge existing notions about the Soviets and their intentions.[31] Departing director of the CIA George H. W. Bush defended the report and warned that the "Soviet Union appears to be driving more than ever toward military superiority" and that the Soviets believed they could win a limited nuclear war because the West would surrender, or that the Soviets could amass so great a conventional military advantage as simply to demand terms of surrender.[32] If even only a portion of the NIE report was true, it painted a frightening picture.

Outgoing defense secretary Donald H. Rumsfeld, in an interview, conceded that the information about the Kremlin had gone to Congress and then to the American people. There would be no cover-up by the Ford Administration on what the Russians were doing militarily. The American people were rightly alarmed, and despite Carter's proposed build-down, they wanted a buildup. Rumsfeld told reporters, "The Soviet Union today is clearly militarily stronger and busier . . . The Soviets continue to press ahead with . . . land-based ballistic missiles and submarine-launched ballistic missiles. The scope of these programs is unprecedented."[33] Carter's new secretary of defense, Harold Brown, was an establishmentarian bean counter who carried out Carter's directive to slash the military budget.

Moral equivalency was the order of the day, as the liberal Brookings Institution published a report that documented that the United States had deployed the military 215 times for political effect since the end of World War II, while the Soviets had done so on only 115 occasions. The study excluded the Korean War and the Vietnam War, and by inference suggested that America had engaged in "gunboat diplomacy" far more often than Moscow. Of the six postwar presidents leading up to Carter, JFK led the way with 13.4 "incidents per year" of saber rattling. Harry Truman, Richard Nixon, and Gerald Ford each had fewer than 5.0 incidents per year.[34]

Leonid Brezhnev was consolidating his power in Moscow, forcing out any potential rival in the Communist Party—which meant that Kremlin photo retouch experts would be forced to work overtime, airbrushing the fallen rivals from photos of official Soviet functions. Jokes about the retouched photos aside, Brezhnev could now spend less time looking over his shoulder and more time plotting Soviet expansionism, which was bad news for America and the West.

While no one snickered about the Soviets' military, everyone did about their propaganda, especially when it came to agriculture. Once again, the Soviets announced "record" grain harvests, and once again, they bought millions of tons of wheat from the United States—using credits provided by the American government, naturally. "The Soviet Agriculture Minister, Valentin Mesyats, said . . . 'the sugar beet crop . . . was below target . . .'" Oh, yes, potatoes, too, well, they were below the targets as well.[35] The status quo of self-deception had ruled in Russia since the fall of the czar. On the possibility of failure, Mesyats boasted, "There can never be a catastrophe. Our socialist economy ensures that."[36]

Testing the new American president, Brezhnev reasserted the status quo in the Warsaw Pact countries, such as Czechoslovakia, where a renewed crackdown was under way and reform leaders, including playwright Václav Havel, were arrested. There would be further moves against democratic reformers inside the Evil Empire.

Meanwhile, the man who would soon bring about the Soviet Union's destruction was rallying the remnants of conservatism to an optimistic vision of a confident and prosperous America. In a memorable speech delivered to the fourth annual meeting of the Conservative Political Action Conference (CPAC), in Washington, DC, in February 1977, Reagan elaborated on his theme of "a new and revitalized second party, raising a banner of no pale pastels, but bold colors," which he had first spoken of just two years before to the same audience.[37]

The earlier speech had been a launching pad for his 1976 challenge to Ford. This follow-up speech would become the launching pad for a complete rebirth of the conservative movement and of his own challenge to the GOP establishment and Carter in 1980.

Aptly titled "The New Republican Party," Reagan's speech would outshine all others, redefining the political platform. In it, he said:

> The time has come to see if it is possible to present a program of action based on political principle that can attract those interested in the so-called "social" issues and those interested in "economic" issues . . . In short, isn't it possible to combine the two major segments of contemporary American conservatism into one politically effective whole? . . . What I envision is not simply a melding together of the two branches of American conservatism into a temporary uneasy alliance, but the creation of a new, lasting majority.[38]

He then laid out a plan of action based upon the notion that it was the common man, not the captain of industry, upon whom the greatness of America was based. Free markets operating by free individuals, he said, were the best way to make the economy work. Communism, he went on, was "the absolute enemy of human freedom," a term that accurately summarized "the ugly reality captured so unforgettably in the writings of Alexander Solzhenitsyn."[39] This was perhaps a particular reference to former president Ford's refusal to see the gulag survivor, much to the chagrin of many in America. It was a smart reference, and immediately emphasized how different Reagan actually was from the one-term president.

Reagan gave a tip of the hat to good old American common sense as well, with his memorable sense of humor. "When a conservative says it is bad for the government to spend more than it takes in, he is simply showing the same common sense that tells him to come in out of the rain."[40] Calling conservatism "the antithesis of the kind of ideological fanaticism that has brought so much horror and destruction to the world," Reagan noted that it simply encapsulates "the common sense and common decency of ordinary men and women, working out their own lives in their own way."[41]

Reagan also noted the (appropriate) white elephant in the room, saying that the GOP must pull in and appeal to black voters. "Look, we offer principles that black Americans can, and do, support," including jobs, education, individual freedom, and liberty.[42] No identity politics was necessary.

This was just the kind of message the long-suffering conservatives in attendance had been longing to hear. Finally, a speech proposing the future of a reformed and potent Republican Party. It could move past the scandals of Watergate, Nixon, Spiro Agnew, and Ford.

Stan Evans, chairman of the American Conservative Union, one of the principal sponsors of the conference, told reporters, "there is a better chance for reshaping the Republican Party in a conservative mold" in the absence of Ford and Rockefeller and that the GOP was "open" for the taking. *National Review* publisher Bill Rusher dissented, saying of the GOP that "it has no more chance of controlling this country than the women's club of Newport, R.I." Rusher had written a groundbreaking book several years earlier that advocated a new conservative party, drawn from both the Democrats and the Republicans, and he was still invested in the idea.[43]

━━━━━━━━━━

Chicago's judiciary was jumping on people as Reagan's famed "Welfare Queen," Linda Taylor, was finally convicted of using multiple aliases and bilking the taxpayers out of thousands of dollars.[44] Reagan had made much of the woman in the 1976 campaign as an example of the "waste, fraud and abuse" that the federal and state welfare agencies engaged in. It was much disputed at the time over exactly how much she stole. About the amount of her excesses, *Human Events*, Reagan's favorite weekly newspaper, said one thing, and some in the media said another. However, the *Washington Post* account verified the conservative's charges about the woman, stating that she'd stolen over $150,000, had twenty-six aliases, three Social Security numbers, addresses at thirty different locations around the city, and "owned a portfolio of stocks and bonds under various names and a garage full of autos including a Cadillac, Lincoln and a Chevy wagon." She also had several dead husbands and was about to head off to Hawaii, presumably to avoid the last bit of the winter of 1977. All her ill-gotten goods were courtesy of the U.S. taxpayer. "Prosecutors say there is no category of public aid—welfare payments, rent subsidies, medical reimbursements, food stamps, transportation allowances, child care expenses, survivors' benefits—that Taylor had neglected 'to apply' for." The *Post* renamed her "The Chutzpa Queen."[45]

Welfare and its attendant costs and the wide range of abuses were very much on the minds of conservative reformers. "Between 1968 and 1976, according to the Congressional Research Service, the benefits rose from $16.1 billion to $62.9 billion—an increase of 289 percent."[46] There were nine major welfare programs and numerous minor ones being run by the government, and none was coordinated with the others. Enforcement by Washington to prevent waste, fraud, and abuse was nonexistent. The problem became generically known as the "welfare mess."[47] From the federal government alone, a welfare recipient could legally get somewhere around $7,000 to $8,000 per year in benefits. Adding state and local benefits on top of this, someone in America could make a good living in 1977 for just loafing. This did not create resentment among the people making $50,000 or more a year, though they clucked their tongues at the fraud, but it did create deep resentment among those blue-collar laborers making $10,000 per year, that is, the core Democratic voter, and they knew the "welfare mess" had been created by liberals in their own party.[48] This building resentment would be significant to the GOP and Reagan's eventual advantage.

Washington was still trying to figure out the Carter clan, including the new First Lady, Rosalynn, who "scandalized" the town when she went with the president to the theater one night wearing the same outfit she'd worn throughout the day. She began her time in the White House cautiously enough, but slowly she began to speak out more, as on the Equal Rights Amendment, and was encouraged by her husband to take on more public issues. The notion that Carter was sharing power with his wife was more grist for the Washington mill, just as it had been when Betty Ford ruled the roost in her husband's White House. In fact, Mrs. Carter made a seven-nation diplomatic tour through Latin America later in the year, meeting with not spouses, but prime ministers and presidents.

Early into his presidency, Carter took a dramatic step to his populist left when he announced his support for universal voting registration, public financing of House and Senate races, and abolishing the Electoral College. Liberals had long championed these initiatives, as the bigger states tended to be Democratic and eliminating the College would reduce the influence of the smaller, more conservative states. Of course, public financing, combined with the perks incumbents used in their bids to be reelected (staff, franking privileges, free travel home, and ready access to the media) could virtually guarantee Democratic dominance of Washington for time immemorial. Some compliant Republicans, including several liberal senators and House members, fell hook, line, and sinker for the bait of "reform." So, too, did the new chairman of the Republican Party, Bill Brock, who curiously said Carter's proposal was "a Republican concept."[49]

The "New Right," however, as the activist conservatives were becoming known, saw the Carter proposal as nothing more than a naked grab for power, and they were already plotting to defeat it. Reagan also saw the power grab for what it was and said as much—namely, that if anyone in the GOP supported it, they were engaging in "a sort of suicide pact."[50]

━━━━━━

Though the 1976 battle between Ford and Reagan was long over, moderates and conservatives were still fighting over it. Congressman Paul "Pete" McCloskey of California, a liberal Republican, had campaigned aggressively against Reagan the year before in New Hampshire and, a year later, was taking credit for Reagan's narrow loss. Jim Roberts, then executive director of the American Conservative Union, took to task McCloskey and the liberals in the party in a column in the *Washington Post*. "The disabling weakness of the 'progressive' Republicans is that theirs is an ersatz liberalism—devoid in most cases of the deep conviction and enthusiasm of the genuine article in the Democratic camp." Of the role of liberals in the Republican Party, Roberts ominously concluded, "Given the precipitous decline of the GOP, their services may not be necessary much longer."[51]

━━━━━━

After Carter was sworn in, Gerald Ford jetted off to California with only two things on his mind: golf that afternoon with Arnold Palmer in the Bing Crosby Pro-Am Tournament, and Ronald Reagan. He told a reporter, after dining on shrimp and steak and a couple of dry martinis on the 707 that had been designated *Air Force One* when he was president, that he was game for another try at the White House in 1980 and that "I don't want anyone to preempt the Republican presidential position."[52] No one needed to guess whom Ford did not want to "preempt" 1980.

Bill Brock had barely settled into his fourth-floor office at the RNC when the issue of the Panama Canal Treaties raised its controversial head yet again. Who was to own the canal, the United States (as stated by the 1903 Hay-Bunau-Varilla Treaty) or Panama? The treaties leaned toward the latter, promising full control of the waterway to the Central American country. During the fall campaign, Ford had backed away from supporting the treaties, as Reagan had pounded Ford and Henry Kissinger mercilessly over it in the primaries.

Senator Jesse Helms of North Carolina called Brock and suggested that Reagan lead the charge on behalf of the party against the treaties. According to columnists Robert Novak and Rowland Evans, Helms also suggested to Brock that he appeal to the networks for "equal time" to respond to Carter's "contemplated fireside chat on the Panama Canal negotiations," and that Reagan act as the party's spokesman.[53] Brock refused to give Reagan such a powerful platform, although the request for equal time intrigued him.

Helms and a group of elected officials that included Senator Strom Thurmond filed suit in DC federal court to block negotiations on the Canal Treaties. Helms told the Associated Press that he'd just returned from a trip to four South American countries, including Brazil and Argentina, and that "top officials in all four countries told him they opposed relinquishing . . . US control over the canal."[54] Later that year, and unbeknownst to Helms, Brock considered funding a tour by Ford and Kissinger to, in fact, support the treaties.

What angered conservatives even more was that they learned of this after Brock had asked Reagan, and Reagan had agreed, to sign a direct-mail fund-raising appeal for the party that opposed the treaties! Millions of dollars cascaded into the RNC as a result, but Brock would not let one dime be used by Reagan. Reagan and Laxalt were furious with Brock over his bait-and-switch ploy.

Carter sent a signal to the Soviets on what his policies on national defense might be when he selected the dovish Paul C. Warnke to head the Arms Control and Disarmament Agency. Carter had previously selected another dove, Cyrus Vance, a former official in the Johnson Administration, as his secretary of state. "Arms control advocates are eager to see the . . . agency once again play a leading role in pushing for arms limitations. Many think their cause was weakened in recent years because outgoing director Fred C. Ikle was less enthusiastic about cutbacks than Secretary of State Henry A. Kissinger," said the *Washington Post*.[55]

Columnists Rowland Evans and Bob Novak scooped the competition when they revealed that Carter was backing away from another campaign pledge to meet with Soviet dissident and Nobel winner Alexander Solzhenitsyn. Carter had defended Russian critics of the Communist regime, including Andrei Sakharov, and Moscow had let the White House know of its displeasure, putting Carter on notice.

Reagan also put Carter on notice, as he blasted the president's proposed cuts in defense spending, in light of "the new reality presented to us by the Russians."[56] He confined his criticisms of Carter to national defense and foreign policy, but this included defending the CIA and once again calling on Carter to refuse to hand over the Panama Canal to the Panamanians. His support of Solzhenitsyn was well known.

———

Reagan was born in a time before lame superficialities and sound bites, when politicians yearned to be great rhetoricians, and any public speaker worth his salt read his Cicero and knew his *McGuffey Readers*. In the early days of the twentieth century, political communication was considered a form of entertainment, and woe to the politician who could not unleash a stem-winder, taunt his opponent, appear sagacious, or quote extensively from the Ancient Greeks and Romans, the Founding Fathers, and the Bible.

Foreign policy dominated much of the debate in Washington for the balance of 1977. Another issue of vital interest to Reagan and conservatives was the U.S. government's stance toward Taiwan. They resented deeply President Nixon's steps toward abandoning the island nation after Nixon began his famous dance with Red China. His Shanghai Communiqué of 1972 did not cease recognition of Taiwan, but it granted public recognition of China, the first time this had happened since Mao led the Communists to power and the pro-Western, National (Kuomintang) government was forced to flee to the island of Taiwan. Reagan and other anti-Communists later learned, to their horror, that Ford had told mainland China that he was in favor of breaking diplomatic relations with the pro-Western Taiwan in favor of recognition of China.[57] The people of Taiwan could only scratch their heads. They were pro-American, they were anti-Communist, and they had a free-market economy. So why were they on the chopping block? Reagan and Taiwan's friends in America were furious at Ford.

What was compelling was how much Reagan figured into nearly every policy and political discussion, both public and private, beginning in 1977. Joe Kraft of the *Washington Post* noted this phenomenon on the "two-China policy" by writing that "delivering these implicit commitments proved politically difficult—especially after Ronald Reagan and the right wing began to challenge President Ford for the nomination."[58]

Carter was also considering the normalization of relations with Cuba, just as Ford had wanted to do during his presidency, but Reagan made this an issue in the campaign and Ford had to back off. Now Carter would have to look over his shoulder, wondering how much political damage he might suffer if he tried to follow Ford's course. That the Administration had sent one of its top diplomats to seek out private citizen Reagan was a testament to their respect for his powers.[59]

The debate over Carter's ideology was coming to a close as far as conservatives were concerned. Most were now in agreement with Reagan and conservative columnist James J. Kilpatrick, who rhetorically wrote, "Is Carter a liberal? Do fish swim? Do dogs bark? Do birds go tweet-tweet-tweet? To the extent Our Leader evidences any political philosophy at all, he is about as liberal as they come."[60]

The Bear in the Room

"We are now free of that inordinate fear of communism"

M iddle age," Ronald Reagan joshed, "is when you're faced with two temptations . . . and you choose the one that will get you home by 9:30." He told the joke to a laughing crowd on February 6, 1977, on his sixty-sixth birthday, in the Virginia home of his and Nancy's old friend Nancy Reynolds.[1]

The event was well attended by former staffers, reporters, and political associates. Included was Reagan's former running mate, Senator Dick Schweiker, along with his wife, Claire.[2] Others who attended included John Sears, Reagan's controversial 1976 campaign manager; Senator Paul Laxalt; Jim Lake, Reagan's former press secretary; and heavyweight reporters such as Tom Pettit of NBC; David Broder, Lou Cannon, and Mary McGrory of the *Washington Post*; and Frank Reynolds of ABC. Also in attendance was a young navy pilot who had only several years before been released from six years of brutal torture in the "Hanoi Hilton," Commander John McCain III. The Reagans were fond of the young hero whose father, retired admiral John McCain II, was an ideological soul mate of Reagan's, especially on foreign policy matters.

Starting in 1977, a conservative consensus, one that had begun to develop in the 1950s with conservative theorist Frank Meyer's brilliant "fusionism," congealed. Up until that time, conservatives had wondered how the economic right, which tended to be libertarian in its view, could work with the traditional, social right, which tended to be more static in its outlook. Meyer correctly theorized that the two could coalesce for two reasons. First, both camps were fervently opposed to communism: the social right because of the -ism's atheistic nature; and the economic right because Karl Marx's theories were antithetical to the free market. Second, both groups shared a minimalist government philosophy. Big government, with its regulations and taxes, was a threat to the business classes, but also to the family, because of the menace it posed to the home, community, and religion. By 1977 the conservatives' antigovernment arguments had come into full flower, as hundreds of examples of abuse, waste, fraud, and silliness proliferated. Conservatives also saw the government as the direct threat to the family—regulations, tax policy, interference, removing religion from civil society. The list goes on.

One favorite poster child for the conservatives was the Occupational Safety and Health Administration (OSHA), which for many was nothing more than a four-letter word. It was the butt of jokes, as it did ridiculous things, such as producing twelve-page reports on the "proper construction of portable wood ladders." The newly elected GOP senator from Wyoming, Malcolm Wallop, effectively used the nonsensical regulations of the agency to his benefit in his 1976 campaign. OSHA had issued a new requirement that agricultural workers, including ranchers, must have a portable toilet facility "within a five-minute walk of each employee's place of work."[3]

The *Washington Post* continued, "The proposal became an instant laughingstock in farm and ranch states . . . He [Wallop] ran a television commercial in which a . . . cowboy was saddling up and riding off to roundup, leading a pack mule and muttering about the omnipresent federal government. 'I think the government is going too far,' the cowboy fumed as he moved off and the mule came into view. Tied to the mule's side was a toilet."[4] Wallop's commercial was a small but vitally important example of how conservatives were beginning to develop winning arguments—and also a sense of humor, which for years they'd been accused of lacking. Even the *Washington Post* editorially attacked OSHA's "bureaucratic paternalism and pettiness."[5]

Reagan did at least one radio commentary blasting OSHA, which included praising an effort by the American Conservative Union to "Stop OSHA."[6]

―――――

As 1977 progressed, Reagan was leaving little to chance. He was unsure about 1980, but he was sure he didn't want to leave the nomination for Gerald Ford to filch, just as Ford was sure he didn't want to hand it over to Reagan unmolested. However, Reagan did far more stumping in 1977 than Ford. Sometimes his staff had to hold him back, such as when he wanted to campaign for the GOP candidate for an open congressional seat in Minnesota, but he was overruled. Evans and Novak said he was "more active nationally than any other Republican politician," but they cited "the enormous odds against him." About his prospects in 1980, they added that "the liabilities are profound."[7]

Still, the perceptive duo found that "critics infuriate Reagan. Anger at his critics, indeed, is just one more sign that Reagan . . . is nourishing a presidential ambition greater than any he had leading up to 1976. At 69, he would be the oldest man ever nominated for President; William Henry Harrison was 68 when he was sworn in and he soon died of pneumonia."[8] The "age issue," as it became known in shorthand by conservatives and Reagan insiders, was becoming more and more a commonly expressed objection to his running a third time for president.

Meanwhile, a group of conservatives, led by *National Review* publisher Bill Rusher, announced the formation of a "shadow government" that would monitor and comment on the actions and policies of the Carter Administration and the various government agencies. Shadow governments are common in parliamentary systems, but this was a novel idea in America, suggested by both Ford and Reagan several months earlier.

Some (and only some) of the members of this "shadow government" were knowledgeable, such as the "shadow secretary of health, education and welfare" Robert Carleson. Carleson had been the welfare director at the Department of Health, Education, and Welfare (HEW) under Richard Nixon, and had directed a state-based welfare agency under Reagan in California.[9] Indeed, Carter's White House was gearing up a plan that would vastly increase the size and scope (and cost) of the welfare state, guaranteeing thousands of dollars, with no strings attached, to the indigent.

The Republican Party itself was too constricted by ideological differences actually to carry out Ford's and Reagan's idea, but since conservatives were far more harmonious and freer to act, the option was one they could exploit. Again, the niceties of party politics such as those in the GOP did not hamstring conservatives. Liberal Republicans made a last-gasp effort to prove they were still relevant with the formation of the New Leadership Fund. A project of the left-leaning Ripon Society, its goal was ostensibly to recruit more moderates into the party.[10] It did not get far.

Conservatives were eager to admonish the liberal Republicans publicly. And if the heresies continued, the offending officeholder might just face an unexpectedly nasty primary, courtesy of the New Right. The latter's growing base of contributors, increasingly middle-class, Catholic, suburban, rural, and evangelical, would prove to be more loyal to the signers of the conservative direct-mail packages than any GOP officeholder, save someone such as conservative senator Jesse Helms of North Carolina.

The New Right also adopted populist tones by challenging the existing order rather than defending it. In this way, it further distinguished itself from the Republican Party. In a special election for a House seat in Minnesota, conservatives raised thousands of dollars and helped produce a moral victory in the form of Arlan Stangeland. He surprisingly won the seat formerly occupied by Democrat Bob Bergland, who had vacated it to join the Carter cabinet as secretary of agriculture. Stangeland had been the district leader for Howard Phillips's Conservative Caucus, a public policy lobbying group.[11]

An alphabet soup of conservative groups was springing forth, all run by dedicated young New Rightists. These included the National

Conservative Political Action Committee (NCPAC, pronounced "Nick-Pack"), the Fund for a Conservative Majority, and the Conservative Victory Fund, along with the older Young Americans for Freedom and its umbrella organization, the Young America's Foundation. New Right leader Howard Phillips told reporters that his organization was "engaged in guerrilla warfare on 435 different fronts," a sly reference to the numbers in the U.S. House, and would eventually hold training seminars in hundreds of locations across the country.[12] If these conservatives were missing anything, it was probably the need for a bit more levity and the ability to laugh at themselves.

Reagan's and Ford's differences continued unabated, virtually over everything. Things came to yet another head in mid-April 1977, when Ford addressed a meeting of California state GOP leaders, telling them that America is "a moderate nation of moderate people . . . a contest within our ranks to prove who is the purer of ideology will not attract the American people."[13] Ford's words were unmistakably aimed at the "extremist" Reagan.

One hundred miles up the Golden State's staggeringly beautiful coastline, Reagan spoke to the conservative California Republican Assembly in scenic Santa Barbara and fired a warning shot across the bow of the GOP, saying, "We hope that those who are in control of the national party apparatus will never force us to look elsewhere . . . to advance the cause of conservatism." He'd also met in private with RNC chairman Bill Brock, and later told people that he was becoming troubled by "Brock's reluctance to criticize President Carter."[14] More and more, reporters and Republicans were slowly concluding that a Ford-Reagan II, like the sequel to *Jaws* or *Rocky*, was in the offing, and were further concerned that this sequel might be as uninspiring as its cinematic counterparts. In fact, both Ford and Reagan were a year or more away from any actual decision to run for the GOP nod in 1980. Yet, as both were highly competitive and each resentful of the other, neither was willing to cede any territory. The skirmishes between the two would drag on, like trench warfare.

———————————

In Washington, at the Republican National Committee, Bill Brock had decided that Reagan's warning shot was enough, and he decided to bring in as his new campaign director Charles Black. Black was an attorney, but more important for Brock, his political skills were well regarded and his conservative credentials were impeccable. The "bad boy" of the New Right, he had worked for Jesse Helms, Helms's Congressional Club, Reagan's '76 challenge to Ford, and as a high official with NCPAC.

Black, a tall, reed-thin southerner and chain smoker with slicked-back jet-black hair, was generally liked and respected in the GOP. He was to Brock what the interpreter Squanto had been to the Pilgrims and the Massachusetts natives: he spoke the languages of both sides of the intraparty disputes and could serve as chief negotiator and arbitrator between those suspicious camps. He could also speak the language of the national media, and this was a bonus for Brock and his wary Republicans. Black brought Neal Peden, an attractive blond divorcée from Mississippi, along as his assistant. Together they had worked for Reagan in 1976, and Peden remembered that the Ford people at the RNC, when she moved over there, were "brusque."[15]

Brock, though he had a moderately conservative voting record in the Senate, had concerned himself too much with the "broaden the base" crowd of moderates, who hated conservatives in the party, and now he needed an intermediary like Black. Black also was one of the only men in politics besides Reagan who could get away with wearing a brown suit. He was replacing another experienced operative, Eddie Mahe, whose task had been even worse than Black's: along with his boss, then-RNC chairman George H. W. Bush, he had had to hold together the party during the Watergate crisis, which nearly destroyed the Republican Party.

Mahe was breaking out on his own to start a political consulting firm, and he hoped to cash in some of the chips he'd collected over the past three years from battle-weary Republicans. He did not look the part of a rescuer: sporting a crew cut, clear-rimmed eyeglasses, short-sleeved white shirts with thirty-seven pens in the pocket, and a no-nonsense demeanor, he bore a striking resemblance to a NASA scientist.

Another NCPAC alumnus was also getting ready to walk through the front door of 310 First Street, SE. While Black was welcomed with flower petals and praises of hosannas by some, others inside the RNC would have liked to pour boiling oil down on the incoming chairman of the Young Republicans, Roger J. Stone. Ironically, these two young men were friends, but they were also like the brothers in *East of Eden*, one the mother's pet and the other the black sheep of the GOP family.

That's not to say that Stone was not good at his business. He was one of the more talented and effective young operatives in either party. His dry wit and knowledge of ethnic and urban policy kept him at the cutting edge of politics in the 1970s—with an eye toward sartorial splendor, he had a flair for high fashion and high jinks—but he derived a daredevil's thrill from edging close (too close for some) to the ethical line in campaigns. Yet he kept coming back, and seemed to have more lives than a cat.[16]

He "was waist-deep in 1972 Nixon-campaign dirty tricks. Stone was never accused of any crime but was cited by the Senate Watergate Committee for sabotage and espionage," wrote Evans and Novak. At the Nixon organization, Stone "assigned an agent named Michael Mc-Minoway to infiltrate Democratic presidential campaigns in several states." He also was directed, in the 1972 New Hampshire primary, to personally "contribute" two hundred dollars to the campaign of Pete McCloskey, a liberal Republican congressman from California, who was making a quixotic challenge to Nixon.[17] He had also been accused of authoring the infamous "Canuck" letter, signed by "a Paul Morrison of Deerfield Beach, Fla.,"[18] which caused Democratic front-runner Senator Ed Muskie of Maine to break down crying in front of the offices of the *Manchester Union Leader* in 1972—a seminal moment that many observers say cost Muskie the nomination by making him look weak. The letter had accused Muskie and his wife of using slurs against French

Canadians, and the *Union Leader*'s William Loeb editorialized against Muskie over the false accusations.

Stone's boarding party of Young Republicans, previously a kind of Junior League for pimply GOP corporate wannabes, was being aided and funded by Richard Viguerie. Stone, despite his controversial and colorful young career, was somewhat bulletproof, as he'd been the youth director at Citizens for Reagan the previous year, so Brock knew that to oppose Stone's election might bring down the wrath of Lyn Nofziger and other members of the Reagan inner circle. Still, when Stone was elected president of the YRs, his defeated opponent led a walkout of more than three hundred delegates to protest his triumph. Backroom finagling was charged, but Stone's pithy response was that those objections were being raised by "Eastern liberals who are too old to be Young Republicans."[19]

Of the coordination between Stone and Viguerie, the columnists concluded, "Even Republicans too insensitive to worry about the impact of Roger Stone's past are getting worried about his future."[20]

Stone, like Black, had also been in on the initial planning of NCPAC, along with two other young Reaganites, Frank Donatelli and John T. "Terry" Dolan. Donatelli was a proud "conservative diaper baby," having been involved in Young Americans for Freedom at an early age, a group with whom the Gipper already had a long, warm relationship. He ran independent campaigns in support of Reagan in 1976, and he was also known, with his fondness for polyester leisure suits, as one of the worst-dressed men in Washington. In 1973, while an undergraduate at Georgetown, Dolan had run for national chairman of the College Republicans but lost to Karl Rove, who in fact had been out of college for several years, never having received a degree. Dolan was a closeted homosexual, and some accused his opponents of using this to discredit him.[21] Rove was a journeyman in Republican politics, working around the country. He was not a "movement conservative" like Dolan, Stone,

Black, and Donatelli; he was a Republican, aligned with then-RNC chairman George H. W. Bush, and less interested in ideology than the game of politics. The Dolan-Rove conservative-moderate fight in 1973 was a metaphor for the larger brawls to come inside the GOP, including the eventual battle between Ford and Reagan in 1976 and, later, between Reagan and the GOP establishment.

The NCPAC foursome were also devoted fans of the Three Stooges, could repeat the dialogue of their favorite shorts word for word, and could impressively reproduce the sounds and sight gags of Moe, Larry, and Curly. Several years earlier, they'd founded the Moe Howard Society. One quiet day, Stone tracked down Mr. Stooge himself, Moe Howard, at a retirement home in New Jersey, and spoke to him at length by phone. He was delighted to discover that Howard was a huge fan of Reagan's and was thrilled to hear from these kids. Correspondence was exchanged for a time between Howard and his youthful fans.[22]

In early 1975 the foursome created its novel political venture. With the new, though bumbling, Federal Election Commission in place, political action committees (PACs) began to flourish, aided by favorable mail rates and creative conservative appeals by Viguerie and his smaller competitors, such as Bruce Eberle, who had handled the direct mail for Citizens for Reagan in 1976.

GOP organizations such as the RNC, the National Republican Congressional Committee (NRCC), and the NRSC attracted more moderate types than the New Right organizations. Temperamentally and culturally, many official party operatives were uncomfortable with the strong ideology and tough tactics of the movement conservatives. Each would make jokes about the others behind their backs, but the groups were slowly unifying—first, along political lines, by their common angst over Jimmy Carter; and later, along ideological lines, when the Republicans saw that issues and ideas counted and that conservative messages against high taxes, communism, and big government, and in favor of traditional family values, could produce winning coalitions in races across the country.

The party was doing its best to right itself. John Sears told the *Wall Street Journal* that "the Republican Party, by itself, can't beat anybody."[23] That was true enough in the early days of 1977, but the Republicans could beat Carter if the Democrats helped the GOP beat their own president and if the Republicans got their act together.

Republicans were feeling the heat and thus seeing the light shone on them by conservatives as Bill Brock took Reagan's criticism to heart and rolled out a strong attack on President Carter's new energy proposal.[24] Carter's plan included additional taxes and fees on oil companies and on the gasoline used by Americans, as well as additional taxes on larger cars, including the family station wagon. It was a "regressive" tax that would hurt mostly those on the middle and lower rungs of the economic ladder.

White House press secretary Jody Powell dismissed concerns about the impact of the tax, telling reporters, "This is a way of making real the future problem," and then astonishingly called it a "psychological weapon" aimed at the American people.[25] Carter also once toyed with the idea of repealing the home mortgage deduction as a way to raise federal revenue, but this notion was junked, as it would have incurred the wrath of millions of homeowners and thrown the economy into a depression of epic proportions.

For the first time in a long time, the Republicans were on offense, not defense. It felt good. Also for the first time in a long time, they had the support of the American people.[26] Indeed, Carter's own pollster, Patrick Caddell, had numbers that showed that over 60 percent of the American people opposed the Carter gas tax. Carter had taken to the airwaves on April 18, 1977, and warned, in apocalyptic language, of "a problem unprecedented in our history."[27] He also told the American people that if they failed to heed him, their country might collapse. He was coming on like an Old Testament prophet when he said the "crisis" was the "moral equivalent of war," but his fire-and-brimstone rhetoric was not sitting well with the American citizenry. He used words such as *unpleasant, overwhelmed, selfish, shrinking,* and *catastrophe.*[28] He also carried with him ten principles of energy conservation. If Moses had his Ten Commandments, then it could be said that, just like Moses, Carter came down from Mount White House and scared the bejesus out of people.

Carter was, ironically, espousing what many liberals, especially in academia, thought at the time about America and the future. Alvin Toffler, in his book *Future Shock,* wrung his hands for the benefit of readers. Barry Commoner, in his 1971 runaway success, *The Closing Circle,* warned that the world faced one of three unpleasant options: nuclear war, overpopulation, or mass starvation—or all three combined. Both books were on the syllabus of every self-respecting left-wing political science professor on every American campus. Apocalyptic pronouncements, mostly from the left, dominated much of the 1970s.

At the very time Carter was imbuing himself with academic dystopia, Reagan was meeting with the scholars at the Hoover Institution on War, Revolution, and Peace, at Stanford University. There, he was listening to men and women talk about lifting the American economy, lifting up the American people, and how and why it was possible to defeat Soviet communism. All in all, Reagan was learning how and why the glass was half full, while Carter was learning how and why it was half empty.

Paradoxically, Carter had come to his views via his religion, and while many academics were atheist, they nonetheless shared his grim outlook. It was stylish to be dour about the future and paranoid about the past, and conspiracy theorists and other snake oil salesmen made a fine living speaking at colleges and universities, spinning myths and half-truths to impressionable young Americans. Invariably, the culprits were business interests, elected officials, racists, the U.S. government, the CIA, the catchall military-industrial complex, and the "greedy" citizens of America.

"Deacon" Carter also took on western U.S. interests, as he attempted to halt dozens of federal water projects designed to bring the much-needed liquid onto arid lands for cattle and farming. Senator Paul Laxalt attended an unpleasant meeting with Carter on the matter and came away distinctly unimpressed. He called his old friend Ron Reagan and excitedly told him, "I just met a one-term president."[29]

Carter was pilloried in newspapers across "Reagan Country." Conservative columnist George F. Will penned, "As Napoleon noted, 'He who says he has made war without making mistakes has made but little war.' And Carter has opened a multi-front war on numerous American practices, habits and mores."[30] The president also took on the home building and home mortgage industry when he proposed that every new home sold have mandatory insulation. If the home did not comply, home buyers would not be eligible for mortgages issued from federally insured banks, which covered just about every lending institution in America.[31]

The RNC launched the first of what would be hundreds of successful campaign management schools, which organized all the particulars of Campaigning 101. The initial sessions were eight weeks long and covered everything from public speaking, press release composition, and the care and feeding of the media, to fund-raising, dealing with the candidate's spouse (frequently a dicey problem), and the correct filing of contribution reports.

The initial school attracted a who's who of guest Republican speakers, including John Sears; F. Clifton White, the legendary conductor of the 1964 Goldwater Express; and Richard Wirthlin, Reagan's pollster since 1966. Thousands of GOP operatives got their start, and made life-long contacts and friends, by attending these schools.[32] Participants paid $35 per course, which included a thick binder with tips and advice on advertising, FEC laws, and fund-raising.

Meanwhile, Bill Brock was moving ahead, bringing the somnolent GOP into the second half of the twentieth century. Brock also convinced the national committee to commit more than $7 million to the 1977 off-year elections in New Jersey, Virginia, and elsewhere.[33] The party had been, since the days of Dwight Eisenhower and through Nixon and

Ford, a "presidential party," but it could not compete at the local level with the Democrats, who had the added advantage of labor unions on their side. So it needed all the new ideas and operatives it could muster. By the summer of 1977, affiliation with the Grand Old Party was still only 20 percent, according to the Gallup organization, which had polled more than nine thousand Americans and found just under 50 percent calling themselves "Democrats," with the balance unaffiliated with either party.[34]

The Republican Party had to fend off GOP members of Congress who tried to foist their otherwise unemployable kids off on the committee. For years, the joke around Washington had been "How many people work at the RNC? Oh, about half of them." These various "Muffys," "Buffys," "Skips," and other offspring, including professional protégés, took up time, space, and money at the RNC, and it was this "preppie" deadwood that Brock was determined to clean out.

━━━━━

Opposition to a Carter election reform bill, which called for the elimination of the Electoral College, minimal identification for voting, and Election Day voter registration, was growing. House GOP minority leader John Rhodes of Arizona had to backtrack. He'd once called Carter's plan a "good idea," but after hearing from angry conservatives, he flipped and called it "a dangerous step toward the federalization of the entire election process."[35]

"Rhodes has been under unremitting attack from party conservatives for his original stand. A recent cartoon in the newsletter of the Citizens for the Republic, the Ronald Reagan political action committee, shows a bleeding GOP elephant that has been stabbed by a figure labeled 'Rhodes' and 'universal registration.' The caption says: 'et tu, Rhode.'"[36]

Brock, too, was forced by conservatives to retreat from an earlier comment he'd made supporting elements of Carter's election reform proposals, after the 162 members of the Republican National Committee devised a statement denouncing the Carter plan. Their appetite for fighting the Democrats was becoming whetted.

At a congressional staff briefing on the Carter plan, sponsored by the American Conservative Union, conservatives walloped the proposals. A young attorney, John Bolton, who had written a book defending the 1939 Hatch Act, which prohibited federal employees from participating in political activities, said the plan would make federal employees subject to coercion from labor unions. Jeffrey Bell, who had worked for Reagan in many capacities, attacked Carter's plan to eliminate the Electoral College. Another former Reagan staffer, Don Devine, a political science professor at the University of Maryland, participated as well. The American Conservative Union promised a full-scale lobbying effort against Carter's reforms.[37]

The Equal Rights Amendment was also in retreat, losing in the Florida legislature. The constitutional amendment, the brainchild of feminists, had been launched in the early seventies with great fanfare, and initially zoomed through many states. A grassroots opposition, led mainly by New Right stalwart Phyllis Schlafly and her legion of "little old ladies in tennis shoes," rallied opposition against the ERA and ground the amendment to a halt. It was just one more example of the growing grassroots might of the conservative movement.[38]

Across the Atlantic, conservatism was staging its comeback in Britain as well, in the surprising form of a grocer's daughter whose public voice got squeaky if she became nervous: Margaret Hilda Thatcher. Though both an attorney and a research chemist by education, she had entered Parliament and risen to the position of leader of the loyal opposition, which at the time were the Tories. She and Reagan shared core principles and values, especially anticommunism. They first met in 1975, and a friendship sprang up instantly.

Her first and only meeting with President Carter went less well. "Carter had a disastrous meeting with her during which she sought to lecture him on foreign policy." The president gave a standing order to his staff never again to schedule a meeting with the minority leaders of foreign governments.[39]

———————

Government was, to many Americans, becoming sillier and sillier. HEW cut off funding for a public school in Connecticut because it sponsored "father-son" and "mother-daughter" light fare evening events. Moreover, horror of horrors, the school had all-boy and all-girl choirs. Someone at the agency finally came to his senses and overturned the ruling, but not before the damage had been done.[40] Every day, bureaucrats were making Reagan's point about government, although the Supreme Court ruled that "corporal punishment" in public schools was not a federal case and that its regulation should be left to the states. Supporters of Federalism were slap happy.

The economy, though, was in fighting shape as it grew at an impressive 5.2 percent in the first quarter of 1977. But inflation reared its ugly head in both March and April, rising over 1 percent each month, for an annualized rate of 13.1 percent, double what it had been the previous year.[41]

Carter also proposed increasing Social Security taxes by raising the "wage base." In 1976 the base had been $16,500, meaning everyone paid into Social Security on incomes up to but not exceeding this level. If enacted, the hike would pull billions out of the economy and into the Social Security Trust Fund, which was beginning to show signs of shrinkage, mostly because pandering politicians kept tinkering with it and giving out benefits that were not intended under the original act.[42]

One person who qualified for Social Security, Gerald Ford, certainly did not act his age. He swam each day, skied, and played tennis and lots of golf. At the end of May 1977 he took time off from his pursuits to go back to Washington for his first visit since leaving there in January. He paid a courtesy call on President Carter, met with a gaggle of reporters, and "came out against 'any old timer' taking control of the Republican Party."[43] He also "offered a veiled criticism of Ronald Reagan" when he told guests at a thousand-dollar-a-plate dinner, "We can divide and conquer ourselves," which was interpreted to point the finger of blame at Reagan.[44]

Ford's verbal slight about Reagan's age (though he himself was only two years younger) and a more overt statement that Reagan should not seek the 1980 GOP nomination were not overlooked in the conservative's camp—especially by Nancy Reagan. She suffered no fools and tolerated no criticism of her husband, particularly by political opponents such as Ford.

In 1977, Nancy was a petite, five-foot-four dynamo of fierce energy, and the most important person in the world to "Ronnie" (and vice versa). Reagan himself was uncertain about 1980, but Mrs. Reagan was not. She may have been even more competitive than her husband, which is saying a lot, because Mike Deaver once called Reagan "the most competitive son of a bitch who ever lived."[45]

Mervin Field, of the respected Field Poll in California, said that it would not serve the interests of the Republicans if either Ford or Reagan stayed in the limelight, "particularly if they continue to battle each other."[46]

What was on Reagan's mind when he sat down with the media in 1977 was not Ford or 1980, but big business and how he saw that the airlines, the trucking industry, and others had a far too cozy relationship with big government. Indeed, these businesses were using government, via their lobbyists, to write rules that squeezed out small competitors. "Ronald Reagan says big business is willing too often to sacrifice free enterprise principles for a comfortable subsidy, tariff or regulation that stifle competition."[47] Reagan, the constitutional-libertarian-federalist-populist was chary about the concentration of power by either government or big business, and when they worked together—well, watch out, little guy. A joke going around at the time went, "When the elephants are fighting, the mice are at risk, but when the elephants are making love, the mice aren't much safer."

Carter, meanwhile, seemed to be softening across the board when it came to the Kremlin. Initially, he'd indicated a willingness to scrap Richard Nixon's détente in favor of a tougher posture, but six months into his Administration, his resolve was waning. Evans and Novak noted that with this new policy, "Moscow may feel politically safe in trying to liqui-date the humiliating agitation for civil and human rights."[48] Indeed, the Soviets issued a new constitution in June that expressly prohibited free speech and any form of opposition inside the Soviet Union, the Warsaw Pact, and the Baltics, while dissidents (who *Pravda* labeled "outcasts" and "moral perverts") were being rounded up anew and shipped off to gulags, labor camps, and "psychiatric hospitals."[49]

The Soviets' newest tough stance did not go unnoticed in the White House. On May 22, 1977, Carter gave the commencement address at Notre Dame. It focused exclusively on foreign policy and human rights while affirming his new support for détente without any strings. Then, before the twelve thousand guests, the president received an honorary de-gree that cited his "generous amnesty to those caught in the misadventure of Vietnam" and for criticism of the "wasteful lifestyle" of the average American.[50] Yet the real news came when he astonishingly told the grad-uating class, the American people, and the world that "we are now free of that inordinate fear of communism which once led us to embrace any dictator who joined us in that fear."[51] From the gulags to the Berlin Wall, from the Pentagon to the British Parliament, anti-Communists across the globe were thunderstruck, dumbstruck, gape-mouthed. Reagan and his conservatives were appalled. President Carter had just canceled the prin-ciple underlying America's foreign policy since the end of World War II, a policy based on an educated fear of the Soviets, given their history of mass murder, betrayal, corruption, treachery, and villainy. The president's remarks caught much of the country off guard and left it in disbelief.

Though Carter received a standing ovation at the elite Catholic school that day, Senator Patrick Moynihan, an anti-Communist Demo-crat from New York, and a Catholic, immediately blasted him. Carterologists tried to figure out what made the guy tick, yet even Carterites didn't know themselves what made Jimmy go.[52]

Significantly, at the very time Carter was telling the American people they had nothing to fear from the Soviets, Reagan was warning them that "the most important reality facing us . . . is that Western influence is shrinking while that of the Soviet Union is expanding."[53]

⸻

Conservatives such as Reagan thought they had Carter figured out: He was more interested in bashing right-wing regimes than left-wing regimes. At the same time that he was denouncing the white-minority, oppressive governments in Rhodesia and South Africa, he was calling for normalization of relations with the repressive Communist regimes in China, Cuba, and Vietnam. And as he moved ahead with plans to withdraw American ground forces from South Korea, it seemed to Reagan as if Carter were retreating on all fronts. The chief of staff of U.S. Armed Forces in South Korea, Major General John K. Singlaub, publicly denounced Carter's plan as one that invited war. Carter, furious, ordered him out of the country and back to Washington, to be reassigned. Singlaub became an instant folk hero to many, and Reagan called Carter's handling of the general "disgraceful."[54] The Singlaub disagreement between Carter and Reagan was the first of many collisions they would have over the next four years.

"The whole Carter handling of the matter shows petulance . . . The President is just plain wrong," Reagan told a GOP crowd in Buffalo.[55] Singlaub was an American hero writ large. Highly decorated, wounded twice in combat, he had parachuted behind enemy lines into Vichy France to work with the French Resistance to help prepare for the D-day invasion. He was also one of the founders of the CIA.

Carter also canceled the B-1 bomber, citing it as costly, technically flawed, and unnecessary.[56] The man who at the beginning of 1977 sounded as if he were going to stand tough and not be rolled by the Soviets, especially on human rights, now seemed more interested in currying their favor than their respect. Early in his Administration, Carter had admirably spoken against the repression of human rights inside the Soviet Union, but when the Russians growled and let him know of their displeasure, he backed off.

On the first day of spring 1977, before a gathering of a trade union in Moscow, Leonid Brezhnev made a major address in which he threw down the gauntlet. He told Carter in no uncertain terms that the Communists would brook no interference in the internal affairs in the Soviet Union and its client states, and that how and to what extent the Communists "handled" their dissidents was the business of the Kremlin and the Kremlin only.

A generous portion of saber rattling was mixed in with Brezhnev's white-hot blast at Carter. The Soviet leader's comments included direct threats to cancel nuclear disarmament talks and warnings of increased trouble in the Middle East if Carter continued to champion the cause of "subversive . . . agents of imperialism" in the Soviet Union.[57] Brezhnev was testing the new American president, just as Khrushchev had done with John F. Kennedy in Vienna in 1961. Kennedy had been badly shaken in their first meeting, and the Soviet dictator came away unimpressed with the young American president, but Kennedy grew quickly in office, and Khrushchev learned, to his misfortune, one year later how much he had underestimated JFK.

The day after Brezhnev's address, the Carter White House weakly responded by merely asking Congress to fund more transmitters for Radio Free Europe and Radio Liberty broadcasts into the Iron Curtin countries. Jody Powell read a carefully crafted statement that indicated to the world that Brezhnev's message had been heard loud and clear by Carter. Now the policy of the United States was "to negotiate seriously in good faith to reduce the burden of the arms race . . . to reduce the real threat of nuclear destruction that now endangers all the people of the world," and human rights in the Soviet Union thereafter disappeared from Carter's foreign policy.[58] A stunned Carter went mostly silent and avoided talk of human rights in the Soviet Union for the next four years. Brezhnev had won round one.

In the last week of March 1977, Secretary of State Cyrus Vance went to Moscow for a meeting on arms control, did not mention human rights, and left with his tail between his legs. The *Washington Post* printed a scalding editorial: "Not since Napoleon had there been a more disorderly retreat from Moscow than that conducted by Secretary of State Vance last weekend."[59] Carter was now publicly supported by Kissinger. The former secretary of state made clear that human rights were, in his opinion, only a small part of the foreign affairs equation.[60] Only a couple of days before, he had implied just the opposite.

Newspapers across the globe filled with stories taking the anti-American side, detailing how the Soviets were displeased with the arms proposals made by Carter, and how they supposedly wanted real arms reductions and real peace. The crush of world opinion helped to put Carter further on the defensive.

The Soviets had taken the measure of Carter and gambled that they could push him around and get the concessions they wanted from the United States. On June 30, Carter asked for a meeting with Brezhnev so they might "get to know one another," but he was humiliatingly rebuffed. His desire for a meeting itself was used as an effective negotiating tool by the Kremlin.[61] The Soviets had very neatly maneuvered him off human rights, a subject for which the president held the moral high ground, and onto nuclear arms, where they could make their "moral equivalency" arguments.

Emboldened, around this time, the Soviets filed a formal protest with the U.S. embassy in Moscow over the showing of two movies inside embassy grounds, *Doctor Zhivago* and *The Girl from Petrovka*, which they labeled "slanderous." The United States finally drew the line and told the Soviets to go review their own movies.[62]

In the late spring of 1977, one of the most underestimated, but hugely successful movies ever made was released. *Star Wars* was an enormous hit in large part because there were no moral ambiguities in the film—

everybody knew who the good guys were and who the bad guys were. It touched the heroic psyche deeply ingrained in American culture, one that Hollywood had been trying to beat out of the country for years, with one movie after another in which casual murder, gratuitous sex, and the triumph of evil were so easily depicted.

The movie was seen as a metaphor for the Cold War. The "Rebels" were fighting "to restore freedom to the galaxy," while the "Empire" was defending the status quo of collectivism. Individuality was repressed in the Empire while it was celebrated within the "Resistance." Before Reagan ever called the Soviets an "Evil Empire," George Lucas, no doubt inadvertently, made sure that everybody in America equated his bad guys with the ones in Moscow. Everybody loved the movie, especially free-spirited conservatives. Unlike most other movies of the era, the bad guys were not illustrious, and the good guys, against all odds, won in the end.

The movie was a welcome diversion for Americans in light of the continuing sad saga of the once-mighty domestic automobile industry. Cars coming out of Detroit were awful, and the Ford Motor Company had to recall cars made since 1974 because it had forgotten to put in adequate holes for lubrication; thus the engines tended to explode, especially in the record heat of the summer of 1977.[63] Also, across the baked country, problems with water supplies and energy demands rippled for several months.

———

In June, Reagan and former senator Eugene McCarthy put in an unusual joint appearance at a gathering of business executives in New Jersey. Both had started out as garden-variety liberals, as Reagan used to quip of himself, but both moved to the right over the years, albeit Reagan more quickly. The two libertarians, with their usual wit and style, zinged both Carter and the vicissitudes of government. It was clear these two old Irishmen enjoyed each other. McCarthy got off the best line, however, when he said, "We can only go 55 miles per hour but the speedometers still go up to 120 miles per hour. The speedometer is a little like Jimmy Carter's lust; it's there, but it wouldn't hurt anybody."[64]

Despite the buffeting Carter was undergoing, his approval numbers continued to hold in the mid-60s.[65] Yet, in another example of the Republicans slowly coming back from the dead, the RNC was given a half-hour time slot by NBC to take on the Carter energy proposal, which even the president's own aides cryptically called his "sky-is-falling speech."[66] At the same time, a study from the Massachusetts Institute of Technology predicted massive oil shortages by the 1980s. The project director of the 291-page book, *Energy: Global Prospects, 1985–2000*, labeled the situation a "time bomb," and the dire forecast was welcomed at the White House.[67] The CIA also issued a report that warned of impending doom, and that, by 1990, gasoline prices would be as high as two dollars a gallon.[68]

Among those featured in the Republican rebuttal on NBC were Bill Brock, Congressman Jack Kemp, Missouri senator John Danforth, and Reagan. While Carter was preaching skepticism and despair for the future, the Republicans, surprisingly, found themselves defending the future.[69] ABC also had granted the Republicans a half hour of free airtime to respond to Carter's three televised broadcasts between April 18 and April 22.[70]

It was a briar patch that Republicans were delighted to be thrown into. Since the days of the New Deal, they'd been the party of fiscal discipline and gloom and doom. "The Republican party showed off its new image as friend of the common man . . . [T]he telecast moved from gas stations to farms to factories to mines . . . questioning workers and consumers," said the *New York Times*.[71] The Carter energy plan called for the establishment of a Department of Energy, conservation, and dramatic new taxes on oil, gas, and coal production. The Republicans attacked the new taxes, not because they'd fallen in love with big business, but because the money taken by the government, they argued, meant less that could be used in the development of new sources of energy, including alternative fuels.[72] And, of course, the new taxes would be passed along to the consumer. Reagan was featured in a four-minute segment standing on a California hillside. The GOP's televised response was a tour de force, and Brock was widely praised for the party's perspicacity.[73]

Despite the Carter Administration's general drift toward the left, it was finding itself under attack from liberals within the party. George McGovern told the Americans for Democratic Action, "We all seem mesmerized by image, taken by symbol. We seem to count the ratings of polls far more than the content of policy."[74] McGovern had been soundly smashed by Nixon in 1972, but he was a man of deep principles, committed to his liberal philosophy.

The words of McGovern carried some weight with the American public, and conservatives might just as easily have invoked them. Curiously, a memo had leaked out of Carter's White House, written by his pollster Pat Caddell, that advocated Carter moving to the right, and stating that the liberal elements of the party were the "easiest to dominate" because they had a "willingness to accept any status quo that provides . . . power and patronage." McGovern and other liberals were not pleased.[75]

Carter had come to Washington indebted to no one and to no institution, but that also meant that since he was indebted to no party or its elders, they would eventually become freer to criticize and oppose him, even publicly. It was a dangerous time for the Democratic Party. Carter was not invested in it, and the party was not invested in him. He came into office with enormous advantages, with his party controlling all the marbles. It wasn't long before his sanctimonious inflexibility caused him to throw it all away.

In the broadcast RNC rebuttal, an effervescent Reagan said of Carter's leadership, "We're not running out of anything except confidence in ourselves." The production of the GOP show was approximately $25,000, but it generated millions in publicity and laid down an important marker in the party's slow comeback.[76]

Of his speeches, Reagan did not charge for all of them, especially not those made for conservative groups or an occasional GOP candidate. Yet, not everybody was on the Reagan bandwagon, particularly liberals in the party. Congressman John Anderson, a nominal Republican from Illinois, pronounced Reagan too old on CBS's *Face the Nation.* Then, perhaps thinking of his own ambitions, he told reporters he believed that the GOP would nominate a "moderate" in 1980.[77]

Another GOP aspirant, the minority leader of the U.S. Senate, Howard Baker of Tennessee, was beginning to stir in his own presidential desires. Baker was one of the most affable men in politics, but he was only a fair public speaker and he had never focused on a significant issue he could use as a platform from which to seek the GOP nomination. Inside Washington, Baker was seen as a formidable GOP contender, but outside the city, of those who did know him, most didn't know much about him. Also, he was going to wait until after the 1978 elections to begin putting together his campaign team. His staff was telling reporters that Reagan was "too old," and they were dismissive of the other candidates as well.[78]

In late August 1977, it was announced that a bill-signing ceremony would take place on September 7 for the Panama Canal Treaties. President Carter himself called Reagan to see if he could persuade him to change his mind, now that some modifications to the treaties had been made. The Carter Administration had wooed Reagan heavily, including a one-hour private briefing with the chief treaty negotiators, special ambassadors Ellsworth Bunker and Sol M. Linowitz.[79]

The pact would still have to pass muster in the U.S. Senate, where the Constitution required that all treaties receive the votes of two-thirds of the senators—in this case, sixty-seven. It was here and in the subsequent weeks and months where a bloody fight would take place. The issue would also be fought over in the news media, across the country, in American Legion halls, over morning coffee, across backyard fences, and within families. No one in America seemed to be without an opinion on the Panama Canal Treaties in 1977 and 1978.

Some in the Pentagon said the canal was "virtually indefensible," but other military brass believed that America had no choice but to defend it. The issue would also be wrangled over in the inner sanctums of the Republican Party. Ford was sick of Reagan's "We built it! We paid for it! It's ours! And we're going to keep it!"[80] He was just as strongly supportive of the treaties as Reagan was opposed to them.

When Reagan had first raised the issue of a "giveaway" of the Panama Canal in the New Hampshire primary in 1976, voters scratched their heads, and the media by and large ignored it. Yet, for weeks, he kept pounding away, day after day, to audiences large and small, and it finally began to sink in and take hold. As Bob Dole quipped, "It's Reagan's issue. Reagan found it, he built it, and he's going to keep it!"[81]

Part of what animated Americans' passion over the canal included the feeling at the time that America's day had passed. Their country had lost a war, had lost so much since the early 1960s. Another motivating factor was that older Americans were heavily and psychologically invested in the canal. Since their childhood, they had been told the tale of its heroic construction and that it was one of the Seven Wonders of the World due to American ingenuity, know-how, and can-do spirit. The building of the canal, to them, represented the finest their country could achieve.

"Are we to be pushed around even by a tinhorn dictator?" lamented a *Wall Street Journal* editorial examining the potency of Reagan's populist argument. Though the newspaper editorially supported the treaties, it rapped Carter hard for giving away more than Ford had originally offered.[82] Like everybody else in America, editorialists and journalists also had opinions about the treaties. A few were calm and reasonable, but most, on both sides, ranted and raved.

Reagan had sown the seeds of this dispute in 1976. It remained to be seen what sort of whirlwind he would reap in 1977—and beyond.

Canal Zone Defense

"The best issue handed a political party in recent times."

Ronald Reagan was out and about as always, giving speeches, flying hither and yon, recording radio commentaries on everything from snail darters to national health insurance, visas and Social Security. In his November 29, 1977, commentary, he spoke favorably of an interesting proposal from Michigan state senator Jack Welborn, a Republican: rather than institute a national day of mourning for Dr. Martin Luther King Jr. for public employees, the government should take the money it cost to pay these individuals and set up scholarships for "disadvantaged students . . . a living memorial" to King.[1] Everybody sought his advice, and he was having a heigh-ho time.

Reagan still hadn't gotten completely used to the idea of flying, having sworn off it for over twenty years after two harrowing flights during his salad days in Hollywood. One flight, in a storm from Catalina to Los Angeles, and another, in a snowstorm over Chicago on a war bond drive, had put him off flying completely. When he did start flying again, in the mid-1960s, new aides would observe him hunched over before takeoff and would invariably nudge Dennis LeBlanc, a former California Highway Patrolman who'd been assigned to Governor Reagan's detail, and ask, "What's he doing?" The answer would come back: "Praying."

On commercial flights, Reagan always flew first class, partly because he liked to greet people as they got on the plane, because of his size, and because he could spread out the papers from his bulging briefcase without inconveniencing anyone. When he flew on smaller charters, he would travel only with Peter Hannaford, Mike Deaver, or Dennis LeBlanc. LeBlanc, the young aide who worked often with Reagan, especially at the Ranch, looked like he was born for the out-of-doors: tall, ruggedly handsome, but also surprisingly shy. Both the Reagans adored him.

In addition, Reagan would often be accompanied by one or two national political reporters, such as Lou Cannon or Bob Novak. Novak recalled that the first time he really got to know Reagan was on a flight in the South, in a terrific thunderstorm, in which they talked alone for over four hours. Reagan nervously told jokes, mostly off-color ones, and opened up to Novak about his life, including the personal details of his divorce from Jane Wyman and his years in between marriages. "I tried to go to bed with every starlet in Hollywood [pause and smile]. And I damn near succeeded. That was before I met Nancy."[2]

———————

In the fall of 1977, voluminous mail and phone calls from angry conservatives to the White House regarding the Panama Canal Treaties were running heavily opposed, as Jimmy Carter's press secretary, the sociable Jody Powell, told the White House Press Corps. One reporter asked the chain-smoking Powell—he dragged on Salem cigarettes as if he were eating breakfast—if opposition to the treaties was running at 87 percent, but Powell acknowledged only that the figure "was a little high."[3]

For conservatives, the Panama Canal Treaties struck a very raw nerve indeed. In their impassioned if not enraged view on the matter, Carter was another quisling and the treaties the culmination of decades of traitorous appeasement that began with Alger Hiss's address to the United Nations in 1945 calling for internationalization of the canal. Some analysts, such as Stefan Leader of the DC-based Center for Defense Information, believed that the Canal Treaties were a great fight for the conservatives.[4]

Reagan was pounding the Carter Administration over the treaties in his daily five-minute radio broadcasts. Because Reagan had to go to producer Harry O'Connor's studio at the corner of Hollywood and Vine in Los Angeles and record as many as fifteen five-minute commentaries in one session, it took a lot of organization and the better part of a day on Reagan's part. Due to the technology of the era, the radio commentaries went out to hundreds of radio stations by mail on reel-to-reel tapes or were impressed on record albums, which were then discarded after the station broadcast Reagan's observations. It also meant picking the right issues so that when the commentaries were broadcast, the issue was still one of national discussion. Reagan hit Carter over the "secrecy" of the negotiations; he hit the *Los Angeles Times* for accusing him of taking on the issue so as to "endear myself to right wing Republicans," and he pointed out the fallacy of the paper's charge, as he claimed that large majorities of "Americans are opposed to giving up the canal."[5] The way the debate over the Canal Treaties was dragging out, no one thought it was going anywhere for a while.

When describing Americans and their senators' attitudes over the contentious canal issue, media reports at the time used terms such as *skeptical*. To be sure, a year before, when Reagan was in hot pursuit of Gerald Ford, thirty-eight senators had signed a resolution circulated by Senator Strom Thurmond of South Carolina opposing the treaties. Senator Jesse Helms of North Carolina released the results of a nationwide poll that showed a vast majority of Americans supported keeping the canal.[6]

Most polling had the two sides of the issue far closer, though, with opponents still ahead of supporters. Senator Barry Goldwater, who had supported the treaties the year before, as he'd supported Gerald Ford over Reagan for the nomination, did a U-turn and was now opposed to them. Carter was surprisingly conciliatory toward the opponents, at least in public, saying that many were "well-educated, very patriotic Americans" and that his job was to "cajole, not to pressure" them into seeing things his way.[7]

This was the fight with Carter that conservatives everywhere, especially Reagan, were spoiling for. The "wigs were on the green," and this would be the first opportunity to draw Carter out into the open.

Turning over the canal represented every frustration they felt as patriotic Americans. For the past several months, there had been several partisan skirmishes over the energy bill and Carter's voter reform proposals, and mounting tensions over his growing softness toward the Kremlin.

Carter's proposal to establish a new Department of Energy that consolidated a number of energy-related federal responsibilities—independent agencies, including the Federal Power Commission, the Federal Energy Administration, and the Energy Research and Development Administration, as well as energy-related responsibilities from the Department of the Interior and other Cabinet departments[8]—under one roof, made it through Congress easily, and he signed the bill into law in August. The department began operations on October 1, but the energy policy proposals Carter had outlined in his much-maligned "Moral Equivalent of War" speech (the MEOW remarks in April) languished on Capitol Hill.

Yet another of the government's alphabet soup welfare agencies, the Community Services Administration, became the *bête noire* of conservatives. A congressional report detailed how the $700 million-per-year agency had no real reason for existence, how public monies were pilfered without prosecution, how employees were hired and never showed up for work, and how meetings were called for which there was no reason.[9]

Congress itself was under scrutiny, as it would, for the first time in history, cost a billion dollars per year just to operate. The lavish salaries and perquisites Congress had voted for itself over the years would become another weapon that conservatives would use against Capitol Hill. The goodies included free flowers, free haircuts, free flights home, and free booze.

After their instrumental role in forcing Richard Nixon from office, a man they had long despised, the media were self-aware of their power as the Fourth Estate. Journalists, once considered rather marginal if not shady characters, took on an air of self-importance. Every kid out of journalism school wanted to become another Woodward or Bernstein. The goal no longer was accuracy and fairness; it was fame, money, and Pulitzers. Consequently, good men got their reputations tarnished and minor scandals were blown out of proportion.

By the time Carter was firmly ensconced in power, the press also wanted to show that it was even-handed, that it could attack Democrats as ferociously as Republicans. The sad result: Bert Lance, a financial sharpie who maybe did (or didn't) cut a few corners in his life as a banker, suddenly found himself as Public Enemy Number One on the front pages of America's newspapers.

The Carter White House had circled the wagons around Lance. After all, he was a charter member of the "Georgia Mafia." Someone who did not get the memo was Carter's outspoken aide, Midge Costanza, who publicly called for Lance to step down. Still, the president's inability to "lance" the affair festered for so long that it robbed Carter of his most important asset, "his reputation for rectitude," as charges of cronyism filled Washington.[10]

During the Lance affair, Carter's public approval rating took a dive and was never to recover to its former heights. Carter's honeymoon was kaput—forever, as it would later turn out.

Carter's Georgians grumbled and drank among themselves, convinced that the snooty sophisticates of Washington had it in for them. An awful pun was coined in the headline of William Safire's column, "Carter's Broken Lance."[11] Safire won a Pulitzer Prize for his musings on the Lance affair. More humorously, though, John Belushi and Dan Aykroyd played Carter and Lance on *Saturday Night Live* in a parody of the American Express "Do you know me?" ads.[12] Lance was finally forced to resign in late September, run out of town by a media mob that had tried and convicted him before he was ever charged. He was eventually cleared of all accusations, but his reputation was in ruins.

Other embarrassments were beginning to crop up when it became known that the head of Carter's anticigarette-smoking task force . . .

smoked cigarettes.[13] More significantly, another instance involved the revelation that of Carter's 488 presidential appointments, only 41 were black, less than 9 percent. This came on the heels of a black official with the Equal Employment Opportunity Commission (a holdover from the Nixon Administration whose term had expired) being ordered out of his office, and in turn calling Carter a "racist."[14] The charge was ludicrous, but this did not stop the Ku Klux Klan from holding a rally in the president's hometown of Plains, Georgia.[15]

Finally, to add another tick to Carter's failures, his election reform proposal died an ignoble death.[16] At the onset, it looked good on paper, but what the Carter folks failed to understand was that the Washington establishment, especially one dominated by Democrats, favored the status quo over uncertainty. Carter's proposal was, in fact, a threat to incumbent Democrats and their city machines. They didn't want just anybody walking in on Election Day, demanding a registration card, and then voting for whomever they wanted. Nor did they want federal campaign officials poking around their political machines in Chicago or Newark or New Orleans or Boston . . . or into their bank accounts, asking uncomfortable questions. Who needed that? And they certainly didn't want a publicly financed federal campaign system, one that would have given money to candidates of both parties, which would have undermined their carefully constructed machine. Why, for heaven's sake, would they want to upset their own apple cart? It was a staggering testament to Carter's naïveté.

———

After six months of detailed negotiations, the representatives of the United States and Panama finally resolved all the minute issues in the two separate treaties that together would be known as the Panama Canal Treaties. The first gave Panama control of the canal starting in 1999, more than two decades in the future. The second gave the United States the right to defend the canal if it determined that its neutrality had been compromised. A signing ceremony, part official duty, part "presumptive close" to pressure two-thirds of the Senate to vote yes in favor of the treaties when it exercised its constitutional duty to "advise and consent," was held on the White House Lawn on September 7. Representatives from twenty-six countries attended, as did President Ford and Henry Kissinger. The president of Panama, Omar Torrijos, a military strongman, had the audacity to quote Abraham Lincoln, lecturing opponents that "a statesman thinks of future generations, while a politician thinks of the coming election."[17]

At a White House dinner, eighteen heads of state attended and "close to half were dictators and several had a sordid background of human rights abuses."[18] Torrijos himself fell into both categories. Disregarded was the fact that, while at Notre Dame, Carter had harshly criticized America's past policies of reaching out to dictators to fight communism. He was now reaching out to dictators to help him give away the Panama Canal.

Reagan's and other conservatives' opposition to the treaties was not just about politics or American morale, or the fact that an American president, Theodore Roosevelt, had conceived and carried out the planning and building of one of the seven marvels of the modern world, which was completed in 1914 at a cost of more than twenty-five thousand lives. Or that an American scientist, Dr. Walter Reed, had come up with a vaccine against the dreaded yellow fever that saved the lives of thousands of Panamanians and Americans working on the canal. The canal was also a part of Americans' national consciousness. From grammar school on up, after the canal was built, it was held up in every public school in America as a wonder of technology and as one of the greatest accomplishments of the American people. (The French had attempted and failed in the 1880s to build it.)

Schoolchildren of the first half of the twentieth century were taught that America did not fail at things. A little boy by the name of Dutch, attending grade school in downstate Illinois, learned this as well. No wonder Reagan said he intended "to do everything he can to generate opposition to the treaty."[19]

Americans did not view themselves as imperialists; it was not part of the American character. It was deeply ingrained in the American experience that the canal was a good thing for the people of Panama, bringing economic opportunity and, with it, physical and mental sustenance.

It was also the practicality of the issue. A ship going from San Francisco to New York via Cape Horn at the tip of South America, in addition to having to negotiate ferocious seas, traveled fourteen thousand miles. But a ship going between the same two cities via the canal had to travel only six thousand miles, and through much calmer waters. It was also the height of the Cold War, and there was a real fear that Panama might "go Communist," like Torrijos's pal in Cuba, Fidel Castro, had done in 1959, which would have precluded the U.S. Navy's use of the canal. The navy each year moved thousands of tons of ships between its chief ports of operation in San Diego, San Francisco, and Pearl Harbor in the Pacific, and Norfolk in the Atlantic. All told, each year, fourteen hundred ships passed peacefully through the canal, a complex maze of several hundred miles through artificial lakes and locks that raised and lowered the water levels.

Along with the fear of Soviet control, there was also the genuine fear that Carter, very quickly, had "gone soft" on communism. In a short matter of months, Carter, who had talked tough about human rights inside the Soviet Union and seemed interested in discarding détente, had reversed his field. Years later, it came to light that a hard-drinking Democratic operative, Bob Beckel, on behalf of the Carter White House, essentially bought the necessary votes with federal contracts and walked around money and women to ensure passage. "Ronald Reagan's 1976 nomination campaign had poisoned public attitudes toward the treat[ies], and most national polls at the time showed most voters opposed to ratification," he later recalled.[20]

Carter Administration officials were so nervous about wavering domestic support for the president's foreign policy that a hastily called meeting was arranged with the Committee on the Present Danger, a group of anti-Communist Democrats and others who would one day become known as "neocons." The private group had first been formed in 1950 to support publicly the buildup of American military forces capable of defeating the Soviet Union, policies outlined in the now-famous but at the time top-secret National Security Council Report 68, authored by then–Assistant Secretary of State Paul Nitze and supported by Secretary of State Dean Acheson, both Democrats who served first in the Franklin D. Roosevelt Administration and subsequently under President Truman. By 1953, many of the group's members were working within the Eisenhower Administration, where they helped implement the policies they had supported, and the organization went into a two-decade period of inactivity.

Paul Nitze served as Lyndon Johnson's secretary of navy, and later as his secretary of defense. Several years later, Nitze and a group of like-minded intellectuals were part of "Team B," a group of outside experts commissioned by CIA director George Bush in 1975 that often produced grim reports on Soviet intentions contrasting sharply with the rosier papers being generated by the bureaucracy.[21]

In March 1976, several of these security hawks, and many who thought like them, resurrected the Committee on the Present Danger, with an eye toward influencing the presidential election that year, bringing to the attention of the public and the candidates the very real military threat the Soviet Union posed to the United States at the time. Carter was trying to reassure these skeptical intellectuals, and through them the American people, that the United States would have free and open access to the canal for eternity.

━━━━━━━

Reagan, with every fiber of his being, disagreed that America owed the Panamanians an apology.[22] As far as he was concerned, all America had done was bring progress, enlightenment, prosperity, and education to the people of Panama, while also aiding in their security for good measure.

The American Conservative Union had bought full-page newspaper ads, which shouted, "There is no Panama Canal! There is an American Canal in Panama!"[23] Senator James McClure, a staunch conservative from Idaho, was the first elected official out with a series of restrictions to the treaties, which included Panamanian forfeiture if their government became Communist, Panamanian support for an American brand of human rights, and the continued presence of U.S. military personnel. Reagan was public and vocal in his opposition to the treaties, but so, too, was Gerald Ford in his support, warning Americans that if their government did not turn over control, the failure to do so "could bring serious international consequences."[24] What consequences, Ford did not spell out.

Reagan spoke in front of the Senate about his opposition. When reporters asked Ford what he thought of this, he icily replied, "[E]very American citizen has the right to petition Congress. I certainly wouldn't want to give him any guidance."[25] The blood between the two men was bad, not thick.

Though Ford's support for the treaties was never in doubt, he called Carter at Camp David in mid-August to tell the president personally. Carter was courting other Republicans to support him on the issue, but so far, only a handful had gone over the wall. Indeed, Carter had initially called Reagan to solicit his support, but Reagan renewed his opposition and held firm, telling the president the alterations to the treaties had not changed enough to allay his concerns.[26]

The battle lines had hardened, with Carter on one side and Reagan on the other, when the Californian announced his formal opposition in a speech to the Young Americans for Freedom annual dinner in New York, which came as a surprise to absolutely nobody. The anti-Reagan smear was on. Carter's White House said Reagan had "helped to mobilize members of the John Birch Society, the Liberty Lobby, the National States Rights Party, and other extremist organizations" against the treaties.[27] The charge was made up out of whole cloth. Reagan had clearly irritated some pro-Carter supporters: a young man and woman were stopped just before they could throw pies in Reagan's face prior to the YAF dinner. At their convention, the "Yaffers," as they had become known, elected a new chairman, twenty-four-year-old John Buckley, a cousin of William F. Buckley Jr.

Reagan's opposition to the treaties was somewhat more measured than it was amid the heat of the 1976 campaign. He was frankly concerned that if the treaties went down to defeat in the Senate and riots broke out in Panama, he could be blamed for them. This did not prevent him from calling Carter's attempt to gain support "a medicine show, a wave of propaganda," though he was forced to admit he hadn't yet read the new language in the treaties.[28]

Supporters for the treaties scoffed at the canal's usefulness, saying that most ships were now too big for it, that its military value was dubious, and that America's continued control represented colonialism. Liberal columnists, such as James "Scotty" Reston, often jeered Reagan, such as when Reston wrote, "There is nothing more depressing in politics than old men crying for a world that is gone and risking wars young men would have to fight."[29] Opponents refuted the charges over usefulness, citing the commercial and naval vessels that used it each year. And since Panamanian military dictator Omar Torrijos was making threats against the United States if the canal were not turned over, wasn't this tantamount to knuckling under to duress? One Panamanian official issued a statement noting that his country would "adopt a course of violence" if the canal were not relinquished.[30]

In August, Republicans scored a big psychological victory with the win of Robert Livingston for an open congressional seat in New Orleans. The seat had long been in Democratic hands. The incumbent, who had defeated Livingston the year before, in 1976, was subsequently sentenced to prison for election fraud, and Livingston won on this second try. It was the third pickup in a special House election for the GOP in 1977, and was welcome news for Bill Brock and the national Republicans. Reagan had campaigned for Livingston, but their first conversation was more utilitarian than ideological. Reagan had flown in as the thirty-four-year-old Livingston waited nervously. When Reagan got off the plane and headed toward Livingston, the first words out of the Gipper's mouth were "Where's the bathroom?"[31]

Livingston had won the wealthier GOP conclaves, but also, surprisingly, had taken the blue-collar, Catholic and heavily Democratic regions of the district from whence he sprung. In addition, he had taken almost 30 percent of the black vote, which was unheard of for a Republican in 1977. One year earlier, he had received just 3 percent of that vote.

The name "Livingston" had a long and storied history in America, and Bob Livingston was a direct descendant of Phillip Livingston, a signer of the Constitution. What's more, his ancestors had long been leaders in Louisiana politics. Livingston, an attorney, was investigating organized crime for the state's attorney general's office. In a bawdy and corrupt town like New Orleans and in a state with a rich history of public gluttony, Livingston had seen it all—especially in politics. From Earl "I Ain't Crazy" Long, to David "KKK" Duke, to Edwin "Vote for the Crook . . . It's Important" Edwards, to Jim "Grassy Knoll" Garrison, to simply flamboyant Harry Connick Sr.—no state in America could rival the Pelican State for colorful characters. The state also had an official canine, a Catahoula leopard dog—whatever that was.

The Livingston coup in this Democratic district was seen as part of an encouraging trend for the GOP. His election was also the first manifestation of the "New Republican Party" Reagan had envisioned earlier that year. As Mrs. Reagan told a reporter, Ronnie's mission was to "revive the Republican Party."[32]

Next door, in Texas, two Republicans, one battle-tested and the other a downy-cheeked kid, were also working to help the GOP. James A. Baker III, who had righted Gerald Ford's delegate operations in the spring of 1976, was getting ready to run for state attorney general. He'd helped other men run for elective office and now he figured it was his turn. The other was George W. Bush, thirty-one, who was preparing to make his first foray into politics, seeking the congressional seat in Lubbock. Baker's campaign was to be handled by Frank Donatelli. Based on Reagan's overwhelming performance in the Texas primary the year before, and how close the presidential race between Carter and Ford had been there as well, Lone Star Republicans were hopeful of actually becoming a "two-party" state, but not in the Dallas or Houston high-society sense.

In the summer of 1977, it was disclosed that the Soviets had been inter-
cepting millions of public and private phone conversations from America
to anyone inside the Soviet Union. As it turned out, the Kremlin had
been doing this for years. America's Fourth Amendment was not something that troubled them. Richard Nixon and then Gerald Ford were
aware of this appalling action, but as the agents of détente, they were not
interested in offending the sensibilities of the Kremlin. Carter also did
nothing to stop the repugnant behavior.

The Soviets also continued to bombard the U.S. embassy in Moscow
with microwave transmissions to prevent CIA agents there from doing
the same in kind to the Soviets. Senator Daniel Moynihan, an anti-
Communist Democrat, told conservative columnist George F. Will that
the United States was engaged in a "disguised retreat" in the Cold War.[33]

Reagan was appalled that Carter had weakened so quickly in the
face of Soviet arrogance. Indeed, an explosive report outlining Soviet
advances was revealed by columnists Rowland Evans and Robert Novak
in the summer of 1977. The report was so hot that government officials
attempted to have its author, John M. Collins, fired.

Reading Evans and Novak's three- to four-times-per-week column
became *de rigueur* in official Washington, as the newshound duo were
the happy recipients of one leak after another out of the government.
One leaked document they reported was an astonishing account that
the Carter Administration would cede a portion of West Germany to
the Soviets if they invaded.[34] This column elicited an odd editorial response from the *Post* itself, not doubting the columnists, but doubting
why someone in the Administration would engage in "mischievous" behavior that could only undermine "Soviet planning."[35]

Because of the column, President Carter was compelled to reaffirm
American military support for West Germany. Months earlier, he had
had to deny a report in Evans and Novak asserting that he had told the
Pentagon to scrap the defenses of the West and prepare a plan, in the
face of thousands of Soviet missiles, using only two hundred submarine-

launched missiles.[36] The Carter White House came to dread waking up the mornings when their column ran.

―――――――

Hanging tough on the canal treaties was becoming of increasing importance for Carter because of the downward arc of his first year in office. The farmers had already rolled him on price supports in the face of huge surpluses and on additional payments for set-side acreage. His energy bill was being picked apart in Congress, and he seemed to waver, day to day, on topic after topic. "Almost any generalization about Mr. Carter is shattered instantly by some contradictory development. Advanced billing pictured him as a hard-nosed executive who'd go to the mat with Congress; then he compromises with congressional leaders on issue after issue." Even his friends said, "There's no consistent philosophy . . . to grab a hold of."[37]

He also was being held up to the usual Washington psychobabble inspection, which concluded that Carter was not a fun guy and that his southern roots made him insecure. A book by Robert Shogan, a diminutive but immensely perceptive journalist for the *Los Angeles Times*, hit the market in October, detailing the first one hundred days of the Carter Administration. Carter's White House had given Shogan unprecedented access, and the book, *Carter's First 100 Days: Promises to Keep*, painted a not very pretty picture of Carter or his people. They were potent but aimless in their goals. Shogan's portrayal was not entirely negative, as he also described Carter's long hours and conscientiousness.[38]

―――――――

Reagan and his opponents to the treaties were now beginning to get help from other folks. One surprising but happy development was the decision of Senator Bob Dole of Kansas to oppose the treaties if they were not heavily altered. Dole had been Ford's running mate, but he was also a conservative who relished a good fight. He would be an effective ally as he immediately proposed eight new amendments to the treaties, dealing with cutting the aid package in half, securing a guarantee from Panama that it would not interfere if America wanted to build another canal in Latin America, and a host of other proposals that would set the Administration back on its heels. The Carter Administration told Americans that, under the treaties, the United States would retain the right to defend the canal, but the Panamanian government embarrassingly disputed this assertion.

All it would take for Reagan's forces to defeat Carter and his treaties would be to land one-third of the Senate, a simple thirty-four senators. Tempers were running high over the fight, and Jesse Helms told reporters that his colleague Howard Baker was "squirming like a worm on a hot brick." Helms estimated that enough senators opposed the treaties, so Majority Leader Robert Byrd of West Virginia pushed the vote back until January, to give Carter a chance to regroup and mount a counteroffensive against Reagan.[39]

In late September, the Senate Foreign Relations Committee invited both Ford and Reagan to testify regarding the matter along with three former secretaries of state, including Henry Kissinger. In the meantime, a plebiscite on the treaties was voted on in Panama; and not surprisingly, it passed by better than two to one. Elections south of the border were considered more or less to be jokes, and the term *banana republic* sprang from this belief.

The Panamanian government had predicted the referendum would pass overwhelmingly, and no one doubted them for a moment, since all knew the vote was rigged. Much of the American media covered the pseudo-vote as if were a real event. Still, one had to wonder why some Panamanians would not want sovereign control of the canal along with a pile of money from U.S. taxpayers—and, as a bonus, the chance to humiliate the gringos running America.

On September 30 at their meeting in New Orleans, the Republican National Committee approved a resolution opposing the treaties. This was a rebuke of Ford, a blow to Bill Brock, and a win for Reagan. Brock had attempted to head off the resolution, to no avail. He was also attempting to navigate the choppy waters separating the two men. "Obviously we are not all united on this," he told the *New York Times*.[40]

The fight with Carter would be a fund-raising bonanza for the party. Anti-canal language was written into the party's platform the year before, in Kansas City, much of it instigated by Reagan supporter Jesse Helms. The California GOP went further in its resolutions at the RNC meeting until cooler heads prevailed. Their original draft labeled the treaties a "dishonorable surrender" and demanded the criminal trial of U.S. negotiator Sol Linowitz. The issue was clearly heating up across the nation.

Senator Paul Laxalt of Nevada spoke at the California GOP convention, held in the first weekend of October. Laxalt was understated, the epitome of class and character. He had friends on both sides of the aisle, but he was also a shrewd politician who could smell opportunity as he told reporters that the treaties were "the best issue handed a political party in recent times."[41] Ford and Reagan had also addressed the gathering, but it was noted that Reagan got a bigger crowd and "at least twice the applause given the former president."[42]

Carter's White House was worried, and lamented the fact that they could not match the growing firepower of the right in its opposition. "The biggest guns include former California Gov. Ronald Reagan and Sen. Barry Goldwater, R-Ariz., opposing the treaty at every opportunity, aided by groups like 'Reagan's Rough Riders.'"[43] Richard Viguerie estimated that his company alone raised more than $1 million for the anti-treaties campaign.[44] Other direct-mail experts were also raising untold dollars. The American Conservative Union began a television campaign with half-hour specials broadcast in targeted states, including Florida, Louisiana, and Texas. The ACU's initial budget was $40,000, but it raised much more and bought more television time as the campaign wore on. Featured in the show were Congressman Phil Crane, chairman of the ACU, Laxalt, Helms, and others, but Reagan did not appear, as the logistics, it was claimed, could not be worked out.[45] Some also suspected that Crane, thinking about the 1980 campaign, did not want Reagan to be the star of the show.

Inside the GOP, the list of presidential aspirants was slowly growing to include (besides Reagan, Ford, Dole, and Baker) Governor Jim Thompson, Senator Chuck Percy, Congressmen John Anderson and Phil Crane, all from Illinois; former treasury secretary John Connally of Texas; Senator Lowell Weicker of Connecticut; and one other Ivy Leaguer, a war hero, oilman, losing U.S. Senate candidate of 1964, former congressman, losing U.S. Senate candidate of 1970, former ambassador to the United Nations, former chairman of the RNC, former envoy to China, former head of the CIA, and all-around utility man for the party: George H. W. Bush of Massachusetts, Connecticut, Texas, Maine (in the summer), and Washington, DC.

Bush could have become vice president several times: once by Nixon in late 1973, when the beleaguered president was looking for a replacement for the opening created after Spiro Agnew resigned; then by Ford in 1974, who instead picked Nelson Rockefeller of New York; and finally in Kansas City in 1976, when Ford went for Dole. In fact, Bush was known to be exceedingly loyal, intelligent, utterly kind, and thoughtful, and had friends and family stretching from one end of the country to the other. "He'd been a bridesmaid too many times" but never the bride, Bush's manager, Jim Baker, told the *New York Times*.[46] It burned inside this extremely competitive man. Bush had been in the same room with presidents and world leaders, sizing them up, and had not found himself wanting. He and Jim Baker flew to California to pay a courtesy call on Reagan and tell him of Bush's intention to seek the 1980 nomination.[47] Bush was going to take matters into his own hands, like his nemesis John Connally.

Someone else taking matters into his own hands was former Minnesota senator Gene McCarthy. He authored an article bashing the Federal Election Commission, which was, surprisingly, published in the newsletter of Reagan's Citizens for the Republic.[48] McCarthy had come up as a traditional liberal, as befitting a politician from the North Star State, and though he'd emerged as the peace candidate in 1968, challenging

Lyndon Johnson in the New Hampshire primary, coming close, and driving Johnson out of the race, his thinking and his beliefs had evolved into a more nuanced libertarian philosophy.

With his rapier-like wit and putdown abilities, McCarthy was the Oscar Wilde of American politics. His article eviscerated the FEC for fining a losing congressional candidate $100 for failing to file a report with the commission for a campaign in which the candidate received no contributions and spent only $150 of his own money. The failed candidate was living in "an old soldier's home in Massachusetts."[49]

———

Pressure on the treaties and the Democrats was growing, and the fight took on a white-hot cast. At a meeting of the Southern Republican Conference in Orlando, Reagan angrily stormed, "Jimmy Carter tells us he is running a tight ship, but it doesn't make a hell of a lot of difference if the ship is sinking." He also told the Republicans gathered at an outdoor luau that over the past year, they had won three out of four special elections for Congress and fifteen seats in the state legislatures while losing only three. "If that doesn't represent a tide, I don't know what does."[50]

Senator Howard Baker would not commit either way on the treaties, and southern conservatives were not amused, pressing him hard at the same event. John Connally also appeared at the confab and did not come out foursquare against Carter over the treaties. The issue had become a litmus test for the right. Any Republican senator who supported the treaties would become suspect for the rest of his natural-born life. At the meeting, the new chairman of the Mississippi GOP, Charles Pickering, made it clear that where presidential candidates stood on this issue would be one of a small cluster of factors that would decide who would be the nominee in 1980.[51]

Meanwhile, Phyllis Schlafly was considering a primary challenge to incumbent moderate Republican Chuck Percy of Illinois, her home state as well, especially if Percy "went South" on the canal. Schlafly was a true grassroots conservative who had always eschewed living and working in

Washington. In the fall of 1977, two competing women's conferences were held in Houston. The first, organized by liberal feminists, attracted two thousand attendees. Schlafly's conference for conservative women attracted eleven thousand.[52]

New Right figures were making noises about primary challenges against other incumbents if they went along with Carter on the canal. A conservative stalwart who did plunge into a GOP Senate primary in 1978, in New Jersey, was Jeff Bell, former Reagan aide and former ACU staffer, who took on the ancient liberal senator Clifford Case. Bell was in his thirties and had been but one year old when Case first went to Congress in 1945. He was a gangly intellectual who had converted to Catholicism and become deeply devoted to his new faith.

Bell had worn a number of different hats for Reagan, drafting speeches and some Reagan columns and radio commentaries and networking the conservative movement on behalf of the Gipper. He had caught hell, unfairly, over the "$90 billion" speech he wrote for Reagan in the '76 campaign, which outlined, without a real plan, a proposal to cut federal spending; it had caused Reagan many headaches in late 1975 and early 1976 and was used against him by the Ford camp in the primaries, yet Bell had earned his spurs and was expecting help from Reagan and Citizens for the Republic as he geared up for his campaign. He was sorely disappointed. Reagan, in preparation for a possible run in 1980, had decided not to support anyone in a primary, even someone as faithful and long-serving as Bell. If Bell was going to do it, it would have to be without the Gipper.

Frustrated with the growing dominance of the anti-canal conservatives in the GOP, a handful of moderate to liberal senators fired off a letter to Brock protesting their counterparts. "The cannibalization of Republicans by Republicans is extremely unwise," said Jacob Javits of New York, one of the most respected men in either chamber.[53] But their numbers were dwindling as the party moved faster and faster to the right. Some Republicans still didn't get it, including former RNC chairman Mary Louise Smith, who attacked treaty opponents, saying they hadn't read the treaties and that their opposition was "emotional." Her brand of "me, too" liberalism inside the GOP was not only the minority; it had become a joke of sorts.[54]

<hr />

From the schoolhouse to the churches to the halls of the state and national governments, American citizens were demanding accountability from their elected leaders. America was moving rightward.[55]

Five days before Christmas, Bill Brock's duplicitous position on the canal treaties leaked out to the media. Though Reagan had raised a ton of money for the RNC on the issue, and though the national committee was on record opposing the treaties, and though the conservative movement and most of the GOP were arrayed against the treaties, Brock still stonewalled, even as the media nailed him for his dishonesty.[56]

The Associated Press story did confirm, however, for the first time, Brock's refusal to give even $50,000 of the money raised by Reagan for a "Panama Canal Truth Squad" to oppose Carter. Brock issued his own statement and mentioned nothing about the controversy, the treaties, or Reagan's request, but patronized Reagan for his support of the Republican Party and its candidates, nothing more.

Paul Laxalt had been stirred up by Lyn Nofziger, who didn't like any of the GOP operatives save his friend Charlie Black, and Laxalt in turn stirred up Reagan. Nofziger especially didn't like Brock's chief of staff, Ben Cotten, who had been the first to reject giving the money to Reagan. Cotten had been the leader of the anti-Reagan clique in the RNC.[57]

Brock's refusal to fund a Panama Canal Truth Squad did nothing to deter the determined conservatives, and a group with a hefty budget embarked on a nationwide five-day tour to spread their anti-treaty message. The tour was marred by an unexpected snowstorm over the Midwest, so stops in Cincinnati and Covington, Kentucky, were canceled. They detoured to St. Louis, where there was no public forum in the works, but this did not daunt the conservatives. Demonstrating their new appreciation for modern campaigning, the group, organized by tour director William Rhatican, who ironically had been a Ford advanceman and was

now in the employ of the New Right, worked the phones from their hotel rooms, talking to print, radio, and television reporters in the targeted
states on the progress of their efforts to forestall the treaties. When they
arrived in Denver, the plan was to hook up with Reagan for a highly
publicized media extravaganza.

In a bow to their growing respect for the conservatives, the Carter
Administration organized counter press conferences at each stop along
the conservatives' way, in an attempt to refute their charges. The cities
selected by Paul Weyrich and others were not chosen willy-nilly; they
were in or adjacent to states where there were U.S. senators who were wavering in their support for the treaties. Florida, West Virginia, Missouri,
Ohio, Kentucky, and Colorado all had senators with their fingers to the
wind, especially if they were up for reelection in 1978.

Still, Brock could look back over the past year with some satisfaction.
The party he'd taken over twelve months earlier was, at the time, on life
support and had been given last rites by nearly all. At the end of 1977 the
party was still on the critical list, but its prognosis was improving. The
Republicans had lost the chance to win an off-year gubernatorial election in New Jersey, but GOP candidate John Dalton's victory in Virginia
provided some of the tonic Brock and his troops needed.

The deadwood had been, by and large, cleared out of the RNC's
building; the Republicans had gone on the offense for the first time
in years; there was still ideological warfare between Reagan's conservatives and Ford's moderates, but some saw this fighting as more evidence
that the party was alive; and money was beginning to flow once again
into the party coffers. Only this time it was not coming from big business interested in buying access and legislation, but instead from grassroots small-dollar donors, motivated by conservative principles. Also,
and maybe most important, conservatives were no longer threatening a
"third party." This was one less nuisance Brock had to worry about as he
headed into the 1978 off-year elections.[58]

The party's leadership was better at coordinating its efforts while
papering over its differences. Howard Baker led only thirty-eight GOP
senators who ranged on the spectrum from Jesse Helms of North Caro-

lina on the right to Jacob Javits of New York on the left, but Baker had done a good job holding his disparate troops together. Even Paul Laxalt acknowledged Baker's yeoman-like efforts as a "miracle."[59]

Other evidence of the resurgence of the GOP came from Ambassador Bush, as he formed a political committee to, ahem, advance the interests of the party. But if it also helped along his national ambitions, well, that was fine, too. In addition, opposition to Carter was helping the GOP to unify itself. In the West, Carter had canceled dozens of federal water projects, necessary for ranching and farming, deeming them wasteful. He had not carried any of these states in 1976, but in cutting these programs, he was helping to ensure he might not carry them in 1980, either.

Though farmers and ranchers were a relatively small part of the western population, everybody was sympathetic to the romantic image of the hardworking outdoorsman and his careworn wife struggling to subdue the land. Both had been heroically portrayed in novels and movies for years. Everybody loved the cowboy, and everybody loved the farmer. Indeed, Carter, in running for president, had portrayed himself as a commonsense, wholesome, honest dirt farmer. "Carter has declared war on the west," said Frank Fahrenkopf, GOP chairman of Nevada and a close confidant of Laxalt's.[60]

———

The meddlesome Federal Election Commission, after months of investigation, after thousands of hours and untold public monies spent, came to the conclusion that two conservative groups that had supported Reagan for president in 1976 were not in violation of FEC statutes that prohibited coordination. A "clean bill of health" was issued to the American Conservative Union and the Conservative Victory Fund, with the ACU having been the far more crucial entity, as it spent over $150,000 assisting Reagan while the CVF had spent only $5,000.[61]

The ACU's efforts in North Carolina alone in 1976, after Reagan had lost the first five primaries to Ford, helped seek out a victory there for the Gipper, to keep him in the campaign and propel him to victories in subsequent primaries, before losing by only a hairsbreadth at the Republican convention. Jim Roberts, the ACU executive who had directed all the work on behalf of Reagan, estimated that he spent dozens more hours answering the FEC's charges than he ever had on the pro-Reagan independent expenditure itself.[62]

As 1977 drew to a close, the Reagans and Paul Laxalt, among others, met for a private dinner to talk through the possibility of another Reagan campaign for president in 1980. Laxalt had always been for Reagan giving it another go, and not even the thought of being embarrassed dissuaded the Reagans. No decision was made, but the purpose of the dinner was shortly leaked to columnists Evans and Novak.[63]

Reagan was not seen as the front-runner by the ivory tower set. In a survey of college professors, Reagan came in a poor third, far behind Howard Baker, who garnered 25 votes, and Ford, who received 15. Reagan himself received only 12 from the learned academics.[64]

But it was Reagan who was selected by PBS to give a refutation to a twelve-part series, which was costing taxpayers $750,000, based on socialist economist John Kenneth Galbraith's book *The Age of Uncertainty*. Reagan, untutored, said,

> Having allowed government to grow to monster size, we have produced a cadre of professional, lifetime politicians, most of whom ply their trade by periodically raising their constituents' hopes of solving the latest problem. Once reconfirmed by the voters, they always follow the same method, dispensing huge amounts of tax money, passing laws that lead to restrictive regulations, then blaming those they regulate when things break down. And then, passing more laws and regulations because things obviously don't work. The ruling class of politicians and bureaucrats we have been breeding for several decades benefits from bigger and bigger government. And the opposite side of that coin is, inevitably, less liberty for individual citizens.[65]

Carter finished his first year in office with more questions surrounding him than when he entered the White House one year earlier. He was in fundamental disagreement with many of the entrenched interests inside his own party. He did not respect the dens of power inside the Democratic Party and certainly not in the nation's capital. He refused to kiss their rings or their asses. He wasn't much for small talk or meetings, preferring to deal with memos alone or with his small circle of fiercely loyal Georgians.

His staff continued to cause him problems with officious Washingtondom. At an embassy reception, his top aide, Hamilton Jordan, drunk, spotted the generous bosom of the wife of the Egyptian ambassador and commented that he "always wanted to see the twin pyramids of the Nile."[66] Within weeks, he was separated from his wife, and just as quickly was in the gossip columns again, accused of spitting his drink on the dress of a woman he had met.[67] Stories and rumors about Carter staffers running amok, smoking pot, and partying until all hours animated the gossip factories in the capital. Beyond the merely titillating, it was a serious issue. For Washington to operate, it was necessary to communicate, sober or otherwise.

The interaction with the rest of Washington was necessary to the successful function of government so as not to produce "group think"— which is what President Kennedy was convinced led to the disaster of the Bay of Pigs early in his Administration. In meeting after meeting, with JFK present, no one had the courage to say no because they were afraid of going against the president.

Carter compartmentalized problems and people. He'd been an engineer at the Naval Academy, and he tried to bring slide-rule computations to the Oval Office. He was also offending members of his own coalition, including blacks, Jews, big labor, and liberals, over a variety of missteps and slights in his first year.[68]

When Carter did deal with people, including congressional members of his own party, he talked down to them and treated most "like children."[69] Congressman Chris Dodd of Connecticut, a Democrat and future senator, was equally blunt as he told Al Hunt of the *Wall Street Journal*, "I hope all their talk about choosing priorities and working better with us is for real because we certainly missed that this year."[70] The economy was also weakening, and this added to Carter's woes.

Jody Powell continued to worry about the "liberal establishment in Washington [that] had no idea in God's earth why Carter had won."[71] Carter's poll numbers were continuing to drift downward, now in the mid-50s according to Gallup, and the year was "ending amid cocktail-party speculation that Mr. Carter may be a one-term president."[72]

Drinking the Kool-Aid

"A new species"

In late January 1978 the Republicans held their annual meeting and were feeling pretty good about things. Fifteen hundred gathered at a fifty-dollar-per-plate dinner in Washington to hear keynote speaker John Connally take it to Jimmy Carter and the Democrats, but the surprise speaker of the evening was the Reverend Jesse Jackson, who was greeted with a standing ovation by the heirs to the party of Lincoln.

Jackson told them, "Black people need the Republican Party to compete for us so that we have real alternatives for meeting our needs. The Republican Party needs black people if it is to ever compete for national office or, in fact, to keep it from becoming an extinct party."[1] Jackson also skewered the Democrats for taking blacks for granted. Some in the GOP questioned the wisdom of reaching out for black votes at the possible expense of lower-income white voters who were opposed to quotas and felt economically threatened by minorities.

Although Jackson made it clear he was not about to change his party registration, he assured beguiled Republicans that he wanted the lines of communication to remain open between the GOP and black America.[2] Before Jackson's advice, however, RNC chairman Bill Brock had already awarded a $250,000 contract to Robert Lee Wright Jr., a self-described "black man and a Republican," whose job was to "introduce the Grand

Old Party to the descendants of the black people Abraham Lincoln emancipated 114 years ago."[3]

Brock's initiatives were not welcomed in all quarters. The *Wall Street Journal* harshly wrote, "Black activists in the party these days tend to be either hustlers, looking to rip off a few bucks, or social climbers, who don't want more blacks in the party because they might be displaced as tokens at the head table."[4] Howell Raines of the *New York Times* referred to Wright and his business partner John L. McNeill Jr. as "hired guns."[5] GOP consultant Eddie Mahe dismissed the insults, saying that the two men were "legit."[6]

Unfortunately for another black Republican, Dr. Gloria Toote, the members of the GOP's national committee were not interested in hearing her message, as they rejected her challenge to Mary Dent Crisp, 118–37, for cochairman of the GOP.[7] Lyn Nofziger had recruited Toote to challenge Crisp, who was a Ronald Reagan basher of long standing. Crisp complained that Reagan's political group, Citizens for the Republic, was taking money away from the RNC. And Reagan was still refusing to raise money for the RNC after he'd been duped by Brock over signing a letter opposing the Panama Canal Treaties several months earlier. Brock had convinced Reagan to sign an anti–Panama Canal Treaties direct-mail letter for the "committee," promising to use the funds raised to help Reagan defeat the Senate vote. When the money cascaded into the RNC, Brock went back on his word, telling Reagan there would be no money, even though Reagan had raised every dime. Paul Laxalt and Reagan burned a red-hot line to Brock and got into a screaming match with the RNC chairman. Reagan vowed never to sign another fund-raising letter for the RNC. (And he did not, relenting only after he won the nomination in 1980.)

The Reaganites lamented Toote's loss, knowing how this tough, conservative black woman would have confounded the Washington liberal establishment. Conservatives saw her defeat as yet another rebuke of Reagan by the party, as was the choice of Connally over Reagan as Brock's preferred speaker.

One crass GOP official, when asked why he preferred Crisp to Toote, replied, "We're getting Jesse Jackson tomorrow. That's enough."[8] Apparently, some in the GOP favored quotas after all.

Meanwhile, in Mississippi, the state GOP chairman, Charles Pickering, a Reaganite, hired two blacks for his staff. This "field" staff, however, was not engaged in the agricultural work that had occupied their great-grandparents a hundred years earlier. Instead, as white-collar, briefcase-toting professionals, their job was to get Republicans elected in the Magnolia State. Indeed, as Jackson had told the Republicans, the "hands that picked cotton in 1966 did pick the president in 1976, and could very well be the difference in 1980."[9]

Electing Republicans in the Deep South suddenly seemed possible for the first time ever. The party was in such pathetic shape in the region that it did not "even bother to field a candidate for statewide office in between 1883 and 1963" in Mississippi.[10] The GOP also believed it had an opening with black Americans. Culturally, they were conservative, and 37 percent of black youths were unemployed in January 1978.[11]

Ronald Reagan addressed the annual meeting of the NAACP in Atlanta instead of the snooty Republicans. Calling Carter's people the "limits to growth" crowd, he told those assembled that "I believe black Americans want what every other kind of American wants: a crack at a decent job, a home, safety in the streets, and a good education for our children."[12]

Several months later, a private lunch was scheduled between the Reverend Jackson and Reagan. Reagan waited patiently, but Jackson never showed up. His staff later said it was a "scheduling error."[13]

———————

Hubert H. Humphrey of Minnesota, the "Happiest Warrior" since Al Smith, succumbed to cancer in January 1979. He'd been a real contender for the 1960 nomination until he ran into the Kennedy family buzz saw. He'd put up with humiliations from Lyndon Johnson while vice president; perhaps the worst insult of all was Johnson's belated and begrudging endorsement of Humphrey in the latter's fiercely contested campaign against Richard Nixon in 1968. Some blamed Johnson's lukewarm support for Humphrey's razor-thin loss to Nixon. Finally, Humphrey had returned to the Senate, where he was the happiest. In his prime, he had more bounce than a Wham-O Superball and he could claim significant legislation to his credit, including the creation of the Peace Corps. From his superb speech at the 1948 Democratic convention right up until his passing, Humphrey had been a tireless and passionate fighter for liberal causes.

Meanwhile, the war over the Panama Canal Treaties continued, especially between Reagan and Carter. Carter told audiences that to "defend the canal if the Senate rejected the treaties . . . It might take 100,000 troops."[14] Still, Washington, and especially Carter, could not ignore another happy warrior, Reagan.

Then, in a nationally televised address, Carter directly challenged Reagan, proclaiming that "we do not own the Panama Canal Zone—we have never had sovereignty over it. We have only had the right to use it."[15] It was a mockery of Reagan's beliefs. Several months earlier, in Hollywood, Reagan had met with television producer Norman Lear, who had the idea of a national debate on the issue between Reagan and someone of equal stature. Ultimately, the idea was never acted upon.[16] The network hosting, CBS, gave an extraordinary amount of time for Reagan to respond to Carter's address.

The Gipper offered an equally strong retort: "That is not quite accurate. What we have . . . are the rights of sovereignty [and this makes] it impossible for a government of Panama to expropriate the canal."[17] Carter had also told the American viewers that it would cost them nothing for the Senate to take this step and pass the treaties, but Reagan refuted this also, itemizing the billions of dollars, direct and indirect, that would accrue against the American taxpayers. He then concluded by shattering Carter's argument that if the canal were not turned over, it would invite terrorist acts against the facilities there.

Carter should have known better. He was inadvertently telling the American people to knuckle under to threats, or else—an attitude that ran counter to the American creed. Some writers ascribed Carter's motives for "giving away" the canal to his channeling Reinhold Niebuhr, because the action would "illustrate Niebuhr's point that a society's highest priority 'must be justice.'"[18] For much of the first part of the twentieth century, Niebuhr had been a prominent American socialist leader, pacifist, and theologian.

Reagan's strong but reasoned and measured opposition to the treaties was not enough for some of the hard-liners of the New Right, the ones he sometimes referred to as the "over-the-cliff-with-all-flags-flying" conservatives.[19]

Terry Dolan, head of NCPAC, pronounced Howard Baker's presidential aspirations as done because Baker had publicly vacillated on the issue and then, after a five-day visit to Panama, came out for the treaties. Dolan's hardball but aboveboard tactics were limited only by his imagination. NCPAC sent letters to fifteen thousand local GOP officials, asking them to forever suspend their support of any GOP senator who voted for the treaties. The New Right was in common agreement that support or opposition to the treaties would be a permanent mark on a senator's record, especially a Republican senator. These conservatives openly warned that any GOP senator who crossed them would never receive any support from them ever again. Such strong-arm tactics made the New Right reviled and even hated in some corners, but the New Rightists didn't care. The only thing that counted was winning or losing on their conservative terms. Ideologues of all stripes, right or left, share two common denominators: passion and certitude.

The mainstream media paid more and more attention to the New Right, and devoted many stories to its rise, documenting and cataloging the various players as if they'd discovered a new species. In a way, they had.

They were on permanent offense. Conservatives were pursuing an alliance with labor on key issues, for one thing. Meanwhile, Stan Evans, former chairman of the American Conservative Union, founded the National Journalism Center, to train a new generation of conservative writers and editors.[20] By setting up new think tanks and training journalists, the New Right was out to win more than an election; it wanted to launch a sociocultural counterrevolution. Historically, the left's domain had been academia, the media, and the other "creative classes." Intellectuals always assumed that rational, thinking people such as they were liberal and that conservatives were unenlightened cavemen. This assumption among the intelligentsia became particularly pronounced after liberalism's successes of the 1960s. But that was all about to change; a backlash was in the works, and liberals soon wouldn't know what hit them. The New Right was outfitting itself for tweed jackets; the culture war was on.

As an early result of these efforts, the *Washington Post*'s David Broder perceptively noted a new phenomenon in Washington: the end of the notion that government could solve every problem. Attitudes in the country had changed, and the idea that a new agency could solve the nation's woes was "blunted by the fear of a meddlesome, bureaucratic big government . . . An era is ending."[21]

In early January, the Panama Canal Truth Squad took wing on a five-state tour to bring pressure on wavering senators. Senator Paul Laxalt led the group of prominent conservatives and opponents of the treaties, which "included Admiral Thomas H. Moorer, former Chairman of the Joint Chiefs of Staff; Admiral John S. McCain Jr., former navy commander in the Pacific; and Lieutenant General Daniel O. Graham, former Director of the Defense Intelligence Agency."[22] Funding for the Truth Squad came from the Conservative Caucus, the Committee for the Survival of a Free Congress, the Young Republicans, the American Conservative Union, and Reagan's Citizens for the Republic. Reagan met up with the Truth Squad in Denver. Their target was the incumbent Democratic senator Floyd Haskell, who was up for reelection later that year.

After their Denver events, Reagan immediately jetted off for a nationally televised debate over the treaties with *National Review* editor William F. Buckley. The debate was part of Buckley's weekly PBS show, *Firing Line*. Moderated by former senator Sam Irvin of North Carolina, Reagan's two "seconds" in the debate were Pat Buchanan and Adm. John McCain II. Buckley's "seconds" were columnist George F. Will and academic James Burnham. Burnham was a former Communist who in 1941 wrote *The Managerial Revolution*, which inspired George Orwell's *1984*. The debate took place in a theater at the University of South Carolina in Columbia. Buckley, true to form, drolly told the live audience of his position on the treaties, though "I fully expect that someday I'll be wrong about something." The audience laughed appreciably.[23]

Buckley also wryly noted the predicament of debating his favorite politician. Reagan spoke and made it clear he did not trust the Panamanian government and that the negotiations of the treaties had begun in 1964, after riots in the streets of Panama. He argued that America should never have been cowed into the negotiations in the first place. Reagan smiled at Buckley and wondered why his old friend was not on his side; Buckley replied, "The force of my illumination would blind you."[24] The audience again laughed.

Reagan would make a thrust; Buckley would parry. Buckley would make a point; Reagan would make an effective counterpoint. It was a serious discussion without vitriol. It was a disagreement without being disagreeable. It was impressive because all the men involved were overachievers and successful in many endeavors, and thus it showcased the best of the conservative movement. These were high-minded men of serious purpose and scholarly thought, and it showed the movement in its best light to millions of viewers.

———

By their estimate, conservatives were one to three Senate votes shy of defeating the treaties. Senator Baker alone received more than twenty-two thousand letters, with 98 percent opposed to the treaties. The American Conservative Union hired an airplane to fly over the University of Tennessee's football stadium with a banner reading "Keep our canal—write Sen. Baker."[25]

The various groups opposing the treaties had, by some estimates, raised more than $1 million, while the forces supporting Carter had raised fewer than $200,000. Conservative groups were purchasing television, radio, and print advertising, but also saw the issue as an opportunity to increase their reach and scope. "I think this issue is a door opener. I view it as a means of enlarging our constituency," Paul Weyrich told the *New York Times*.[26] Conservatives also organized a "Keep Our Canal Day" timed for Washington's birthday. Motorists who opposed the treaties were urged to turn on their headlights. But the Carter White House's political operation was fully engaged, supervised by Hamilton Jordan. Jordan approached the task as he did Carter's startling campaign in 1976, with creativity and thorough planning.

Vice President Mondale remembered years later that one day he had received a phone call from Republican senator S. I. "Sam" Hayakawa of California. He told Mondale he'd been elected running against the treaties but that he would be willing to support them "if Carter would start taking advice from him about the Presidency . . . I said, let's call the president. I think he'd need that. And Carter said, 'Yeah, that's exactly what I need . . . I'm down here all alone . . . and if you'd come up once in a while to help me I'd really appreciate it.'" Hayakawa suggested they meet every two weeks, and Carter replied that he might have to see Hayakawa even more often than that. "And we got his vote; I don't think they ever saw each other again," Mondale recalled, howling.[27] Mondale also knew that the protracted fight was weakening Carter politically.

———————

Reagan's Achilles' heel had always been his lack of foreign travel. He had traveled for a couple of trade missions for President Nixon, had been to Taiwan in 1972, and had vacationed in Mexico and France, but that was about the limit of his foreign affairs, not counting the time after his divorce when he went to England to film *The Hasty Heart* with Patricia Neal, where, she claimed, Reagan made a pass at her.

Dick Allen, Reagan's principal foreign policy adviser, and Peter Hannaford, his principal speechwriter, had been urging Reagan for a time to go on an extended, low-profile tour of the world later in the year. Eventually, the Allens and the Hannafords journeyed with the Reagans. Stops included Cairo, Moscow, Tokyo, London, Paris, and other destinations. While in London, Reagan met with Tory leader Margaret Thatcher.[28] But in all his years and all his travels, Reagan never went to the Middle East, including Israel.

———

At the end of February, it was Bob Dole's turn to create a political organization, Campaign America, to help the GOP and advance the White House aspirations of Robert Joseph Dole.[29] Several former Reagan men, including John Sears, were involved, and some took this to mean that Reagan would not run in 1980 or that Sears had signed up with a younger star.

Dole signed on, in an unpaid position, none other than William Miller of Buffalo, a former congressman and the running mate on the ill-fated Barry Goldwater ticket in 1964.[30] Goldwater had returned to the Senate, but Miller opted not to seek public office again. Still, he would be heard from as he briefly gained new fame starring in American Express commercials.

Dole was one of the funniest men on either side of the aisle. Once standing before a group of farmers, he wisecracked, "Can you hear me in the back all right?" After a pause, he added, "Just in case I should say something?"[31]

Shortly afterward, Dole went to New Hampshire for a media tour. His inexperienced aides scheduled him to attend, and squirm through, a nearly two-hour local GOP town meeting, the highlight being a slide show. Those who knew how impatient Dole could be could only giggle in private. The savvy Granite Staters knew how to leverage their unique position of being the first-in-the-nation primary, and they would exact from every presidential candidate their required pound of flesh or more, to serve their own purposes. In Dole's case, he was trying to shake his image as a "hatchet man." One critic, a moderate GOP senator, said the Kansan was so hard to get along with "he couldn't sell beer on a troopship."[32]

Charles Brock, in late March, announced an impressive plan that, if successful, would net the GOP millions of dollars. Fifteen gigantic fund-raising dinners would take place simultaneously across the country, linked by closed-circuit television, and each dinner would have GOP stars as attractions. It was a logistical nightmare but also an exciting extravaganza. Dinners would take place in Los Angeles, Chicago, Boston, Washington, Philadelphia, and elsewhere. Gerald Ford and John Connally would be the draws in Los Angeles, with Reagan and Bush as the featured attractions in Chicago. Other GOP luminaries tapped for the auspicious evening included Jack Kemp, Barry Goldwater, Henry Kissinger, Howard Baker, and Dole, along with lesser lights. "The dinners will be a major fundraising effort for this year. We intend to show our candidates the type of financial, organizational and volunteer support they can expect from our party in November," Brock told reporters.[33] Goldwater had fortunately gone back on his promise of one year earlier that he would never raise money for the GOP again, as had Reagan after being badly used in the fall of 1977 by Brock.

The night of the multicity GOP affair, former president Ford, in Los Angeles, grabbed the headlines when he told the thousands of Republicans via closed-circuit television that any of the contenders for 1980 could do a better job for the country than Jimmy Carter was doing. Ford included himself in that estimation. But he also named names, including Howard Baker, Bob Dole, George Bush, and "Ron Reagan."[34] In the end, the Republicans had fallen somewhat short of the goal of $4 million; nevertheless, Brock and his team had successfully stitched together a fourteen-city event, sold more than 4,600 tickets at $500 or $1,000 per ticket, depending on the city, grossed over $3 million, boosted GOP morale, and wowed the political establishment.[35]

The National Republican Congressional Committee (NRCC), too, was expecting a big year in 1978, after a record fund-raising year in 1977, in which it raised over $9 million. An ambitious goal of $14 million was announced, and why not?[36] The committee direct-mail impresario, Wyatt Stewart, had developed a "house file" of four hundred thousand proven repeat givers to the committee.[37]

After months of a titanic struggle, the Panama Canal Treaties barely passed, but Carter got very little political benefit, and most of it melted away within weeks.

The first treaty was ratified by the Senate on a 68–32 vote in March, just one more than the two-thirds required.[38] It took another month before the Senate ratified the second treaty by the same 68–32 margin.[39] Ten Democrats crossed the aisle to join the majority of Republicans who opposed ratification. But almost half of the Republican senators (sixteen) crossed the aisle to vote in favor of ratification.

Reagan was traveling in Japan at the time, part of his foreign tour. Before facing the media in Tokyo, "Reagan was turning the air blue in the limousine," angry at the treaties' passage, according to Peter Hannaford.[40] Carter had put everything on the line. In the 1976 campaign, he seemed at times to have taken a harder line in opposing the canal treaties than even Reagan, but that changed after he was elected. He also was forced to do something he swore he'd never do if elected: deal for votes. His success in getting the treaties passed arguably had less to do with high-minded geopolitical issues and more to do with sewer projects and dams and other patronage. Winning the treaties was costly to Carter's squeaky-clean image.

For a moment, conservatives were crestfallen. In the end, they learned there is little that a senator can deny the president of his own party if that senator wants to continue being invited to White House Christmas parties and getting pork and largesse for his home state.

Reagan recorded a hot radio commentary, denoting all the largesse spread among the Panamanians over the years, including housing developments, giant hospitals, "trawlers with full equipment—$30 mil." Overall, the American taxpayers had sent "almost a bil. $" to the Central American country. Reagan was of course refuting Carter's moralizing that Americans should somehow feel guilty as "treaty proponents droned on & on about how we'd taken advantage of this country."[41]

The conservatives had spent untold time and millions of dollars to defeat the treaties.[42] But when the dust cleared and they took stock of their work, they came to realize that, though losing, it had been one of the most impressive populist uprisings in American history. Their long efforts would portend greater successes in the months ahead. Columnist Tom Wicker, in a long article, perceptively wrote, "By 1980, it's entirely possible—some would say probable—that the third Republican to be elected president since World War II could be preparing to take office."[43]

Wicker wondered if it would be Reagan. "Reagan is trimmer and more vigorous than many men 20 years his junior, but advanced age is an increasing handicap in American politics." Reagan "is already a two-time loser in Presidential nomination battles, and might understandably have no taste for the . . . battle that almost surely will be waged."[44] On this last point, Wicker was way off the mark. Reagan relished a good fight. The day after losing the vote, Reagan addressed the annual Conservative Political Action Conference, telling conservatives to cheer up and that, though they'd lost, they brought down the "smug assumptions" of Carter's foreign policy.[45]

———————

Reagan was also still getting encouragement to run, if not always from his own men. Each time he went on the road, people implored him to run. When he and Nancy took in a show on Broadway, *Ain't Misbehavin'*, word spread among the audience that he was there. "People spontaneously stood and clapped when they discovered the Reagans in their midst. Many came over to shake hands, ask for autographs, and wish them well."[46]

During this time, he was meeting regularly for a private dinner in New York with leading conservative and neoconservative intellectuals, organized by *American Spectator* publisher Bob Tyrrell. Also, he met every so often with his inner circle of advisers at his home in Pacific Palisades, to take their temperature on 1980, but oftentimes the reply was less than enthusiastic.

Conservatives were waxing and liberals were waning inside the GOP. The Wednesday Club, that group of moderate to liberal GOP senators, was losing ground to the new Senate Steering Committee, an organization of conservatives. The *Wall Street Journal* noted this: "Washington may be controlled by the Democrats, but liberalism oddly seems to be running out of new leaders and new ideas. The lethargy on the left contrasts sharply with a resurgence on the right, at least in the Senate. The New Right, dressed up in cowboy boots and Western belts, is spoiling for a fight."[47]

A study released by the Brookings Institution said that Carter had accomplished little in the first year and a half of his presidency.[48] The president was catching hell from all sides, it seemed. According to a poll by the Gallup organization, his standing in his party was the lowest since Harry Truman's in the spring of 1952, just before he decided not to seek another term. This poll, however, did have Carter beating both Ford and Reagan in a general election matchup, 51–45 and 50–46, respectively. The poll also showed Ford defeating Reagan decisively among Republicans in a two-candidate GOP contest, 54–42.[49] The *Washington Post* said of Reagan, "[H]e seems at this point to have a slim chance for the presidency in 1980. Age is considered his greatest handicap."[50]

In April, the *New York Times* began the first of a lengthy three-part series on fraud in the federal system, which conservatives had warned about for years. The paper reported that the annual rip-off amounted to over $12 billion a year, and more painfully, there was no system in place in most government agencies to detect the fact that they were being ripped off. Most of the fraud was in programs to assist the poor, including food stamps, housing assistance, and medical aid. It was not just the penny-ante "welfare queens," but also white-collar professionals who had government contracts to provide services. One employee had defrauded the Department of Transportation to the tune of $856,000.[51]

"Asked by a judge how this could happen, the Federal prosecutor replied, 'Your honor, he posed as a subway system.'"[52] Another example was a doctor who billed the government for seven tonsillectomies he performed on the same individual. Another was the daughter of a Civil War veteran who continued to receive her mother's widow's benefits even though her mother had been dead for twenty years. Sometimes the fraud was downright sad. One dentist, under a federal welfare program, had unnecessarily pulled all but three teeth of a thirteen-year-old girl because he was compensated on a per-extraction basis.[53] One creative fellow in California signed up people for a fraudulent health care plan by telling them the documents they were signing were impeachment proceedings against then-governor Reagan.[54]

In June, what many conservative revolutionaries had threatened became a fact, when the baby-faced, thirty-four-year-old Jeff Bell upset Senator Clifford Case in the GOP primary, 51–49 percent.[55] The young Reaganite's defeat of the liberal Case, a courtly gentleman who'd been a fixture in New Jersey politics for over forty years, sent shock waves across the political spectrum and squeals of delight across the New Right. Bell based his campaign mainly on tax cuts, but also on the Panama Canal and social issues. Bell was a tall, rail-thin, clean-cut Ivy Leaguer who had served in the army and saw action in the Mekong Delta. He was a first-generation purebred conservative. He worked for the American Conservative Union, wrote for *National Review*, and had worked for Reagan for several years. Though Reagan did not campaign for his young disciple, Jack Kemp did—and a lifetime friendship began.

There had been a fight inside the Reagan operation over whether to support Bell openly. Lyn Nofziger, head of Reagan's Citizens for the Republic, along with Reagan's 1976 campaign manager, John Sears, unlikely allies, argued for supporting Bell in any way possible. Paul Laxalt disagreed, according to Evans and Novak, insisting that it was not the time to go "head-hunting" for Republicans. Laxalt and Case served together in the Senate. Laxalt won the day, and Reagan stayed out of New Jersey. Nofziger sent fifty dollars out of his own pocket to his friend Bell,[56] but not before Nancy Reagan upbraided Bill Simon at a dinner party when Simon openly questioned why Reagan was not supporting Bell. Art Laffer was one of the dinner guests and remembered the rest of the evening as "quite stressed and . . . unpleasant."[57]

Case was shocked that he lost, and blamed it correctly on voters' disaffection with big government. Bell's win could not be underestimated, as ideology was becoming more important in the Republican Party than experience. He'd out-raised and out-spent Case, $500,000 to around $100,000, and had also out-hustled him and clearly out-"conservative'd" him.[58]

―――――――――

Helping to shape the new conservative movement's thinking was *A Time for Truth*, a scholarly work by William E. Simon. Simon was a Renaissance man and, it appears, a Midas man: his touch seemed to turn everything to gold. He was a successful entrepreneur, a former Cabinet secretary, a confidant to the rich and famous, and on every board and foundation imaginable. He was also president of the U.S. Olympic Foundation. Simon was an admirer of Reagan's, though he flirted briefly with running for president himself in 1980. His book, a national bestseller, called for an *Atlas Shrugged*–like movement in which the wealth in America went only to those institutions that were distinctly pro-freedom. David Keene, formerly of the Nixon campaign in 1968 and the Reagan campaign in 1976, was playing footsie with Simon as he mulled a 1980 campaign for the GOP nomination.

The national media were truly mystified by this new, grassroots antigovernment movement, and NBC went so far as to bring its "prime-time special news reports" out of retirement to attempt to dissect the phenomenon. The special, called *Mad as Hell*, was hosted by David Brinkley and featured interviews with revolutionary tax-cutting acolyte Howard Jarvis and Reagan.[59] Not to be outdone, CBS also broadcast its own special, with Walter Cronkite anchoring, called *The Angry Taxpayer*.[60]

———

Though Reagan had not slowed a bit in crisscrossing the country, the "age issue" continued to surface. Tom Wicker of the *New York Times* led a column in which he pointed out to Reagan that he'd be seventy years old if inaugurated in 1981. "Put this proposition to him and you get a rueful grin together with proof positive that he's been pondering the matter—a practiced recital of how many United States senators are past 65, and a quip that in China he'd be considered just a kid."[61] If he sounded repetitive to reporters, they must have sounded like a broken record to him.

The issue of age came up in a different forum, this one lighter. He and Nancy had gone to Dixon, Illinois, for his high school reunion. Some of the locals wanted a street renamed for Reagan, but that idea was rejected. Then they wanted the local high school renamed for the Gipper, but the principal shot that down because it would mean reprinting uniforms and stationery. So then it was suggested that the auditorium be named after the town's most famous son, but that didn't go over so well, as no one knew if Reagan had ever actually been on the stage there. So they finally settled on renaming a local bridge after Reagan.[62] When the school and the town surprised Reagan with the news, Reagan quipped, "Thanks, I'm greatly honored, but you've called it the Ronald Reagan Memorial Bridge and I'm still alive!"[63]

Dave Broder of the *Washington Post* had seen the unrelenting hostility between Ford and Reagan over the years up close, and he speculated that the reason the two were both eyeing the White House in 1980 was because "neither . . . seems willing to quit until the other keels over."[64] Reagan was trailing Ford in most surveys of GOP voters. In another poll, congressional Democrats were trouncing their GOP rivals 49–29 in a ballot test. Republicans were getting smeared among Catholic voters 50–25 and among blacks 77–9, despite Brock's best efforts.[65]

Congressman Jack Kemp, who represented Buffalo, New York— where he had been the starting quarterback for the Buffalo Bills during two championship seasons in the old American Football League in the 1960s, before the NFL merger—was a rising GOP star for the New Right. With an unexpectedly scratchy, high-pitched voice, he had latched onto tax cuts and economic growth as his signature issues. Don Rothberg of the Associated Press wrote that cutting "taxes are to politicians, what the long, touchdown pass was to quarterback Kemp—a dramatic move sure to bring the crowd to its feet to cheer its hero."[66] Kemp read books, attended lectures on fiscal policies, and omnivorously devoured the white papers coming out of the growing number of conservative think tanks. And he talked, a lot: sometimes about running for president in 1980.

In fact, the entire group of potential GOP 1980 candidates, from Reagan to Ford to George H. W. Bush to Bob Dole, Phil Crane, John Connally, Howard Baker, John Anderson, and Jack Kemp, was quite impressive, as all of them had been either standouts in other fields besides politics (business, acting, sports, academia) or, in the cases of Bush, Ford, and Dole, genuine war heroes. Also, Bush had been an All-American baseball player at Yale, and Ford was a talented-enough football player at Michigan to be drafted by the Green Bay Packers. Dole had been an outstanding high school track star. Several were polished public speakers, especially Connally, Anderson, and Reagan.

Not Ford or Bush, however. All the potential candidates were over-achievers, almost none took themselves too seriously, all had above-average intelligence. By any standard, it was one of the most impressive groups of candidates ever to seek the nomination of their party for president. The group, not including Ford, would eventually produce two presidents, a presidential nominee, a vice-presidential nominee, an ambassador—and a number of books. All enjoyed the company of women, and women enjoyed their company. Before his first marriage (and before his second), Reagan had been known around Hollywood as quite the ladies' man, and was seen often out on the town with an attractive woman on his arm. Dole, after having divorced his first wife and before he married the attractive Elizabeth Hanford, a Ford White House aide, had for a time dated Ann Dore, an equally striking Washington lobbyist. Lyn Nofziger, never at a loss for a quip or pun, jokingly referred to her as "Bob Dole's swinging Dore."[67]

Another group of men was assembled around this same time, though they were not celebrated national political stars. Three hundred former prisoners of war met in Los Angeles for a reunion of sorts, the idea Reagan's and Texas billionaire H. Ross Perot's. They gathered at the home of Richard Nixon for a late-afternoon reception, where he gave each a personally inscribed copy of his memoirs. In turn, they gave him a painting of a POW uniform, inscribed to Nixon, thanking him for bombing North Vietnam and thus hastening their release from captivity. One attending was navy lieutenant commander John McCain III. During his six years in the hell of the "Hanoi Hilton," he'd been beaten, starved, and had bones deliberately broken, but his captors could not break his spirit. One of his fellow prisoners had had his arm broken so badly that the bones jutted out of the flesh. McCain took what little rags he had and fashioned a makeshift bandage and sling for his fellow POW.

When he was asked about being a hero, McCain drolly said, "It doesn't take a helluva lot of talent to get shot down."[68] Reagan, in meeting with the men and their wives, told them that America should never go to war again "unless we intend to win."[69]

With all the activity on the right, the national media should have taken better note, and in fact usually did, but on one occasion, the *New York Times* oddly wrote that "the conservatives are in imminent danger of having no active, charismatic national leader."[70] Nothing could have been further from the truth. Reagan defined the conservative standard, and no one, except maybe Phil Crane, could get to his right. Almost all the GOP candidates, with the exception of John Anderson, were right of center, had their followings, and were considered charismatic, with the exception of Gerald Ford. Still, Ford was mulling another run in 1980, so he swallowed his pride and endorsed Jesse Helms's reelection in North Carolina. Helms had been a passionate supporter of Reagan's and was passionately opposed to Ford in 1976. By 1978, however, Helms was talking down a Reagan run in 1980, and some saw this as evidence that the Tar Heel senator was eyeing his own run for the nomination.[71]

Ford hired a political operative, Charles Greenleaf, to coordinate his political activities. It was only one hire (a man who had worked on the 1976 campaign and hailed from Michigan), but the national media treated it as a seismic event, indicating that Ford was moving toward a 1980 candidacy. Ford also stepped up his pace: before the fall elections, he put in more than one hundred appearances across the country for GOP candidates and organizations. Ford had a full-time staff of twelve, yet the National Republican Congressional Committee would pay for Greenleaf's salary, and this raised eyebrows. The justification was that Ford was raising money for the NRCC.[72] The irony was lost on the party chieftains because, just the year before, Reagan had been turned down flat by the RNC when he wanted to use some of the money he'd raised for the party to oppose the canal treaties. Reagan was shot down cold. Even out of office, Ford was more equal than Reagan inside the party committees.

In late April, Ford and Reagan were not invited to attend the Tidewater Conference in eastern Maryland, an ad hoc group of Republicans gathered to address issues facing the GOP as a whole. Attendees were spread among a number of tables, each named after a Republican president. Three Republican presidents had no table named after them: Richard Nixon, Herbert Hoover, and Gerald Ford. The Republicans addressed a variety of issues, including tax cuts, and on this issue they were in near-total agreement.[73]

Bill Brock saw the potency of the tax-cutting issue, and with the full support of the RNC, the "Kemp-Roth" tax-cut plan took off. "Roth" was Senator Bill Roth of Delaware. The rumor at the time was that when Jack Kemp was developing the issue of the 33 percent across-the-board cut in federal income taxes, he went shopping for a Senate partner, and first approached Dole, but the Kansan wanted it to be called "Dole-Kemp," which Kemp would have none of. Roth was more amenable to having his name go second, and the marriage was consummated. Brock had the RNC produce bumper stickers and seminars, so GOP candidates could become disciples of the way of the tax cut. A film was also prepared by the RNC, but it was a Democrat, John Kennedy, who got top billing in it, due to his 1962 tax cuts.

Tax cutting had picked up even more steam, and Republican candidates across the land, many of whom had made a career as favoring balanced budgets and opposing tax cuts as fiscally unsound, became converts. The supply-side evangelist Kemp led millions of budget-balancing Republican souls down the sawdust trail to tax-cutting political salvation. There was an important difference between the tax cutting of Jarvis and that of Kemp. With Jarvis's plan, with its cuts in property taxes, there would be accompanying real cuts in government services. Kemp's were far different, as his supply-side tax cuts were theorized to increase, not decrease, revenue for government treasuries. This meant no cuts in services. It was an important doctrinal difference, but to the unschooled and less sophisticated, it was just one big, happy tax-cutting party.

Kemp's idea was not a new one, but Washington's creed rejected it as irresponsible, even though JFK had done almost precisely the same thing fifteen years earlier. Speaking for Carter, his treasury secretary Michael Blumenthal sniffed, "Whatever benefits might be envisioned would be quickly negated by the rise in prices and in interest rates."[74] Bills to cut capital gains and personal income taxes had been introduced in Congress, and by opposing them, the Democrats fell into the GOP's trap.

Kemp played professional football for thirteen years, but he was now quarterbacking the tax issue, and doing so just as effectively as he'd done on the gridiron. Playing football, Kemp broke his ankle twice, broke a knee, twice broke a shoulder, and sustained eleven concussions and untold broken fingers. But he was a winner, a perennial All-Star, a league Most Valuable Player, and the leader of championship teams. Politics, Kemp knew, could also be a full-contact sport, but sometimes you had to take the hits if you wanted to end up with the trophy.

Liberals met the Republicans' love for tax cuts in heated (more than hot peppers) opposition. The Carter Administration swung at the GOP's pitch in the dirt and rolled out its top economists to attack its tax cuts. The parties had gone through a role reversal, and hell, it appeared, had frozen over.

With the growing plethora of issues and potential GOP standard-bearers for 1980, Reagan began to drop his coyness, and by the summer of 1978 he started talking more openly about the possibility of running. He saw the momentum the others were gathering and that 1980 was potentially becoming more favorable for the GOP nominee, as Carter weakened in the polls and the economy and the world weakened along with him. Some aides were even franker; Lyn Nofziger said, "I expect that Reagan will run."[75] Reagan also moved quickly to make sure everyone knew that tax cuts were his issue. Before a cheering audience of seven hundred in Philadelphia, he told them he wanted Proposition 13 to unleash a "prairie fire" that would lead to "a renewed America, dedicated to limited government and unlimited freedom, a renewed America so strong we would have no need to fear foreign tyrants or domestic agitators."[76] Even Ford was caught up in tax-cutting fever and said that the movement that started in California was indicative of a "nationwide groundswell against excessive government."[77]

Only several months earlier, Moscow had ordered MiG fighters to shoot up a Korean Air Lines plane that had wandered into Soviet airspace. Two passengers were killed and more were wounded.[78] Somehow, the plane barely landed safely and the Kremlin made it clear that the pilots had been ordered to down the plane.[79] The Soviets wanted everybody to know they were the toughest kids on the block and that the rules simply did not apply to them. Reagan and a handful of hard-liners seemed to understand this about Moscow and called for swift and decisive measures against the Soviet Union, but the wheels of the U.S. government never budged.

President Jimmy Carter seemed intimidated by Moscow. He manfully gave one speech at Wake Forest by telling the graduating class that America would stand up to the Soviets over the attack on the civilian plane, but no presidential action followed, and the incident was quickly forgotten after the Kremlin told Carter that his policies were "the main obstacle on the path to détente."[80] Carter backed off.

Vestiges of an aggressive posture toward the Soviets were spiraling downward when Carter made it clear in a later graduation speech, this time at the Naval Academy, that there would be no linkage between SALT II (the follow-up to the Strategic Arms Limitations Treaty) and Soviet behavior in any way, shape, or form. Carter issued a blank check to the Soviets to run amok, which they were delighted to cash. In thirty-one minutes, Carter, the first president to graduate from the academy, received only four rounds of applause from his brethren.

In late June, Carter attempted one last time to aim some tough rhetoric at the Soviet Union on a swing through Texas, but by now he was on the defensive as anticommunism grew in America. The *Washington Post* took note, in an above-the-fold headline: "Carter Under Pressure to Respond to Soviets," which was accompanied by two major stories.[81] The Soviets also held a show trial for Anatoly ("Natan") Sharansky and Alexander Ginzburg, two Jewish refuseniks, and the Administration squawked a bit, but the trials went forward, and both men were of course convicted and sentenced to labor camps.[82]

"He's working harder than he ever has," noted one Reaganite.[83] But Reagan stepped up the pace and began to campaign even harder for House and Senate candidates nationwide, doing so full-time by the fall, covering as many as thirty states. Nofziger also stepped up contributions to GOP candidates from Citizens for the Republic. CFTR was running ahead of all other conservative organizations in terms of fund-raising, according to the FEC, with more than $2 million raised by October.[84] The original plan had been more money to fewer candidates, but it was decided to spread the wealth, and by early July, sixty-two GOP office seekers had received at least one contribution from the Reagan group. Some received as little as one hundred dollars, but the impact of the check and the letter that came from Reagan's organization could not be underestimated.[85] Each and every candidate put out press releases, issued statements, and did interviews trumpeting their support from the Gipper. Reagan's blessings thrown down from the balcony and upon the masses of GOP candidates were enlivening for their morale and their souls in 1978.

Reagan had been extremely busy with CFTR business throughout the year, speaking at conferences and meeting in six appearances in Texas, Los Angeles, Atlanta, Philadelphia, and other jam-packed locations. For many, he was still their Moses and they were waiting for him to lead them to the Promised Land. Still, Reagan had to refute a story in the *Christian Science Monitor* that said he'd already created a campaign committee and that a plan had been written for a 1980 effort. By the end of the election, Reagan's CFTR had contributed greatly to more than four hundred Republican candidates.[86]

The base of the party and especially the grassroots conservatives had much invested, emotionally and otherwise, in Reagan, but some of the leaders of many of the New Right groups had much less devotion to

Reagan, including Paul Weyrich, whom Phil Crane brought aboard to handle his campaign management.

Complaints were cropping up against Reagan, if behind the scenes. They ranged from the fact that he wasn't campaigning hard enough, or at all, for conservative primary challengers taking on incumbent moderate Republicans to the attitude and beliefs of some of his staff. Indeed, House minority leader John Rhodes of Arizona, appearing on CBS's *Face the Nation*, was asked about the growing conservatism of the party, and he referred to "a younger strain of Republicans who identify themselves with conservative ideas," saying, "they are going to be quite a force in the party."[87]

The Democratic leadership in Washington continued to underestimate Reagan. John C. White, the former Texas commissioner of agriculture and a protégé of Lyndon Johnson, whom Carter had tapped at the beginning of the year to become chairman of the Democratic National Committee, was but one of many in that group. White appeared on NBC's *Meet the Press* and called on his fellow Democrats to fall in line behind President Carter. He whistled past the graveyard, though, as he forecast that neither Ted Kennedy nor Jerry Brown would challenge Carter in 1980. He also predicted, maybe hopefully, that the GOP would nominate Reagan.[88] Though the Carter White House was focused on its own day-to-day problems, when they did let themselves daydream about 1980, visions of pummeling Reagan into the ground danced in their heads.

To a person, they all wrote Reagan off as a dumb bunny, a not very good actor who was old, rigid, and an extreme right-wing kook. They all bought into the caricature of Reagan and, frankly, thought Carter would decimate him in 1980 if the GOP was foolish enough to nominate him. Reagan, as in most circumstances, was once again being underestimated. This opinion came not only from the Carter White House, or Harvard or Georgetown or Manhattan, but also from many quarters inside the Republican Party and the New Right. Frankly, many did not see in Reagan what the American people saw in him, which explained why such a large group of GOP candidates was forming. To a man, they all were positioning themselves to be the beneficiary if or when the sixty-seven-year-old Reagan stumbled, through a misstep, rhetorical or physical. As Howard Baker bluntly said, nobody "has a headlock on the nomination right now."[89]

Reagan was not going to lie down for Crane or anybody else. When Crane announced, he was peppered with questions about whether he was a stalking horse for Reagan. He repeatedly denied it and told the assembled media that while Ford was "encouraging," Reagan was "not discouraging."[90]

Bill Brock was encouraged as he announced in August that the GOP would undertake a national blitzkrieg to rally support for the Kemp-Roth bill. "Not since abolition has one issue so united the Republican Party," he told reporters. The plan was for a charter jet to leave Los Angeles to barnstorm the country, including stops in New York, Philadelphia, and Detroit. At each city, politicians would speechify and then spread out to do local media interviews.[91] Kemp-Roth had been defeated in the House, but that only added fuel to Brock's fire. Further helping the Republicans was Carter's pledge to veto the bill if it ever got to his desk. But some Republicans didn't get the memo about the new populist image of the GOP. At the tax-cut rally in Chicago, local GOP officials arrived in a black limousine.

The populist antitax uprising that Brock had taken note of, and hoped to channel to the greater good of the GOP, came to the nation's capital when none other than Howard Jarvis, father of Proposition 13, held a press conference on the steps of the Internal Revenue Service headquarters. He was there to announce the formation of the American Tax Reduction Movement and an immediate campaign to cut $100 billion in federal spending over four years, a $50 billion cut in income taxes over the same period and a 2 percent cut in the national debt, which was approximately $750 billion in the summer of 1978.[92] Jarvis, as always, was plainspoken and never at a loss for words. He told reporters that if elected officials signed on to spending cuts, they would get his support. If they opposed them, he would work for their defeat.

The resurgent Republican Party had developed not only an issue advantage over the Democrats, but also a financial one. The National Republican Senatorial Committee was now handing out to incumbent and challenger Republicans checks totaling millions. Rod Smith's hard work as the NRSC's finance director beginning in January 1977 was paying off in dollars. The Federal Election Commission allowed the national committees to hand out "coordinated" money, the amount based roughly on the population of the state: for example, $200,000 to Senator John Tower of Texas, $110,000 to Senator Jesse Helms of North Carolina. According to FEC reports, the three GOP national committees had raised to date a total of $49.6 million, as compared with only $14.4 million for their Democratic counterparts.[93] The beauty, too, was that much of it was raised from small donors, who were not in it for access but for blood.

Other conservative PACs, besides CFTR, showed healthy financial numbers, including the National Conservative Political Action Committee, at $2 million; and Paul Weyrich's Committee for the Survival of a Free Congress, at $1.5 million.[94] Though the Democrats' money problems were not as severe as it appeared, due to the millions they got from big labor, it was an impressive period of renewal for the GOP. One Republican Senate nominee who had a tussle with the NRSC before he got his money was Gordon Humphrey.

Humphrey was a copilot for Allegheny Airlines who had worked his way up from cleaning privies to wearing a blue uniform with two gold braids. He was based in Boston but was on the road constantly, so he chose the no-tax state of New Hampshire as his residence. His only political experience was as the state coordinator for the Conservative Caucus, and he collected thousands of signatures for a petition opposing the Panama Canal Treaties. Humphrey got a taste for politics and decided to run for office himself, setting his sights on the three-term incumbent, Democrat Thomas McIntyre. Humphrey was only in his thirties and had lived in the state for just four years.[95] When he sat down with the temperamentally tough curmudgeon of Granite State politics, William Loeb, publisher of the *Manchester Union-Leader*, Loeb assumed Humphrey was talking about running for the state senate, not the U.S.

Senate. Loeb counseled Humphrey to lower his sights. Nothing doing. Humphrey took a leave of absence from the airline, and his future wife, Patty Greene, acted as his campaign manager for a time.

The stuffy New Hampshire GOP establishment saw Humphrey as something of a flake, and it and the national committees got behind the candidacy of Jim Masiello, former mayor of Keene. There were several other candidates in the field, but Humphrey whipped them all in the GOP primary in September, not just winning but also taking over 50 percent of the vote. Simply put, Humphrey was more conservative and had worked harder than had all the other candidates combined. He turned down no speaking invitation, no matter how small the town or the crowd. He was awkward in a one-on-one setting, but the flinty New Hampshirites, who themselves could be off-putting, liked the passion of this impetuous young man. Though he won his primary convincingly, the Senatorial Committee refused to give him the more than $60,000 in coordinated funds the nominee should have routinely received.

They bought the line of the country-club Republicans in New Hampshire that Humphrey was a kamikaze and thus would be easily defeated by McIntyre. "Save your money," they said. The establishment also just didn't like Humphrey. He wasn't old money, he wasn't Dartmouth, and he wasn't a Cabot or a Lodge. He had not attended Phillips Exeter Academy. His father had been a laborer in Connecticut, and Humphrey was tough and unrelenting.

Before Humphrey could score the biggest upset of 1978, he would have one unforgettable encounter with Ambassador George Bush, one that would sear into Humphrey and animate his later relations with Bush to no good, to Bush's misfortune. In October 1978, Bush had been invited to address the Manchester Women's Republican Club. The scions who ran it thought Humphrey tacky and had never invited him to speak, even after he won the nomination. Humphrey knew there were several hundred Republican votes in the room, and he needed them, so he swallowed his pride and crashed the event, where he was badly snubbed by Bush. A photo appeared in the *Union-Leader* showing Bush with his back rudely turned toward Humphrey.

Briefed badly by his staff, Bush regarded Humphrey as some sort of nut. Humphrey's pride burned. The upper-class kid from Connecticut who went to a private school in a limousine had snubbed the blue-collar kid from Connecticut who walked to a public school. Humphrey would never forget it and would, over time, exact his revenge on Bush.

Reagan came to New Hampshire in the fall of 1978 to stump for his friend, the peculiar governor Meldrim Thomson Jr., but he also cut a commercial for Humphrey at the old New Hampshire Highway Hotel in Concord. Reagan arrived with only a few aides, and when the camera crew was late, he sat down in the lobby for a few minutes with Humphrey's nervous twenty-one-year-old press secretary.

The young man had nothing to offer the Gipper, but that did not matter. Reagan was kind and genuinely interested in what he had to say. They talked about the weather—they both didn't like the cold—where they were from, sports they played in college. When the camera crew finally arrived, Reagan studied the thirty-second script for a moment, looked into the camera, and got it perfect on the first take.

Dole also campaigned for Humphrey on several occasions, and during the fall of 1978, Humphrey grew fond of the Kansan as well.

───────

Reagan briefly went off the campaign trail and appeared on *Face the Nation* to assail the Carter Social Security taxes and the upcoming talks about SALT II with the Soviets, but he also took some time to defend former president Nixon on the show.[96] Reagan had not known that Nixon once described him as "strange" to his White House staff, but it would not have mattered.[97] Reagan was loyal and compassionate, and he felt for the humiliated ex-president. He also knew Nixon got to the presidency by stumping aggressively for GOP candidates from 1964 until his second try for the White House in 1968, and if Reagan was going to take another bite at the apple, he'd have to continue his blazing pace of the last two years. To Reagan's misfortune, John Sears had other plans for the Californian in 1979.

Reagan was not the only potential 1980 GOP presidential candidate delivering speeches across the hustings. The hustling George Bush, who visited forty states, and Bob Dole, who visited nearly as many, despite a busy Senate schedule, bested the Gipper in a number of pre-announcement political events.

Both were also stalking New Hampshire early. Dole became reacquainted with his ex-wife and former mother-in-law, who were both Granite State natives. Bush, too, was seeking support, but he was beginning to earn the scorn of conservatives in the state. Former governor Hugh Gregg, who had been Reagan's honcho in New Hampshire, was edging away from Reagan, as the year before he'd declined to join the board of CFTR. He said with spite, "Thomson and [Jerry] Carmen and Loeb will run his campaign."[98]

By late September 1978, it was becoming more and more obvious that not only was there a political upheaval especially over taxes, but that Americans were swinging to the right on a number of issues. According to a new poll out by NBC News and the Associated Press, 31 percent said they had become more conservative "in recent years," while only 17 percent said they'd become more liberal. Yet the needle hadn't moved on GOP party affiliation. Only 22 percent said they were Republican, while 36 percent said they were Democratic. Democratic affiliation, however, had fallen more than 10 percent in a year's time.[99] In another poll released at the time, by Lou Harris for ABC News, the Democrats' advantage in the November elections was only 49–40 percent in 104 competitive congressional races. A CBS News poll had congressional disapproval at 51 percent, but it also showed that nearly half thought their own representative should be reelected.[100]

━━━━━

Reagan and Ford, the two rivals who went down to the wire in their fight for the 1976 GOP nomination, remained publicly cordial as they circled each other for a possible 1980 rematch. In mid-September the two made a joint appearance in Texas to campaign for GOP gubernatorial nominee Bill Clements. They appeared gracious and charming toward each other. In confidence to columnist Bob Novak, however, Reagan told of his concern about the growing chasm between the two men. Reagan naïvely thought the fault lay with the "anti-Reagan" folks around Ford more than with Ford himself. In Dallas, Reagan said with a straight face to thirteen hundred Republicans that "history will record that Gerald Ford healed our land and by his example reminded us that this nation deserves our love and demands our courage." Ford turned to a friend and whispered, "Why didn't Reagan say that in 1976?"[101] Two of the nicest men in politics could not stand each other.

Publicly, Reagan was still maintaining a "we will see" public stance about running for 1980, though his frenetic schedule belied this. In private, he told at least one person that he was making the plunge . . . one more time. State senator Fred Eckert of Rochester, New York, a devoted Reaganite, said that Reagan told him, "Fred, we're going." Reagan had appeared at a fund-raiser for Eckert in upstate New York. He also told Eckert, "We'll be there" when the state senator told Reagan he'd see him in New Hampshire.[102]

Eckert was passionately for Reagan. The Soviet news agency *Pravda* was anything but, as it decried the state of the Republican Party, saying it did not "have much to count on," and the Kremlin predicted that the Democrats would pick up seats in the off-year elections. They singled out Reagan as the principal leader of the Republicans and said, "In his speeches on radio and television, the former actor who has become a politician criticizes the unions, attacks détente and calls for stepping up the arms race."[103] Reagan was getting under the Soviets' skin, and it must have pleased him mightily.

Despite the displeasure of the Soviets, the Republican tide was running in their favor, but they were letting expectations get out of control on how many seats they would pick up in the House and the Senate. It was becoming accepted that the GOP would have a net pickup of thirty or more in the House and three or more in the Senate. In fact, Howard Baker was having troubles in his own reelection. He was facing a stiffer-than-expected challenge from Jane Eskind, a wealthy liberal who, shock and surprise, opposed the Panama Canal Treaties, as they were hugely unpopular in the state. Other Republicans around the country were struggling as well.

On Election Day 1978, Republicans picked up 3 net seats in the Senate, improving to a 58–41 deficit, and 15 net seats in the House, where Democrats still enjoyed an overwhelming 277–158 edge. Democrats held more than 60 percent of the seats in the lower House, though Speaker Tip O'Neill would have to get by with fewer than the two-thirds support he had in the previous Congress.

Two of those three net Senate victories were in Iowa and New Hampshire, which the GOP national committees had by and large written off. The nominees were unreconstructed Reaganites, and it took all the browbeating they could muster to get anything from the RNC and the NRSC. Moderate Republican challengers had received far more help from the national committees. The party committees had the resources. Due to Rod Smith's prowess, the Republican Senatorial Campaign Committee had raised millions more than its Democratic counterpart had.[104] In Mississippi, Thad Cochran won the Senate contest, becoming the first Republican senator since Reconstruction to win there.

In Kansas, Nancy Landon Kassebaum, the diminutive daughter of the Sunflower State's popular former governor Alf Landon, who had been decimated as the GOP candidate in 1936 by FDR, also won a Senate seat. Conservatives were disappointed that another one of their own, Jeff Bell, had lost in New Jersey to Bill Bradley by twelve points.[105] The contest had been spirited and admirable, as both men simply debated the issues with no mud thrown. In fact, the two men debated twenty-one times, and the *New York Times* was forced to concede that Bell "emerged as the better campaigner. He has been articulate, earnest and witty."[106] Both were Ivy Leaguers, and Bell had once been a sports announcer at Columbia, calling games when Bradley was an All-American forward at Princeton while it was routinely crushing Columbia, and everyone else in the Ivy League. Bell never did end up asking Reagan to campaign for him personally and instead opted to bring in Ford, who had carried New Jersey in 1976.

Gordon Humphrey won in New Hampshire, scoring the biggest upset in the country. Nobody saw it coming. The Friday before the election, Eddie Mahe took Humphrey's seventh campaign manager, a pocket-size ne'er-do-well named Jim Murphy, and the campaign's press secretary, who had set much of the theme and tone of the campaign, out for a cup of coffee in Concord, near the campaign headquarters. Mahe stopped by the campaign once a week to advise and consult, as per his role with the national GOP Senate committee. Mahe asked the two young aides, "Well, who's gonna tell Patty [Humphrey's wife]?" The aides, befuddled, looked at Mahe and said, "Who's gonna tell Patty what?" Mahe said, "That Gordon's going to lose next Tuesday." Murphy and the Humphrey press aide told the older man, "Look, Eddie, maybe Gordon is going to lose, but something is going on out there. Gordon can feel it, and look at the way the mail, the volunteers and the phones have been popping for the last couple of weeks."

Meaning well, Mahe took it upon himself to tell Patty Humphrey that her husband would in all likelihood lose the election. When Patty returned to the office, she was a blubbering basket case, offering anyone in the office a bet on whether Gordon would win the following Tuesday. No one took her up on the offer, and fearing her outburst would hurt the morale of the campaign volunteers, Humphrey's press secretary took her aside, sloughed off another bet challenge, and instead gave her his handkerchief, which she used abundantly. She offered the handkerchief back, and the aide demurred. The next day on his desk was a gift-wrapped package from her; inside were a dozen silk handkerchiefs.

One other Humphrey tale must be told as a lesson to all those know-it-alls in American politics. Humphrey's Manchester city cocoordinator was a tall, striking Nordic woman by the name of Enid MacKenzie. She was a recent émigrée to America and had never worked in politics before, but she was a conservative. She ran the Manchester phone banks her way, and despite pleas to Humphrey by so-called experienced hands such as the hired gun Murphy to tell her how to run it correctly, Humphrey sided with MacKenzie.

In 1978, Rule 101 in politics was to "work the base," meaning identify potential supporters inside your own party and then move on to identify supporters among the registered independents. It was not taught to call your opponents' presumed voters. This is precisely what MacKenzie did. Rather than having her phone bank volunteers call Republicans and independents, she had them calling registered Democrats, in an attempt to convince them to change their minds and vote for Humphrey. Humphrey carried the overwhelmingly Democratic city of Manchester by five hundred votes. Six years earlier, the Republican nominee lost the city by six thousand. Who knew?

The pro-life movement was credited with the win in Iowa of Roger Jepsen over incumbent liberal Dick Clark.[107] Proposition 13–inspired referendums won in thirteen states. Tax cuts had been credited with Republican wins in many elections, but the fight was on in earnest inside the party; would the GOP emphasize image or issues? Tax cuts or good taste?

In other interesting campaign results, the Republicans lost a house seat in Florida when, before the election, the sixty-five-year-old J. Herbert Burke was arrested outside a strip bar for drunk and disorderly conduct, thus proving that an old dog can also learn a new trick.[108] Over in Arkansas, the youngest governor in the nation, a stripling at thirty-two, was elected: Bill Clinton. One minor candidate running in Illinois won his election for county auditor, only to learn that voters had abolished the job in a referendum the same day.[109]

In Texas, young George W. Bush lost his bid to win an open House seat to conservative Democrat Kent Hance. Bush had won a bruising primary over Reaganite Jimmy Reese, but never recovered for the fall election. The primary had become nasty and personal, as Reagan had personally campaigned for Reese, and Lyn Nofziger had sent Reese

$3,000 from CFTR.[110] A phone call between the elder Bush and Reagan was strained, to say the least, and other conservatives pounded the young Bush for his father's association with the Trilateral Commission and the Council on Foreign Relations, two internationalist groups that some paranoid conservatives suspected of seeking world domination. During the primary, Ambassador Bush complained, "I'm not interested in getting into an argument with Reagan . . . but I am surprised about what he is doing here, in my state . . . They are making a real effort to defeat George," his father had lamented.[111]

Two other conservatives running for the House who won that night were Newt Gingrich, on his third try in Georgia, and Dick Cheney, on his first try in Wyoming. Another prize for conservatives in New York was the election of William Carney in the First Congressional District.

A Democratic congressman, Leo Ryan, tragically met with a fate worse than losing an election: he lost his life when he went to Guyana to investigate rumors of a strange cult that had begun in his hometown of San Francisco. The cult, called the People's Temple and led by a man named Jim Jones, had established a camp in the sparsely populated South American country. While investigating, Ryan was shot and killed by Jones's guards. There, more than nine hundred followers of Jones drank or were forced to drink cyanide-laced grape Kool-Aid, or were injected with the deadly poison. Most of the nine hundred were Americans, and many were black. Mothers and fathers made their children drink the deadly mixture, and the macabre and horrible mass suicide gave rise to a tired and overused cliché.

The week before the elections, President Carter had gone to California, where he addressed raucous and disrespectful crowds. At one stop in Sacramento, protesters held up a banner opposing U.S. military involvement in Iran; and at another, they chanted against U.S. support for the Shah of Iran. The Shah, a strong ally of America's for many years, was implementing Western-style reforms in his country, including many that would benefit women. The Carter Administration urged the reforms on the Shah. Fundamentalist Muslims were becoming increasingly troublesome about these initiatives. Hundreds of people had been killed in rioting by the fanatical followers of Islam, with what the *Washington Post* was referring to as "snowballing unrest," the growing power of the militants, and the waning power of the Shah.[112]

The rioters were being egged on by the Ayatollah Ruhollah Khomeini, whose life the Shah had spared, ironically, and who was at the time in exile in Paris. Iran requested riot-control gear from the U.S. State Department—the same State Department that had urged the Shah to modernize in the first place, which had led to the rioting. U.S. government officials rejected Iran's request for the equipment. At the urging of the United States, the Shah was also releasing political prisoners from jail in the hope of placating his opponents, but this only emboldened them.

Carter openly speculated that the Shah would fall, and this was seen around the world as a clear sign that the United States would do nothing to help its old anti-Communist ally of many years. Carter's main focus seemed to be on "an absence of violence and bloodshed."[113]

That was it for Americans living and working in Iran. If their own government could not offer words of support for the Shah, they knew that a revolution was just around the corner, and they did not want to be in between the Iranians when they went at one another. A slow trickle of Americans leaving Iran became a flood. Sensible Iranians were also leaving their homeland.

In December 1978 there was upheaval in Afghanistan, as a Soviet-backed coup ousted the previous government, which had been less hospitable to Moscow. The Soviets angrily denied they were behind the ousting, but quickly signed a "treaty" with the new regime. Historians had known that the Russians and later the Soviets had coveted a warm-water port via Afghanistan and Pakistan. Of course, the route would have to go through the Khyber Pass. One journalist wrote that "some analysts wonder if the United States and the Soviet Union are involved in a regional trade-off—Moscow's continued good behavior in Iran swapped for similar American forbearance here."[114]

The Soviets' puppet in Kabul, Hafizullah Amin, had his strings pulled and said with a straight face, "Do you know any Soviet expert in the world who after World War II has intervened in the internal affairs of any country?"[115] The new socialist regime in Kabul announced plans for the collectivization of farming and manufacturing, with the assistance of thousands of Soviet advisers. They were going to keep doing it until they got it right.

Carter had made human rights the centerpiece of his foreign policy, especially in dealing with the Soviets, a staple of his early presidential rhetoric. But when Brezhnev fought back, Carter had backed away. In early December 1978, he gave a mild speech on human rights, but the biggest abuser of human rights on the globe, the Soviet Union, was lumped in with six other countries that, with the exception of Cambodia, were pikers in comparison to the Soviets when it came to running roughshod over their citizens: "Cambodia, Chile, Uganda, South Africa, Nicaragua and Ethiopia."[116] China was not even mentioned, as the president would shortly reveal why.

President Carter attempted to repair the damage of his previous statements that undermined the Shah, but it was too late. The genie was out of the bottle, and hundreds of thousands of Iranians, encouraged by Soviet agents, took to the streets calling for the overthrow of the Shah. Carter meekly warned the Soviets not to interfere; he could not even bring himself to mention the Russians by name in doing so.

Carter was fending off complaints from Republicans about the presence of Russian-made jet fighter planes in Cuba. They said the planes were in violation of the 1962 agreement between Kennedy and Khrushchev, but Carter worried that the public would make a linkage between the Soviets' latest move against America and his beloved SALT II. Anti-Communists worried about this newest chess move in Cuba by the Russians. Columnist George Will lacerated Carter over the arms pact, knocking down all the arguments the Administration was using, which were identical to those used by Nixon: the president would be "crippled"; it is "better than no deal"; and the old Washington soft shoe, "[W]e'll do better next time." Will, who two years before won the Pulitzer Prize for writing, showed why in his conclusion: "This time, as last time, the most remarkable aspect of the SALT debate is American readiness to justify Soviet behavior. And when supporters of the SALT II deal say things will be better next time, opponents reply: This is the next time."[117] (Many months earlier, Reagan, too, had made clear his disdain for the treaty, saying that Carter simply wanted a treaty, "good or bad.")[118]

Carter stunned the nation on Friday, December 15, when he announced that he was abrogating America's mutual defense treaty with Taiwan in favor of full recognition of the Communist government of mainland China. All official contact with Taiwan would be severed, including the Mutual Defense Treaty, and a friendship that had lasted since 1949 ended. Since the beginning of the Cold War and the "Who lost China?" debate that swept across America, all sensible anti-Communist politicians paid testimony to their loyalty to the Republic of China on Taiwan.

Gerald Ford was the only leading Republican who supported Carter on his decision to recognize the government in Peking. Across the board, every other leading Republican, from Ronald Reagan to Barry Goldwater to George Bush, eviscerated Carter for his actions. Goldwater called it "one of the most cowardly acts ever performed by a president," and Bush said, "[W]e gave all and got nothing" from the Communist Chinese. Some Democrats, including Senator John Glenn of Ohio, condemned Carter. Carter's statement said the only reason for the move was "peace," but it contained no mention of "freedom."[119]

Carter claimed he first received assurances from the Communist government that it would not move militarily against Taiwan, but when reporters pressed, "[A]dministration officials who briefed reporters following the speech provided little additional information."[120] Conservatives weren't going to take this lying down. A court case was prepared, challenging whether Carter could abrogate a treaty of the United States without Senate approval. Again, as in so many cases over the previous three years, from the Panama Canal Treaties to Proposition 13 to the ERA to SALT II and tax cuts, conservatives would maximize issues to their fullest advantage. George Will picked up the verbal equivalent of a thirty-two-ounce Louisville Slugger and walloped Carter and the State Department. He chalked up the move to the "ideology of the State Department's seventh floor" and said that Carter had done "more to jeopardize the rights of the Taiwanese than he has done to enhance the rights of any other people."[121] Anti-American protesters streamed into the streets of Taipei by the thousands; and one man, protesting Carter's action, doused himself in gasoline and set himself ablaze. Others stomped on peanuts to protest the president.

———

While in West Germany on a world tour, Reagan was struck by the evilness of the Berlin Wall. He later wrote a column about the complexity of it, the monstrous meaning of it and how it represented the great struggle between East and West.

"The bottom line is still human freedom," Reagan said.[122]

———

Jimmy Carter's numbers accelerated downward, and he appeared to be in a political freefall. In early November, an NBC/Associated Press poll found that Carter's presidency was now viewed as good or excellent by only 41 percent, while 57 percent said it was either fair or poor. Only 23 percent of self-described conservatives gave him favorable ratings. The survey was monstrous, as almost thirty-five thousand people were queried.[123] The bad shape of the economy was cited by many as their reason for disapproving of Carter's performance. Near the end of the year, the Soviets and the United States announced they had reached an agreement over strategic arms, but it still had to be approved by the U.S. Senate. The *Post* was skeptical, writing bitingly that "the real SALT issue is whether the Senate thinks Jimmy Carter is a fit guardian of the nation's security."[124] Carter and Leonid Brezhnev were photographed bussing each other, European-style, in what would later be a widely ridiculed embrace.

John Sears and labor leader George Meany spoke to the Public Affairs Council in Washington, and both had grim takes on Carter and 1980. Bluntly, Sears said, "If I were Carter, I'd be worried to death." Meany, eighty-four, knew the American worker and pointedly told the crowd of business leaders that no one should underestimate Ronald Reagan in 1980.[125] Carter was losing his base of support, people who thought he was different from other politicians. Tom Braden, writing for the *Los Angeles Times*, cited an example of Carter firing someone in the government who happened to be a friend of Tip O'Neill's. O'Neill pitched a fit, and Carter subsequently gave the man another job in the government.[126]

———

Republicans had reason to be joyful as 1978 came to a close. The air had been cleared of the stench of Watergate. Many of, if not all, the old internecine ideological fights had been swept under the rug. The band was back together.

A new, self-confident, freshly minted Republican Party was stepping out, champing at the bit to take on Carter and the Democrats. The Republicans (so recently the political equivalent of a ninety-eight-pound weakling) had bulked up on ideas, self-confidence, and money. The conservative movement was bursting with new initiatives on economic, social, national defense, and foreign policy matters. New conservative publications and columnists were venturing forth. Attractive and articulate young conservative leaders were coming to the fore as well. They vowed to win the pretty girl, or in this case the voters, back from the bully Democrats. The *Washington Post* was impressed with the GOP's gains and editorialized about the "Republican rejuvenation."[127]

Now it was the Republicans' turn to kick sand in the Democrats' faces.

Reagan on Ice

"Reagan is a lot older than I think he ought to be."

S ince, and even before, the 1976 convention in Kansas City, the "age issue" had off and on dogged Ronald Reagan. However, the resistance to one more campaign by the Gipper was now also coming from some of the men and women across America, for whose votes he would eventually ask, in the 1980 GOP primaries. If he ran again. The complaints were also coming more and more openly from the national media and the chattering classes in Washington, whose main industry was speculation and gossip. Questions about his abilities also came from some of Reagan's own supporters.[1] The media harped on it constantly, and in Jack Germond and Jules Witcover's political column, they complained about "the spots and veins of his hands."[2] "Reagan can't overcome that age thing," his former Georgia coordinator Betty Jones, who had switched to Phil Crane, told *Newsweek*.[3] More and more, he found himself queried in forums by voters about his age, and he repeatedly said, "I feel just fine."[4] But as 1979 wore on, Reagan was on the road less and less, which only added more and more to the notion that he was not up to the job of president.

It was not how Reagan looked or felt that they were hung up on, but the fact that he would be seventy years old in February 1981, just one month after being sworn in. All knew the mantra by this point: "If inaugurated, Reagan would be the oldest president in American history." Part of this idle chatter included the historical fact that the oldest president up to that point had been William Henry Harrison, a sixty-eight-year-old codger elected in a decade in the first half of the nineteenth century when the average life expectancy for an American male was barely forty, and who had been foolish enough to give a two-hour inaugural address in the cold, pouring rain in March 1841. Harrison caught pneumonia and died one month later.

The unfortunate William Henry Harrison, however, was not the proper comparison when it came to the age issue for Reagan, as a cursory study of history by his staff would have shown. Instead, Calvin Coolidge, one of Reagan's favorite presidents, and Dwight Eisenhower were more appropriate. Had Reagan won in 1976, he would have been three years younger on his inauguration, at sixty-six, than the average male life expectancy of sixty-nine that year, little different from Coolidge, who was fifty-three when inaugurated to his first full term in 1925, four years younger than the average male life expectancy that year of fifty-seven.[5]

Concern about presidential age was a relatively recent political phenomenon, historically. Every president, covering twenty-nine inaugural ceremonies from George Washington in 1789 to William McKinley in 1901, was significantly older than the median life expectancy of his era. In those 112 years, the average life expectancy for an American male was 41.6 years, but the average age of all the presidents during that time was 56.0.[6]

In the twentieth century, 60 and older became the new 41.6, and that fifteen-year age differential between the president's age and average male life expectancy of the nineteenth century shrank dramatically. In fact, all the presidents of the twentieth century after Woodrow Wilson were younger than the average male life expectancy when they were first inaugurated.

President Carter's aides confidently predicted that the Californian would "be denied the nomination because he is too old," and that would be the end of Reagan, once and for all. Carter, they believed, would be facing a younger GOP contender.[7] Not that they were happy about it. They saw Reagan as easily beatable. The doubters of the national media, the GOP, disgruntled former Reagan staffers, and the other Republican aspirants were out for Reagan. Anti-Reagan conspirators made deals against Reagan to destroy his candidacy and with it his nettlesome conservatism.

But he soldiered on in early 1979, going to Mexico for a meeting with President José López Portillo. He then gave a major policy speech at Pepperdine on trade matters, taking American businesses to task for not trying to sell their products more aggressively, as he'd been told by Japanese officials. He also spelled out his concerns about the growing superiority of the Soviets' nuclear capability.[8]

Despite stumbles, Reagan's hard work for the past many years had paid off by the dawn of 1979. Since Gerald Ford lost narrowly to Jimmy Carter in November 1976, Ford had continuously beaten Reagan in trial heat polling among Republicans until November 1978, when, in a *New York Times*/CBS poll, Reagan became the preferred standard-bearer over Ford, 37–22.[9]

The Californian knew where he was going and the consequences of the times he lived in, and his speaking style was muscular. Yet, as always, he was underestimated. *U.S. News & World Report* came out with its annual ranking of the thirty most influential Americans as judged by nearly fifteen hundred influential Americans. Reagan came in a distant twenty-second, behind Ralph Nader and just ahead of the First Lady, Rosalynn Carter. Howard Baker was the only other Republican on the list. The rest were media figures, Democratic officials, industrialists, and labor leaders.[10]

The complaints about his age notwithstanding, Reagan neither looked nor acted like a man deep into his sixties. He seemed to all at least ten or more years younger. Tom Wicker of the *New York Times* said as much.[11] His wrinkles became more pronounced around his jawline, but he was also over six feet tall, broad-shouldered, blue-eyed, with a winning smile and winsome ways, and he always looked handsome, regardless of what he was wearing.

He always tied his tie in a Windsor fashion and wore shirts with a spread collar. Mike Deaver once shook his head and said that somewhere along the way, some Hollywood type had told Reagan that his head was too small for his body, and Reagan had believed him. To fix this, the entertainment operative said, a little visual magic, including the even-sided, triangle-shaped knot and shirt gimmick, would work. Many men in politics at the time wore preppy button-down shirts and tied their ties in an even preppier four-in-hand, or "schoolboy," uneven knot fashion. Not Reagan.[12]

UPI reporter Don Lambro went to Reagan's ranch for an interview in the late 1970s, and upon his arrival, he found Reagan high in a tree clad only in his bathing suit, cutting limbs.[13] He was vigorous, loved the outdoors, and enjoyed a good joke, the more ribald the better. "He is remarkably fit and moves through 12-hour days during speechmaking tours with the bounce of a man in his prime. He seems to thrive on work, using spare time on airplanes and in hotel rooms to write commentaries for his syndicated radio program and newspaper column. 'I wasn't built for a rocking chair,' he says."[14]

Still, more than a few people would express their concern about his age: "Reagan is a lot older than I think he ought to be."[15] It was a scab that the media; GOP officials; Reagan's primary opponents, including Congressman Phil Crane; and some of Reagan's own people kept picking at, and the wound kept opening more and more. Crane was a "Reagan without wrinkles," according to one of his supporters.[16] Reagan shrugged it off with good humor in public, but in private, he often fumed. Gamely, he said, "You know, I was in the Orient last year. They thought I was too young."[17]

He tried to do his best to deflect the growing criticism by saying voters would judge if he was "physically competent" and if he had all his "faculties," but he was being more cautious. If he made a verbal slip in public, "Reagan takes pains to correct it, something he rarely bothered to do when he was running for governor of California."[18] The *Los Angeles Herald-Examiner* gamely came to his defense, saying that a recent editorial cartoon it had run, with Reagan's smile depicted as dentures on a bedside nightstand, was grotesquely unfair. The editorial said that all the harping on Reagan's age was out of control, "obscene," "irrelevant," and not helpful: "It is possible for a president to die in office for reasons having nothing to do with age."[19]

One of the few columnists who did not harp on Reagan's age was George F. Will of *Newsweek*. "From Jerry Brown's father to Jerry Ford, many people have paid a high price for underestimating Reagan."[20] Will had once been a Doubting Tory when it came to populist conservatism, but as the established liberal order on both sides of the aisle in Washington reeled under the weight of incompetence and corruption, Will became more and more repulsed by what he saw, and hence much more of a Washington critic and Reagan champion.

As with much of the Republican establishment, Reagan was also not held in great esteem by the new crop of moderate GOP governors, ironically by many for whom he'd campaigned. A "Stop Reagan" effort was floated, although it could not decide on whom to get behind other than Gerald Ford—and recently he'd been giving off more and more signals that he was not going to make the race. David Broder of the *Post* wrote of Reagan and ex-Democrat John Connally that "neither . . . is widely revered among the GOP executives."[21] Broder, while acknowledging that Reagan was the presumed front-runner, echoed the sentiments of all the other candidates: "The threat to Reagan is a misstep or blunder that might cost him victory."[22] Jim Thompson, the new governor of Illinois, was one of the more vocal anti-Reagan ringleaders.[23]

In late January, Reagan journeyed to Washington for a number of meetings with his top supporters and to woo some new friends, organized by Senator Paul Laxalt. Laxalt was reaching out to more moderate Republicans, but he maladroitly highlighted Reagan's two main weaknesses: "You're not talking about a right wing nut with horns out of his ears . . . [Y]ou're talking about a responsible conservative whose age may be a problem."[24] Reagan and Laxalt hosted a private dinner with a group of GOP senators in the U.S. Capitol the evening of President Carter's State of the Union address. "A Gallup poll released today shows that Carter has a 57–35 percent lead over Reagan. The survey . . . indicates a steadily widening margin for the incumbent over the Californian." Since the previous spring, Carter had gone up 6 percent and Reagan had fallen 10 percent in trial heats.[25] Reagan had been dropping steadily in head-to-head polling. Carter was getting a second look from the voters, but Reagan wasn't getting any looks as he was pulled back more and more from public appearances as the year wore on.

John Sears wanted to conserve Reagan's strengths, not knowing that Reagan's strength came from being out and about, speaking to and meeting with people. This decision by Sears to "ice Reagan" would be his most disastrous, as it nearly cost Reagan the GOP nomination in 1980. On Sears, Reagan wrote to William Loeb, publisher for the *Manchester Union-Leader*: "We are well aware of his shortcomings as well as his talents . . . But the head man is definitely Paul Laxalt, and John will be doing those things we think he does well, but not a solo job with all of us wondering what happens next."[26]

Carter's strength relative to Reagan's gave the White House a lot of confidence that their policies were correct, including relations with mainland China. Carter had invited the Chinese premier, Deng Xiaoping, to Washington for a series of high-profile meetings.[27] The new senator from New Hampshire, Gordon Humphrey, was invited to a command performance at the Kennedy Center. It was not a request. A command performance from the president meant you must go. Humphrey had already signed on to the conservative lawsuit against Carter over Taiwan and was not enthralled with the niceties of Washington. He rejected the invitation from the White House, scribbling, "So solly! And I don't need a damned reason, either!"[28] Deng was seventy-four years old when he assumed power in China and eighty-eight when he stepped down, yet at no time did any reporter in Washington speculate that Deng was too old to be leader of China.

———

Carter decided to set an example for the American people by having solar panels installed at the White House to heat water. But the project went way over budget.[29] Carter also proposed federal subsidies for wood-burning stoves. A favorite bumper sticker of the era proclaimed, "Split Wood, Not Atoms."[30]

Reagan had mellowed, and while the fire inside had never waned, he used more intellectual humor to get his point across. "We're so used to talking billions, I wonder, does anyone realize what a single billion is? A billion minutes ago, Christ was walking on this earth. A billion hours ago, our ancestors lived in caves. And a billion dollars ago was 19 hours ago in Washington, D.C."[31]

━━━━━━━

The Gallup poll in January 1979 had Reagan with a name identification of 95 percent among all Americans. Within the GOP, he was leading Gerald Ford 40 percent to 24 percent, although it was much closer among independent voters, 26 to 25. No one else was even close. Senator Howard Baker was at 9 percent, John Connally was at 6 percent, and George Bush wasn't even an asterisk.[32] Internecine warfare within the Reagan camp had not broken out yet, and they were supremely confident, at least on the surface. Yet rather than working hard to hold or extend his lead, Reagan was working a light schedule, and consequently his lead in the polls was slowly beginning to diminish. "What is also implicit, however, is a recognition that any signs of slippage early in the campaign are going to increase the pressure on the age issue. If Reagan loses his image as a winner, the actuarial tables are suddenly going to become far more important politically than the polls suggest today."[33]

Reagan, through it all, remained serene—at least publicly. Privately, he was concerned.[34] He was still one of the most persuasive politicians in America, in front of large groups and small. However, one who in fact could top him, especially in a small setting, jumped into the fray. John B. Connally declared himself available for the GOP nomination at the National Press Club in Washington in late January. Connally was a smooth operator and put in an impressive showing before a packed media house. Connally had signed as his campaign manager the soft-spoken party technician Eddie Mahe.[35] Mahe would have preferred to work for Reagan, but John Sears would never have allowed him any actual power. Plus, he'd been in an ongoing feud with Lyn Nofziger about the neutrality of the RNC in the 1976 campaign. Mahe was offered the number-two slot answering to Jim Baker in the nascent Bush campaign, but Mahe took a pass. A year and a half with the vacillating Bush at the RNC had been enough for Mahe.[36]

Earlier in the month, another Texan—though Connally disputed this—Ambassador George H. W. Bush filed papers with the Federal Election Commission that left him just short of a formal declaration of candidacy; but there was little, if any, doubt that he, too, was spoiling to take on Reagan et al., and win the nomination. Bush started with meager assets out of an eight-room office in Houston, but the phone number was easy to remember: GOP-1980. His formal announcement that he was in the race would not come until May 1.[37]

Plans were under way for a "Reagan for President" headquarters to open shortly, to be followed later in the year with an official announcement of candidacy by "The Governor." Washington columnists and politicos, including Democratic operatives inside and outside the Carter White House, were frankly flabbergasted that the front-runner for the GOP nomination was this old, right-wing extremist. They could scarcely believe their good fortune. One Carter operative snidely said of Reagan, "Is it possible that one man could make Richard Nixon and Gerry Ford look good?"[38] First, however, Jimmy Carter had to get past Ted Kennedy if he should decide to jump into the primaries.

Carter was waning before the waxing Ted Kennedy–for–President groundswell. Carter's weakness against Kennedy was showing up in the polls, especially in the West and, surprisingly, in his native South.[39] Although Carter was losing to Kennedy among Democrats, he was whopping both Ford and Reagan among all voters, 53–39 and 57–35, respectively.[40] Carter's White House team believed that if they could get past Kennedy in the primaries, they could coast to reelection, despite the president's attendant problems, which of course were offset by Kennedy's own land rush of problems.

Reagan's initial campaign office was in a small building at 809 Cameron Street, in historic Old Town Alexandria. The building also housed a psychiatrist, a handy happenstance for the colorful operatives who were, to put it charitably, idiosyncratic. Newly hired Bush political aide David Keene, along with a group of friends including Lyn Nofziger, Roger Stone, and Paul Russo, all with the Reagan campaign, and Tom Winter, editor of the conservative journal *Human Events*, owned the building.

Though Keene had recently left Reagan, unpleasantly, the Gipper's political organization was still paying Keene and his group in the form of a $1,200-per-month rental payment. Washington was a small town.[41] Russo had taken the position as Reagan's liaison to Capitol Hill that Keene had turned down. Nofziger tweaked Keene, saying with tongue in cheek, "We think we owed it to him after all the nice things he has said about us since he went to work for George Bush."[42] It was all very "inside baseball" stuff, but whenever Reagan came to Washington, he almost always called or met for a cup of coffee with Winter and his esteemed colleague Allan Ryskind. Ryskind was the son of famed Hollywood screenwriter Morrie Ryskind, of Marx Brothers fame.

Bush and Reagan were already competing for political staff. Sometimes old friendships and alliances were damaged as the participants jockeyed for position, power, and money. Keene, a Reagan staffer from the '76 campaign, was one of the first players in this inside-baseball drama. He'd expected a promotion from the field staff for 1980, but when Sears offered him a job dealing with interest groups, he blanched. The idea of working with "Serbs for Reagan" or "Poles for Reagan" or with some complaining congressman was distinctly unappealing to the habitually impatient Keene. Either job would also mean a great deal of travel, and Keene was a ten-to-four guy who liked to go hunting and fishing as much as possible on the weekends and have dinner with friends during the week. Reagan had once quipped about his own work habits, "It's true [that] hard work never killed anybody, but I figure, why take the chance?"[43] Keene utterly agreed on that point, but the comparisons ended there. Keene had a low opinion of the Reagans and often made jokes about both. Keene was a good talker, but no one could ever really figure out what it was he did—other than talk.

Perhaps sensing something amiss about Keene, Reagan wouldn't pull the trigger, however, and did not specifically ask the heavyset apparatchik to stay with his campaign even after being told by Nofziger that Keene would stay if asked. Keene specified that if he did leave for the Bush campaign, which was putting the heavy woo on him, he must receive exactly the same salary ($50,000 per year) that he anticipated receiving from Reagan. He claimed he didn't want anyone saying that he'd sold out on Reagan for money.[44] He was running a bluff, but this was Reagan, who had outmaneuvered the fearsome Hollywood moguls. It was akin to pitting George Washington against George of the Jungle. Also factoring in was Nancy Reagan's and Mike Deaver's concerns about Keene.

Keene was not sure if he wanted to accept the firm offer from the Bush campaign or the anticipated offer, still somewhat vague on the details, from the Reagan campaign. Even then, he wavered until some in the Reagan camp, including Mike Deaver, told reporters he'd sold out on Reagan for money. Reagan, when asked by a reporter, also seemed to suggest that Keene had left over a salary dispute. Keene burned a hot phone line to the Reagan campaign and Deaver's attack stopped, but that was it. Keene now had the excuse he was waiting for. He was going with Bush.[45] The Reagan campaign heaved a huge sigh of relief.

———

As battle lines were starting to be drawn in DC, Vietnam and Cambodia moved toward war. The Communist country of Vietnam had exterminated untold numbers, and Cambodia's Communist leader, Pol Pot, and his Khmer Rouge snuffed out the lives of millions. Pol Pot was eventually overthrown by the Vietnamese, but not before millions of innocents died in the bloody chaos.

Pope John Paul II had his own message for Communists, especially those in the Kremlin and those occupying his beloved Poland. Communist officials in Warsaw had jammed his Christmas radio broadcast, so several days later, he ordered the powerful Vatican Radio to give it all it had, and he broadcast yet another speech, this one ripping Marxism. The UPI reported that "his voice often trembled with indignation."[46]

At the halfway mark of the Carter presidency, George Will evaluated the president's performance and found it wanting, to say the least. Two years earlier, Will had praised Carter in several columns, because the president was moving toward policies that would be tougher on the Soviets than Gerald Ford and Richard Nixon had been.[47] But Carter reversed his field, and Will raked him over the coals. Other transgressions were not overlooked, either. For instance, while running for office, Carter had supported both deregulation of natural gas and aid to parents with children in parochial schools, but after becoming president, he actively opposed both initiatives.[48]

Carter was racking up a record as a flip-flopper and vacillator, managing to alienate groups who ordinarily would have been in his camp. Notably, during his 1976 campaign, he had promised full amnesty for Vietnam War draft evaders; once in office, he partially reneged, thus pleasing nobody.[49] He pushed for intermediate missiles in Europe, twisting the arms of skeptical allies such as Helmut Schmidt, prime minister of West Germany, and getting them to embrace the missile deployment despite domestic opposition in their countries. Then he abruptly shelved the idea, causing allies who had gone out on a limb for him to tear their hair out in exasperation. The name "Carter" was becoming a synonym for "incompetent."

The president's personal life offered no refuge. His woes with his brother, Billy, mounted, to the point where the sibling rivalry became the stuff of tragicomedy. Billy's newest antic was to give an interview to *Penthouse* magazine in which he said Jody Powell "should be running a farm" and dismissed Hamilton Jordan with a curse word. Sometimes he could be not only outrageous but also deeply hurtful and cruel. In 1977, Billy Carter was at an event featuring a black candidate in Oakland, California, by the name of Carter Gilmore. "When Gilmore made a passing reference to their common name, Billy commented: 'We all left a n— in the woodpile somewhere.'"[50] Billy also had signed a potentially illegal and certainly embarrassing lobbying deal with Libya.

In another incident, at the Atlanta airport, Billy Carter emerged from his limousine, in full view of reporters and an Arab diplomat, opened his trousers, and relieved himself. He was an equal-opportunity annoyer. At a roast for Phil Niekro, the Atlanta Braves pitcher, Billy said to him, "I didn't know you were a Polack; I thought you were a bastardized Jew."[51] As a justification for taking money from Libya, Fred Barnes of the *Washington Star* reported that Billy said, "There's a hell of a lot more Arabians than there is Jews."[52]

Not surprisingly, American Jews and their supporters were growing increasingly critical of the Carter Administration. The Russians had a long history of anti-Semitism that went back to the pogroms of the czar, but Soviet policies were growing particularly harsh in the 1970s, drawing international outrage on the plight of Soviet Jews. Carter's SALT II did not address this issue, but it did not mean the U.S. Senate would not ignore their treatment or oppression. Conservatives urged "linkage" of a myriad of issues for the Senate to consider when taking up the arms pact, including Soviet Jewry, Cuba, Afghanistan, and other Soviet military adventures. As Howard Baker put it, "Linkage is a fact of life."[53]

New Hampshire was still the first real contest in each presidential season, at least as far as the political intelligentsia was concerned, although the Iowa caucuses were moving up in importance. New Hampshire was the first primary, and reporters and politicos looked forward to the quadrennial event. It was their chance to watch the candidates, laugh at their antics, make sport of the quirky New Englanders behind their backs, and of course eat and drink excessively with their fellow "knights of the keyboard" at all their old haunts, especially in and around Manchester and at Concord's New Hampshire Highway Hotel. The Granite State's first-in-the-nation primary was political junkie heaven.

One of those quirky New Hampshirites, Gordon Humphrey, had won the previous November, scoring a huge upset, but Humphrey was faced with $100,000 in campaign debts.[54] If you are the incoming freshman senator from New Hampshire a little over a year before the presidential primary and you are shopping for a candidate to support for president, what's the first thing you do?

You throw a big shindig in your honor, get all the 1980 GOP would-be presidents in one room with the national media, charge the locals a sum they can afford, but still enough to eliminate your debt, and—you have a good time. Bob Dole, Phil Crane, Howard Baker, Jack Kemp, and Senator Orrin Hatch of Utah, representing Reagan, all trundled up to snowy New Hampshire for the festivities. Reagan and John Connally could not attend due to scheduling conflicts, but they sent telegrams of congratulations to the incoming conservative senator.[55]

George Bush also wanted to attend. Humphrey, remembering the nasty slight by Bush several months earlier, wouldn't hear of it. Each time someone from Bush's staff called, the request to attend was relayed to the senator-elect, and each time his answer was an emphatic no. Bush's staff was becoming frantic, and young aide Karl Rove was dispatched to New Hampshire to see if he could pry loose an invitation for Ambassador Bush. He met with one of Humphrey's aides, whom he'd known casually in the College Republicans. The two young operatives got down to cases quickly, and the Humphrey aide told Rove, in no uncertain terms, that Ambassador Bush would not be invited.

Rove pressed, and in exasperation, the aide said, "Karl, you don't have to convince the rest of us—it's Gordon you've got to convince!" Suffice it to say, Bush was shut out of the big event, which was swarming with national media and hundreds of influential New Hampshire Republicans, warmed that chilly night by Humphrey's victory, good cheer, and free-flowing booze.[56]

Humphrey had really not decided on whom to support. He was personally fond of Dole, respecting the war hero, but he also liked Crane, admiring his intellect. He was truly focused on getting his offices in Washington and New Hampshire organized, choosing his committee assignments, and hiring staff. The media, however, were not interested in this mundane nonsense. What they wanted to know was who Humphrey, the giant killer, was going to support for president. The stakes had risen as the media focused on the Humphrey operation and the lists of volunteers, contributors, and operatives it had collected. Humphrey, they knew, wouldn't be coming to the game without his bat and ball.

The stakes got even higher when a reporter for the Associated Press asked Humphrey's inexperienced press secretary about the lineup and he blurted out, "[T]he batting order reads Crane, Reagan, Dole"[57]—on the record. The wire story ran across the nation the next day and escaped no one's attention.

That did it. All hell broke loose for the poor, unsuspecting Humphrey as he now heard it from all sides, including Loeb and the state GOP chairman, Jerry Carmen, who had helped Humphrey at a time when few gave him a lick's worth of a chance. Carmen had already committed to Reagan, and it would have been most embarrassing for him not to bring along the new, freshman Republican senator from his home state.

Humphrey did not order his press secretary to "drink the Kool-Aid," as hundreds of followers of the Reverend Jim Jones had done in Guyana several months earlier in committing mass suicide, but the thought must have crossed Humphrey's mind. A series of hurried meetings and phone calls eventually quelled the situation, even though the young aide, in upping the ante, had also raised Humphrey's importance and influence.[58]

Carmen won reelection narrowly as state party chair. His challenge had come from a stalking horse for George Bush, James Masiello, whom Humphrey had defeated in the GOP primary the previous year. It was the newest version of the same old state power struggle, usually with conservatives losing. However, Carmen, with Humphrey's explicit and Reagan's quiet support, prevailed 198–186.[59] Humphrey eventually endorsed Reagan and, in time, would assume a small stagehand role in one of the biggest dramas in American political theater.

One of the first hires of the nascent Reagan campaign, as yet neither announced nor formalized by the establishment of a presidential campaign committee, in December '78, was Charlie Black, late of the Republican National Committee. This was Black's second tour with Reagan, and his work four years earlier, plus the rave reviews he'd received as political director at the RNC, meant he was due for a promotion. In 1976 he'd been a regional political director with meager resources. This time around, he would be national political director with a projected pot full of money to help the front-runner sew up an early nomination. The last thing anyone wanted was another long and bruising battle down to the wire as in 1976. The goal was to swathe Reagan in gauze to prevent him from getting bloodied in the early nomination process and then wrap up the nomination quickly, cleanly, and efficiently. Black received a generous pay raise from the last time he'd worked on a Reagan for President campaign.[60]

By January 1979, many people had resolved that America was not much fun as a place to live or work—if one could find work. Hollywood had taken note of the sour mood of the nation and produced a low-budget film called *The Late, Great Planet Earth*, based on a bestselling book of the same name. The author of both the book and the film, millenarian fanatic Hal Lindsey, prophesied the imminent demise of the planet, though he could not decide if it would come from killer bees, floods, famine, nuclear war, pollution, or the Rev. Sun Myung Moon. The movie also tried to use a computer to determine if Carter, Kennedy, or Reagan was the anti-Christ.

The Republicans appeared, on the surface at least, to be more united than at any time in recent memory. Many had coalesced around Reagan's opposition to the Panama Canal Treaties (Tennesseans Bill Brock and Howard Baker being the two notable exceptions), and they were becoming unified on tax cuts, government spending, regulations, a strong national defense, and anticommunism. Although they were less unified on abortion, the Equal Rights Amendment, and other social issues, they were together 100 percent in their opposition to Jimmy Carter.

Opposition to "big government" was one of the tenets of the new GOP. On big government, it depended on who was beholding it, but Reagan knew it when he saw it and made clear his disdain for it. In a rare appearance before the American Farm Bureau's annual convention in Miami Beach, he called for a new alliance of labor, agriculture, and small business to stand up to the "arrogance" of big government. He also told the nine thousand cheering farmers that they need not operate with government subsidies, but could and should operate in the free market, where they could fetch a fair price for their goods. Of doing business with the government, Reagan told the tittering crowd, "If you get into bed with government, you're going to get more than a good night's rest."[61]

The new GOP was moving forward, but the unity was sometimes hard to maintain in practice. Opposition to Carter was one unifier, and the party had most definitely moved rightward on many issues, but there were still deep cultural fissures in it, the fight over the site of the 1980 convention being one good example.

Reagan's men wanted the Republican convention in Dallas, center of operations in the New West and the name of a new, racy, popular Friday-night television series on CBS, but after pulling the party from the morass it had been in since January 1977, RNC chairman Bill Brock was feeling his oats and opposed the selection of Dallas. He wanted instead for the convention to take place in, of all cities, Detroit, ground zero for organized labor, the muscle of the Democratic Party.

It had been twenty years since the party journeyed north of the Mason-Dixon Line or east of the Mississippi River for its national convention. Brock's opponents fought him on the grounds that Republicans would encounter massive protests, that Detroit was an economic dystopia, that it was overwhelmingly black, and that the city itself looked like a war zone from years of race riots, neglect, and decay.[62] It was also a city with nearly 100 percent Democratic voter registration. Republicans were almost as scarce as jobs in Detroit in 1979. One member of the Republican National Committee complained that not only was Detroit a "Democratic stronghold," but it was also a "crummy city."[63]

They all missed the point. The symbolism of picking Motown was what Brock was considering. For two years, he had assiduously wooed black, labor, and urban voters for the GOP, not by specifically trimming the party's ideological sails, but by simply talking to their leaders, making overtures, and keeping the lines of communication open. Although he may not have increased appreciably the number of blacks registering Republican, Brock was, just by reaching out, winning the plaudits of the media, lowering the long-standing hostilities between the GOP and urban America, and, most important, soothing the concerns of other Republicans about how the party went about reaching out beyond the country club and the Sun Belt. Brock was having a "dialogue" long before the term became fashionable.

GOP conventions, up until 1960, had been selected not for symbolism but for practicality, including ease of rail travel, which is why almost half the party's conventions had been held in Chicago, the last being in 1960, before the explosion in commercial jet travel. In addition, as the party of Lincoln, which was built on destroying slavery, it could not hold conventions in the segregationist South without fear of "massive resistance." No one considered Miami Beach, Nixon's preference in 1968 and 1972, as part of the South. Republicans had chosen Philadelphia in 1948 in part because it was in the North, it was easy to get to, and it had a rudimentary television system.

Conventions had grown in size and scope over the years. What used to be a weeklong collection of party bosses and delegates had grown into a two-week media extravaganza that involved tens of thousands of people and hundreds of millions of dollars. Cities hired lobbyists and influence peddlers to try to win national conventions for their municipalities, because the conventions poured tons of money into the local economy.

Delegates in both parties, though, were steadily diminishing in influence. For years, men and women from all walks of life (housewives, small businessmen, local elected officials, church leaders) came together after they were elected or selected locally and took their responsibilities very seriously, including writing the all-important platform. For the week before the actual convention, these patriotic people would listen to testimony from senators, Cabinet officials, captains of industry, philosophers, military leaders, labor leaders, and other nationally important or influential individuals. From this testimony and endless discussion over endless cigarettes and coffee would come impressive documents, known as "platforms," stating the party's goals, dreams, and aspirations.

———————

Despite their election successes, all was not sweetness and light at the Republican National Committee heading into the Christmas season, as nearly sixty employees were laid off. The committee was half a million dollars in debt, having used even more than the record $20 million it had raised that year. Contributions slowed after the elections, and the staff was reduced to around two hundred. Their counterparts at the Democratic National Committee were going through the same, only in spades. Their staff had gone from two hundred individuals to about fifty.[64] Trench politics could also be hazardous duty.

Nevertheless, the growth of the political consulting classes and the increasing importance of primaries would signal the demise of the functional convention delegate. High school physics didn't teach that Sir Isaac Newton's theories also applied to politics; power cannot be destroyed, but only recast and moved around, and the power of the delegates, unfortunately, was draining away, to be sucked up by the political consultants. These operators feared that anything beyond their control, such as a delegate with an opinion of his or her own or an idea that was not generated by the consultant, was dangerous and therefore had to be muzzled or destroyed.

Brock also had popular opinion to consider. The fight over the Equal Rights Amendment was in full force. The party had a tradition of supporting universal suffrage. But this controversial amendment had mixed support, and not just among women within the GOP. The elites of both parties and of the media, the academy, and entertainment, supported the ERA. However, a small but growing vocal and determined band of women, led by the estimable Phyllis Schlafly, thought something else, namely that the ERA would lead to needless lawsuits and a breakdown in traditional family roles. An unwritten "law" of the Republican Party at that time was that it could not hold its national convention in any state whose legislature had not ratified the ERA. Michigan had done so, but Texas had not. Brock had enough sway over the voting members of the RNC to prevail, albeit by only one vote. The party was headed to the Motor City, and it was "a vindication of Mr. Brock and the kind of leadership he has tried to give the party . . . that the party must reach out, not merely try to further please those voters it pleases already."[65]

In mid-January, Reagan met in Los Angeles with approximately fifty-seven members of the steering committee of CFTR and told them he could not foresee any circumstances that would prevent him from running for president.[66] Yet he was still saying publicly that he'd not decided whether to throw his cowboy hat into the ring. Members of the steering committee had already been involved in the 1976 campaign and in Citizens for the Republic, Reagan's PAC. These included old friends such as Ray Barnhart and Ernie Angelo from Texas, Billy Mounger from Mississippi, and Dick Richards from Utah, but also new friends such as the downy-faced California lieutenant governor Mike Curb and Senator Jake Garn of Utah.

Another potential presidential candidate, one who was doing his best to imitate Hamlet, was Jack Kemp. Some conservatives who had abandoned Reagan saw Kemp as the next hero of the movement, according to his chief aide, David Smick.[67] Others were supporting Kemp for president for other reasons; for instance, New Jersey's Jeff Bell was still bitter at Reagan for having failed to support his primary challenge the year before. Running for president was on Kemp's mind.

━━━━━━

Fresh off his Detroit selection victory, Bill Brock met with party leaders at their now annual Tidewater Conference on the eastern shore of Maryland to debate an amendment to the Constitution to balance the federal budget.[68] Federal spending had grown out of control, beginning with Lyndon Johnson and continuing through Nixon and Ford. It was on everybody's lips in the GOP by 1979. Jimmy Carter, in fact, was at loggerheads with liberals in his own party who felt he had thrown over the party credo of "tax and tax, spend and spend, elect and elect," and that he was far too concerned with holding the line on spending. The Republicans, on the other side, were of course bashing him also. No presumed 1980 aspirant attended Tidewater save Bob Dole, but the conference nevertheless provided a good snapshot of GOP concerns.

After a long day, the evenings were punctuated by well-lubricated sing-alongs featuring the men of the GOP. All in the party appeared to be in harmony, at least on issues.

Although the Republican conferees did back away from a hard-line balanced-budget amendment, they did strike a hard line against communism, especially on the United States' dealings with the Soviets. They decided formally to oppose Carter's SALT II. This was in the face of overwhelming public support for the treaty. One Associated Press/NBC News poll showed that 81 percent of Americans agreed with the measure.[69] Support had actually risen over the preceding month, but a "Stop SALT" grassroots campaign was mounting, organized by the American Security Council and led by retired military brass, including Admiral Thomas Moorer, former head of the Joint Chiefs, and Bill Middendorf, former secretary of the navy.[70]

Conservatives were also singing from the same sheet as they gathered at the sixth annual CPAC in Washington in February 1979. Only two of the 1980 candidates, Reagan and Phil Crane, were invited to speak—which was a mistake in that nearly all the candidates were warbling the conservative tune. All, including Dole and Connally, would have been comfortable in the conservative surroundings. Two specially honored guests were newly minted senators Gordon Humphrey and Roger Jepsen of Iowa.[71] Humphrey gave a thoughtful address in which he admonished conservatives not to resort to character attacks in their political campaigns but instead to rely on ideas and persuasion.

Reagan also gave a compelling speech, as he examined the Carter Administration's recent abrogation of the Mutual Defense Treaty with Taiwan to favor relations with Communist China. He then restated his "no pale pastels" speech, updated for 1979: "I call for a long range program . . . which will earn and retain the support of the American people and which will help to restore the trust and confidence of the world in an America which once again conducts itself in accordance with its own high ideals."[72] A month before, Reagan had been less measured in his response. Then he assailed Carter because, he said, he "bleats about human rights in a moralistic and highly selective manner," and then he dropped the anvil when he said the sellout of Taiwan was a "shabby, needless blow, hasty and ill-timed."[73]

"Single-issue groups" on the right came under the scrutiny of the national media. Groups that worked on abortion, prayer in school, the ERA, and right-to-work (i.e., anti-union) laws all earned the newfound fascination of journalists. It was just one more piece of evidence that there was a growing populist movement, disgusted with the status quo. Pro-life groups came in for much of the media inquiry.

———

On February 6, Reagan's sixty-eighth birthday, the *Los Angeles Times* trumpeted in a headline, "Reagan to Form Campaign Unit."[74] He hadn't really decided once and for all yet, but this third try was not going to be like the late-starting efforts in 1968 and 1976. Reagan's forces seized outright control of the California GOP and extinguished an attempt to alter its "winner-take-all" delegate-selection process. A new, pro-Reagan chairman, one Truman Campbell, was installed.[75] Paul Haerle, who had once been Reagan's appointments secretary in Sacramento, who Reagan later made state chairman, and who betrayed Reagan by supporting Ford in 1976, had been banished from Planet Reagan. Haerle had also committed the unpardonable sin of dating the boss's daughter, Maureen. Reagan was of course all for the winner-take-all formulation, but John Connally, George Bush, and the others mobilized against it. After a protracted fight, it was momentarily settled, in Reagan's favor.

If Reagan was less sure about the staffing of his political campaign, he was far surer of himself as he aggressively took on the Carter Administration on the issue that he was most passionate about: anticommunism, specifically Soviet communism. At a rare speech, he told an Iowa GOP crowd that Carter's policies reminded him of "the sorry tapping of Neville Chamberlain's umbrella on the cobblestones of Munich. He, too, talked of peace in our time."[76]

The others, including Connally and Bush, followed Reagan's lead, but on this issue he could not be trumped or overtaken. Reagan made anticommunism his issue long before it became fashionable to be anti-Communist. He'd been anti-Communist when it was fashionable in some Hollywood circles to be pro-Communist. He became anti-Communist long before he switched parties, when there were many Democrats who were repelled by communism at the time, when liberals in Hollywood were making paeans to communism during World War II, including *Song of Russia* and *Mission to Moscow*.

Reagan's anticommunism was deep and firm; and he'd forced the Republican Party to discard its policy of détente initiated and carried out under Nixon and Ford and became much more confrontational, which meant increasing the military budget and opposing any new arms agreements. Reagan was firmly convinced that those agreements were putting the United States at a strategic disadvantage.

Across the board, from the recognition of mainland China to trade with the Soviets to the Vietnamese invasion of Cambodia, Reagan had staked out the most dramatic foreign policy contrasts with Carter, far more so than had the other GOP presidential candidates. If he was to run, he was going to make sure that foreign policy was front and center as an issue, and that he would own the lion's share of those American voters worried about their country's future in the world. "The Republican definition has a decidedly macho ring. It plays to that rising frustration among Americans who feel that the period of retreat and retrenchment after Vietnam has gone far enough, that it is time to stop letting others 'push America around.'"[77]

The problem was that, as Reagan was speaking less and less, he was leaving more and more of his issues on the table for the others to abscond with.

Jimmy Carter's despair continued. The economy was spiraling downward, as were his poll numbers, even in his home state, where, according to a Darden poll of Carter's fellow Georgians, he was getting only a 44 percent overall approval rating. The pollster Claibourne Darden said it was a "disaster" for Carter.[78]

The growing criticism of Carter was beginning to wear on his team. His press secretary, the normally genial Jody Powell, barked at one reporter, complaining about "people with more access to publicity than responsibility."[79] Powell could have been talking about any number of Carter critics, not just those from the Republican Party. A healthy number of Democrats openly wondered if Carter was up to the job.

Nationally, Carter's approval had dropped again, now at only 37 percent, and Democrats outside the White House were in a sour mood as well. Carter had been in Washington for two years and still was a complete stranger to the doyens, scions, and hostesses of Washington, though they were all Democrats. They were not invited to the very few state dinners at the White House or any of the other rare social functions in the Executive Mansion. It had been only fifteen years since the assassination of President Kennedy, and most of Washington high society remembered the sweet and happy parties that always seemed to be going on when Camelot was in full flower. If the party was not at the White House, it was at Hickory Hill, home of Bobby and Ethel Kennedy.

Official Washington had waited a long time to laugh and drink again on the South Lawn or the Truman Balcony or in the East Room or, even better, in the "Residence." Carter was not giving them what they wanted. He had done nothing to cultivate them, and they returned the scorn. The natives were restless. Party leaders, however, were holding their breath. Breaking with one's president was not something one took lightly, unless of course one didn't mind risking one's entire political career by making the wrong choice and thus losing one's friends, power, and access.

———

Reagan was tied with Ford at 31 percent apiece in Iowa, but the former president was showing no inclination to trade the golden warmth and comfort of Southern California for the cold and muddy campaign trail that led through the Hawkeye and Granite States. Carter himself had trundled through Iowa for, all told, about one hundred days in winning the 1976 primary; and even then, he took only 28 percent of the Democratic primary votes.[80]

Presidential primaries had changed, as had the campaign industry, over the years becoming more sophisticated and more dominated by pollsters and consultants. There were "general" consultants, but there were also media consultants; fund-raising consultants (and even here, specialties would break down into specific categories, with event, direct-mail, and high-dollar consultants); speech coaches; research consultants, including the dark-side "opposition research consultant"; wardrobe consultants—everything except, it seemed, consultants on substance, ideology, or principles. That would have required something out of the grasp of most. Many consultants were not especially bad, and some were quite good, but some were downright stupid or corrupt. As NCPAC head Terry Dolan used to tell candidates at the National Conservative Foundation's campaign schools, "never forget the first three letters of the word 'consultant.'"[81]

It was all about winning, about sail trimming to follow the crowd rather than lead it. Political pandering and patronage had always been a staple of American campaigns and government, but it was mostly a side show to the more important but mundane tasks of governance, making difficult decisions and maintaining the balance of compromise, this being the great genius of the American experiment.

With the explosion of emphasis on winning Iowa and New Hampshire, the stress was put on organization and control. No longer would people run as a delegate in the caucuses in Iowa or in the primaries as "uncommitted." It was just too dangerous for the candidates. The last thing the political pros wanted were delegates going to the national convention with minds and opinions of their own. Among the GOP candidates, Reagan was probably the most unimpressed with political consultants. Yet, for the time being, he listened to the advice of his own top political consultant, John Sears, who kept him off the road for much of 1979.

Reagan knew what he stood for, and knew he could lead. All he needed was a chance to speak to the people. He'd listen to briefings on poll results and agree sometimes that one issue needed greater emphasis over another with a particular crowd or state, but it was a fool's errand to try to get him to change his mind on something about which he felt deeply. He could be as pragmatic as the next politician about little things, but on the big stuff—forget it. He was far more interested in a briefing on the mood, temperament, and makeup of a given crowd he was about to address than in the results of some abstract poll.

On March 7, Reagan submitted to the FEC his plan for his own campaign committee. Paul Laxalt, who was to be its chairman, hosted a press conference in Washington, where a twenty-three-page list was released of 365 prominent supporters of the "exploratory" Reagan for President committee.[82] Reagan did not attend, but he sent a letter to Laxalt acknowledging the committee and thanking Laxalt for his hard work. "Dear Paul . . . The work of your committee will be of great help to me when I make the final decision."[83]

At the press conference, a list of senior staffers was released including Lyn Nofziger, Mike Deaver, Martin Anderson, Loren Smith as chief counsel, and Angela "Bay" Buchanan as the campaign's treasurer. By the end of 1979, almost all would be gone, all because of Sears. Also joining the nascent Reagan staff was Helene von Damm, to work initially on scheduling for Reagan as she'd done for him for years with Cindy Tapscott as her assistant.[84]

The Reagan direct mailings started in earnest, and an initial prospect letter went out to five hundred thousand people, asking them to join Reagan's roundup. Bruce Eberle, who had so adroitly handled the fund-raising mail for Reagan in 1976, had been shunted aside for 1980, and a company named Integrated Communication System was handling the all-too-important effort. Reagan needed to get his message out, and Laxalt believed it would be difficult to do so through the mainstream media. "The press, which is largely liberal, will be less than friendly . . . They do not want a conservative President."[85]

Bread and Circuses

*The slick, the shrewd, the smart, the hustlers, had never
believed in government promises.*

Consumers were bombarded with mindless advertising in the 1970s,
their brainpans pounded over and over with further rationaliza-
tions for being self-absorbed American citizens. Sober college students of
Psychology 101 would remember this phenomenon as Sigmund Freud's
fully unconscious id. It was a decade of mawkish movies, where they
turned the old Hollywood and publishing formula on its head: introduction, conflict, resolution; or, in the Tinseltown vernacular, "Boy meets
girl, boy loses girl, boy gets girl back." In the end, in these movies, the
boy lost the girl. So it was in the culture of the '70s.

It was also a culture where men's leisure suits (indeed, all clothing in
the 1970s) seemed to be woven from a synthetic fabric, polyester, made
entirely from oil. Other apparel available to men included the "4-way,"
which matched a jacket with a reversible vest and two pairs of pants,
one matching the jacket and one in contrast. Ads of dorky male models displaying the nauseating sartorial disaster for Bond's men's store
in Washington offered the getup for only $129. As comedian Dennis
Miller would say years later, "Those suits had lapels the size of hang
gliders. Stores would offer 'two-for-one' sales. If the store really wanted
to screw you, you'd get three suits." Lapels and ties were clownishly wide
and loud. One might have thought that had Americans stuck with natu-

ral fibers, the oil crisis that plagued the decade might have been averted. The "oil depletion allowance" took on a whole new meaning, and it was suggested that each sleazy disco suit and slinky dress come with the warning label "Do not wear near open flame."

Great American novelists, notably John Updike, John Cheever, Philip Roth, and Tom Wolfe, chronicled the decadence of the 1970s in sardonic, lacerating prose. From Rabbit Angstrom to Alexander Portnoy, everyone had a complaint. Years later, author Bruce Reed would write for *Slate*, "Parents want to relive the '50's; liberals want to relive the '60's; conservatives want to relive the '80's . . . but nobody wants to relive the '70's."[1]

Like most other parents of the 1970s, Ron and Nancy Reagan struggled with their children, especially young Ron, "Skipper" to his father—Ron Jr. hated the nickname—and Patti, who had rejected the name "Reagan" in favor of her mother's maiden name "Davis." Mike and Maureen Reagan were having their on-again, off-again struggles as well, but nothing like their younger siblings. Mike went through a series of jobs, racing speedboats, working in alternative fuels. Maureen moved east, married a cop, divorced the cop, moved back west, and jumped from job to job.

Patti was "shacking up" (in the vernacular of the '70s) with a member of the rock band the Eagles, Bernie Leadon, and generally doing whatever she could to infuriate or embarrass her parents. Rumors of casual drug use surfaced from time to time. Son Ron was also rumored to have had an affair, while still a teenager, with a married woman in her thirties. Reagan did what he could to play Father Knows Best, but he tried less to control their lives and more to be there to listen and write letters of advice and admonition. Most of the parenting responsibilities for the younger two children fell to Nancy.

————

Nancy Reagan had given a rare speech in Grosse Pointe, Michigan, the Fords' home court, about traditional values, just about the time Betty

Ford was in the White House telling *60 Minutes* that the abortion, marijuana smoking, and affairs her daughter Susan might have were all right with her. The Reagans were struggling with their children against the onslaught of the 1970s culture, made doubly hard by living in California, ground zero for every new fad, every nonsensical idea, every new hedonism. Tellingly, most of Reagan's commentaries and speeches of the era did not touch on the seamy culture, sticking mostly to public policy issues.

With every stimulus, there is a response, and the decade saw the beginning of the rise of the "Religious Right" with the advent of *The 700 Club*, hosted by the Reverend Pat Robertson on a small station in Virginia Beach, and the ascent of the Reverend Jerry Falwell, up the road in Lynchburg. Falwell had been a hell-raiser as a teenager, but turned to the cloth, founding the Thomas Road Baptist Church in the mid-1950s, opening his house of worship in a former bottling plant. Falwell's following grew quickly, aided by his *Old Time Gospel Hour* broadcasts. He inspired a series of successful "I Love America" rallies and, later, the "Clean Up America" campaign, aimed at addressing the growing squalidness of the culture.

In 1979, appalled at the downward-spiraling culture and determined to bring traditional values four-square into American politics, Falwell created the Moral Majority, through which he could conduct permanent political activities. Most of these Southern Baptists had supported the "foot-washing" Jimmy Carter over Gerald Ford in 1976, attracted to Carter's born-again message. But Falwell and others became incensed over a proposal floated by Carter's chief of the Internal Revenue Service, which threatened to revoke the tax-exempt status of every Christian school created after the 1954 *Brown v. Board of Education* decision.

The IRS arrogantly assumed that each of these schools was created to avoid integration, even as black Christian schools were also growing. When the Religious Right fought back successfully against the proposal and the Administration backed down, it developed a taste for political combat. Falwell, Robertson, and others had been driven away from Ford in 1976 in part because of Betty Ford's aggressive support of abortion and the Equal Rights Amendment. Also, there was the simple fact that they were inheritors of the southern, "yellow-dog Democrat" tradition, and the hangover from the Reconstruction era made the Republican Party repugnant. By the late 1970s, however, the national Democrats were beginning to embrace "secular humanism" over traditional Christian values, and this helped herd religiously motivated voters into the eagerly awaiting arms of the GOP.

Contrary to urban legend, ministers and religious leaders had always been involved in American politics. From a certain point of view, the American Revolution was a "religious war," as the Colonies, and especially the South, fought off the authoritarian Church of England. Religious leaders helped with the Underground Railroad prior to the Civil War, and the Catholic Church organized nationally to help elect the first Catholic president in 1960, but this new development by Falwell and other lesser figures in the Religious Right was a departure from the past—or a glimpse of the future.

They would marry traditional values with technology in an unprecedented fashion, using the mail, computers, the airwaves, and the pulpit to organize Christian voters into a potent new force in American politics. "Not only did evangelical churches add members faster throughout the 1970's than non-evangelical . . . but the more evangelical the church, the more likely its members were to attend services. Through the 1970's, the Baptists consistently showed the highest turnout of any Protestant denomination; Episcopalians the lowest," wrote David Frum in his definitive book on the 1970s, *How We Got Here*.[2] Another group, Christian Voice, was also emerging as a political force, led by a charismatic minister, Dr. Robert Grant. Grant hired as his Washington lobbyist Gary Jarmin, who had previously been the executive director of the American Conservative Union. Interlocking directorates and cross-pollination were not unusual in the New Right. CV, as it became known, was a fervently pro-Reagan operation.

The Democratic Party was going through its own introspection. Since the time of the New Deal, it had been the party of government activism, but government activism was no longer working or popular.

Ted Kennedy saw this in part when he led the fight to deregulate the trucking and airline industries in the late 1970s. Kennedy's motivation was pro-union but also antibusiness (which was enjoying government protection and a guaranteed market), not antigovernment, but he came to the same conclusion that Reagan had years before, at least on this issue, which was that the concentration of power, by either businesses or government, threatened the American people and the entrepreneur. For Teddy, it was at least a twofer. He could stick it to business and to the Teamsters, which supported trucking regulation. The Teamsters had been a burr under the Kennedy family saddle for years.

The good news for Democrats was that the party had controlled Congress uninterrupted since 1954, and it had a two-to-one lead in registration over the Republicans, though the Republicans were beginning to outraise them financially. The Democrats had been the party of the workingman, except that in the 1970s, the workingman was often not working, and if he was, his paycheck was being eaten up by spiraling inflation and rising taxes. Food costs in some areas of the East Coast had gone up by more than 7 percent in just the first three months of 1979.[3]

The workingman—the forgotten man—and now woman in many homes needed two paychecks just to get by. They usually sent their children to public schools, but these schools, when not torn by race riots, were not teaching their kids. The failing teachers, in the minds of many American families, were associated with the Democratic Party. American workers also went to church, or had gone to church as children, and though their attendance may have become sporadic, they did believe in God and Christ. Yet, increasingly, their party was attacking organized religion. Working people also tended to be pro-life, especially Catholic Democrats, but their party was becoming militantly pro-choice. They were also union men and women, and believed in unions, but corruption and arrogance on the part of many of their leaders had left them wondering if they were being faithfully represented anymore.

Carter and the Democrats were growing out of step with blue-collar American values. "He has a right-wing domestic policy and a left-wing foreign policy, and that's a sure way to lose an election," Senator Pat Moynihan told David Broder of the *Georgian*.[4]

———————

America's ambassador to Afghanistan, Adolph Dubs, was kidnapped and killed in mid-February, the fifth American envoy to die within ten years.[5] Afghanistan was going through a civil war, although the revolutionaries were being supported by the Soviets, who wanted a more Kremlin-friendly government installed. Iran was going through internal turmoil, but the Kremlin told Americans it was none of their concern. They'd handle it, since they had a border with Iran that extended over twelve hundred miles. Kremlinologists in the West knew the Russian bear was hungry once again. The Soviets were implicated by U.S. officials in a report about the death of Dubs, but the Carter Administration did little to signal its displeasure with the Russians murdering American ambassadors. Indeed, State Department officials tried to quash the report, to avoid embarrassing the Soviets.[6]

Conservatives were becoming increasingly dismayed over Carter's softening policies toward the Soviets. Columnists Evans and Novak referred to "Carter's submissive policy of turning the other cheek every time Moscow fouls Uncle Sam."[7] George Will was equally dismissive, referring to Carter's government as "the surprised administration."[8]

In a brilliant two-part series for the *Washington Post*, the much-esteemed David Broder examined at length the internal turmoil inside the party of Andrew Jackson. He interviewed everybody, from Vice President Walter Mondale to senators and House members. Words such as *adrift* and *vacuum* animated the story. Democrats were open in their criticism of Carter. To Republicans, the question over whether Carter was a "closet Republican" was nonsense, but inside the Democratic Party, it was a hotly contested debate—in large part because Carter ran in 1976 as an outsider, a populist and as someone who spoke vaguely about a strong national defense.

After he became president, he moved to the left, though not fast enough for many liberals in his party. Pat Caddell had written Carter a memo in early 1977 in which he urged the new president to pick Republican issues that would create a new governing majority. "We have the opportunity to co-opt many of their issue positions and take away large chunks of their presidential coalition by the right actions in government. Unfortunately, it is those same actions that are likely to cause rumblings from the left of the Democratic Party."[9] The party was going through more than rumblings, but out of this discontent would come a new and potent political force in American politics by 1980: the Reagan Democrat.

In his 1979 State of the Union address, President Carter yet again told his fellow Americans that there were limits to what government and they could do at home and around the world, and he called for a "New Foundation." Carter was attempting to jump-start his presidency with a new slogan, but more of the same policies. Obviously, his team wanted it to take its place alongside "New Deal," "New Frontier," and "Great Society," but what they misunderstood was that the American people were not focused on the adjective but the noun. And *foundation* could have been about anything, including women's undergarments. "Not since Howard R. Hughes designed that cantilevered strapless superstructure for the formidable Jane Russell has the term taken on such importance in conceptual reform."[10] It could also be segued into metaphors about unstable structures.

In preparing for the speech, Carter met at length with his senior staff to discuss the previous two years. Carter said it was more difficult to lead than at any other time. The times were different, more complex; it was difficult to get the attention of the American people, and they were too suspicious of power, he told his team. The story of the intimate meeting was leaked to the *Post*, but that was no surprise. The Carter White House leaked like a sieve from day one.[11]

The Republicans prepared a weeklong counteroffensive against the speech, but one who didn't understand what was happening to the GOP and its evolution was former Ford adviser Alan Greenspan, who said the president's speech was "as Republican as I can imagine."[12] In fact, as Carter was remaking the Democratic Party the home of limited horizons, Reagan was embarking on a course to make the Republican Party the home of limitless horizons. Greenspan had lost his moorings years ago as a devotee of Ayn Rand, who wrote and preached of objectivism and the superiority of the individual over the state.

Carter's New Foundation of 1979 was stillborn, as were his earlier energy proposals, none of which he even mentioned in the speech. The decade also saw gas lines and gas rationing, the loss of Southeast Asia, the resignations of Spiro Agnew and Richard Nixon following the Watergate scandal, high unemployment, high interest rates, and spiraling inflation—all of which sapped the confidence of the American consumer.

Americans also were coming to believe that government could not solve all the ills that afflicted society. Like Ronald Reagan, they, too, were moving away from the notion that government, absent a strong national defense, was not the panacea they'd been told by every Democrat and many Republicans since the 1930s. It was a revelation in that, though government had not solved the Great Depression, the New Deal nonetheless gave Americans hope. Also, they saw their government defeat the empire of Japan and Nazi Germany in World War II, institute the GI Bill for returning veterans to attend college, build the interstate highway system, and generally govern in a beneficial and mostly benign manner. Yet, beginning with the assassination of John Kennedy on a Dallas street in 1963, Americans would begin to discover that government, despite the promises of elected officials, could not protect or save them from all the vagaries of life. Nor could government keep Senator Robert Kennedy and civil rights leader Dr. Martin Luther King Jr. from also being gunned down. Nor could government win the Vietnam War; losing a war was unthinkable to the American people. Respect for government bottomed out in the 1970s, with the ignoble and humiliating departure of Nixon—though, ironically, the Constitution ensured the smooth transition of power to his vice president, Gerald Ford, without riots, revolution, or a shot being fired.

The problems of the '70s mirrored the past. Rampant inflation and food shortages, coupled with vast unemployment, had destroyed the German people's will and that of their government, which helped ease Adolf Hitler's march into power. "What the great inflation had done, by 1979, had been to separate America into layers of different but resentful (or greedy) people . . . The slick, the shrewd, the smart, the hustlers, had never believed in government promises, but those who had believed those promises . . . all those had been cheated," wrote the great historian Theodore H. White in his book *America in Search of Itself.*[13] The Carter recession was in full flower.

The phrase "a conservative is a liberal who was mugged by reality" came into vogue in the late 1970s. The American people had not only been mugged, but also shot, beaten, raped, and left for dead by their own government. They were all too ready for an alternate reality. "More than any other period of American history, the decade of the 70's is the time when the seat of the Federal Government has become the national center for the petition of grievances."[14]

Americans had always been skeptical of government, and this attitude was deeply ingrained in the American experiment. After all, the term *public servant* was supposed to mean just that—a person who worked in government served at the pleasure of the people, and the people ruled.

Washington had become Rome: corrupt, complete with its own form of bread and circuses. Government bureaucrats drank deeply from the public trough. They were exempt from nearly every law imposed on the American people. They were exempt from participating in the Social Security system into which all wage-earning Americans were required to pay. They had an unparalleled health care plan and could retire after only twenty years of service, on half pay plus "bennies" for the rest of their lives. They could then go work somewhere else in government, put in another twenty years, and get a second pension, thus becoming "double dippers." The only thing missing was the sprinkles.

Washington's household income was an astounding 33 percent above the national average. The average home in the DC metro area sold for tens of thousands above the price of houses in the rest of the country. Carter's attempts to bring to heel the self-centered, self-satisfied, self-important, and self-absorbed citizens of the city-state had come to naught. They'd seen his type come and go.

The national government as envisioned by James Madison was unrecognizable, a great deal larger and more comprehensive in its scope. America in 1787 was a do-it-yourself enterprise. The notion that the government must eliminate despair from the American experience was unknown, and if someone had suggested it, he would have been laughed out of Philadelphia. Now, what had grown exponentially was a second unofficial government that fed like a parasite pilot fish off its host, a giant white shark. Often, corrupt interests claimed to want to help others, but only first by helping themselves.

Washington had had its attendant lobbyists and hangers-on since Lincoln's time, but this was far different and far more dangerous. The manipulation of money and regulations had led to too-cozy relations between politicians, their aides, and the special interests. In the DC telephone book, there were more than "200 listings under 'National Association.'"[15] Compounding this mess were the "consultants" who floated in a limbo-like state, neither fish nor fowl, as they worked for the government and themselves. In the early days, President Carter had asked the agencies to answer one simple question: "How many consultants do you guys have?" He never did get an answer.

Some estimates put the budget for consultants at more than $2 billion per year. Sometimes consultants were brought in not because the personnel at a given agency could not handle the task but because they simply did not want to do it. Corruption abounded. The Nuclear Regulatory Agency attempted to conceal the number of consultants it had under contract.[16] Tens of thousands of federal workers commuted into Washington each day, where none of them had to pay for their parking, unlike those in the private sector who also worked in the nation's capital. The American taxpayer subsidized it all.

Idiotically and insanely, also paid by the American taxpayers were college students on Capitol Hill whose job was to keep moving the cars of congressional employees so that "3,300 cars" could "be crammed into 2,465 spaces." It was musical cars, as there were more cars than parking spaces, so day in and day out, these college students drove cars to nowhere. The cost of paying twenty-two college students to shuffle the cars of the pampered staffers was more than $110,000 annually. These staffers, of course, parked for free, and when Senator Chuck Percy offered a timid proposal that Hill staffers pay for their parking, they howled, bawled, stomped their feet, and threw temper tantrums, and the suggestion was shot down.[17]

———

Carter was dwindling in the polls. In desperation, he went to the scene of his 1976 breakthrough, Iowa, in an attempt to reconnect with the voters. The welcoming band played "Happy Days Are Here Again," the theme song of the Democrats since FDR. "I feel like I'm part of you," he told the crowd. But one person who did not feel a part of things was incumbent senator John Culver, who begged off on accompanying the president to his home state. Culver was an ally of Ted Kennedy.[18]

While some Democrats were hoping Kennedy would challenge Carter for the party's 1980 Democratic nomination, Republicans had their choice of Reagan challengers. Phil Crane, the former professor of history and Republican congressman from suburban Chicago, had his presidential campaign up and running, but it soon hit some speed bumps. Charges of stolen lists and fraudulent stock deals and micromanagement by Crane went round and round the rumor mill. As more and more people left, often due to Mrs. Crane, the black humor of the remaining staffers had them calling it "the coup of the Week Club."[19] Crane manager Rich Williamson also resigned after clashing with Richard Viguerie over the direct mail and its attendant costs. Arlene Crane referred to the departed staff as "the Rat Patrol."[20]

While some conservatives such as Viguerie had signed on to the Crane campaign, most still had their hearts set on Ronnie, including the Gipper's number one fan, William Loeb, publisher of the *Manchester Union-Leader*. Loeb, seventy-four, had not lost a step or his acid pen.

Loeb was furious with Crane for what he saw as Crane's betrayal of

Reagan. The stories alone that Loeb published, which alleged a wide range of immoral behavior against Crane, ran more than one hundred column inches and charged that the couple had "mutual and separate sex lives."[21] Crane had promised to back Reagan again but reneged, earning the ire of Loeb and other Reaganites. Commenting on the faltering Crane campaign in late April, Reagan told Adam Clymer of the *New York Times*, "You'd think I was the one who got in the race early and had been out there trying and now the bubble had burst."[22]

Howard Baker stepped forward and suggested some sort of mediating intervention organization to sort out these types of problems inside the GOP, but the Gipper's folks and Crane's team were having none of it. John Sears's above-it-all strategy for Reagan was not working, as Reagan was getting involved in things he had no business being involved in, such as personally denying some ridiculous charge Crane made against him about being the source of Loeb's hatchet job. Also, Reagan was being held back by Sears, and missing events he should have been involved in, such as the all-important Midwestern Republican Conference in Indianapolis.[23]

Party leaders and the national media all planned to attend, as would all the GOP contenders, except, of course, the Gipper. John Connally had an especially good day, as he was rated the best speaker by the three hundred influential Republicans, with 53 percent (to only 35 for the absentee Reagan). Again, Reagan was raked over the coals by the media and GOP officials for ducking yet another event. In fact, he instead was giving a speech at a seminar in Miami. The complaints against Reagan made the rounds over cocktails in Indianapolis. The Republicans were not happy with his snub.[24]

Of the field of GOP would-be presidents, virtually all were more or less accomplished, sane, sober, and confident men, save one: the senator from Connecticut, Lowell P. Weicker. Weicker had said when he announced his candidacy in March, "I'm told that all this talk of investment for a future during times of economic difficulty runs against a tide of conservatism. That's right. That's what I'm going to be all about."[25] A rumored sexual harasser of women on his Senate staff, Weicker was a bully who rubbed almost everybody the wrong way. The *New York Times*, in covering his announcement, wrote, "Senator Weicker said today that no major Republican Party official had encouraged his campaign."[26]

———

Reagan was not exactly marching in place, though he had postponed a long-standing overseas tour to Moscow and the Middle East. His campaign organization was continuing to reach beyond its traditional base of conservative supporters, and even signed up the former Republican governor of Connecticut, John Davis Lodge, who had held two ambassadorial posts under Eisenhower. For hundreds of years, New Englanders knew from childhood that among certain snobby descendants of the *Mayflower*, the "Lodges talk only to the Cabots and the Cabots talk only to God." Indeed, Joe Kennedy, a lace-curtain Irishman, had moved his children away from the Boston Brahmins to Cape Cod to escape their kind.

Finally, one of their descendants was talking to a real, live man of the people: Ronald Reagan. On the other hand, this Lodge was not one of the Boston Lodges. He was scandalously married to a beautiful Italian actress who was not an Episcopalian; nor did she have blue eyes or blond hair, and these factors may have explained his apostasy.

Republican candidates, operatives, and journalists were swarming like bees over New Hampshire in search of the ubiquitous primary voter. Seeing the financial opportunity, the state party announced plans for a giant fund-raising dinner and identified fourteen potential GOP presidential candidates to appear at the event. They even invited Harold Stassen, along with George Bush, Jim Baker, John Connally, Phil Crane, Lowell Weicker, and Bob Dole. They all attended, except for the most popular conservative in America.[27]

Of course, as per John Sears's strategy, Reagan did not attend, but Nancy Reagan did in his stead. Grumbling was heard from Reagan supporters in the Granite State. Senator Gordon Humphrey fretted that "some of [Reagan's] support is already going to his challengers and if he waits too long, he'll have some serious trouble."[28] Mrs. Reagan was none too happy, either, as she noted that "Ronnie's" announcement for president, scheduled for the fall, was "plenty early enough" for her. She elaborated that people, the media, and the candidates were getting worn out from politics, and on this she had an excellent point. It was fully a year before the New Hampshire primary, and the November election of 1980 was a year and a half away. Many shared her opinion.[29]

═══════

The greatest nuclear accident—but not a meltdown—in American history occurred in early April 1979 at the Three Mile Island facility in Harrisburg, Pennsylvania. The surrounding area was evacuated as some hysterical leaders feared a "China Syndrome," meaning that the reactor core would super-heat and the facility would literally melt all the way through the earth. Public opinion shifted with lightning-like speed, from supporting nuclear power to opposing it. Yet liberals had made support for nuclear power a centerpiece of the Port Huron Statement, a document written by college students in the early 1960s, including activist Tom Hayden, as a statement of ultraliberal principles. Now, Hayden's wife, actress Jane Fonda, was starring in a new movie called, serendipitously, *The China Syndrome*, and she and producer Michael Douglas squeezed as much out of the crisis as possible, making untold dollars for themselves as Americans streamed to the movie that advanced an antinuclear message. Greed was good for Douglas and Fonda, but leftist actors often had a hard time being taken seriously. Robert Redford had given a recent speech on solar power. The first question he fielded was "Did you really jump off that cliff in *Butch Cassidy*, or was that a stand-in?"[30]

An "antinuke" rally was held at the U.S. Capitol in May. "No More Harrisburgs" was the slogan. Fonda, sporting a tube top, spoke to the crowd, as did Jerry Brown. The "usual suspects" of liberal groups participated, including those organized by Ralph Nader, along with Dick Gregory, Kurt Vonnegut, Barry Commoner, and of course Hayden. District police estimated the crowd at sixty-five thousand.[31]

The burgeoning crop of GOP candidates wasn't waiting for Reagan. All were moving ahead at various speeds to scoop up supporters and workers in not only Iowa and New Hampshire but in other early primaries, including South Carolina, Florida, Alabama, and Illinois. They were also brazenly bringing up Reagan's age, most especially John Connally and Phil Crane, by portraying the Gipper as "yesterday's hero."[32] A cruel joke going around the drinking establishments and GOP circles in Washington was "there is no age issue. Reagan is a shoo-in, if he lives."[33]

In fact, Reagan's age had been somewhat of an issue when he first ran for governor, in 1966. After several nasty exchanges with reporters on the campaign trail, "downtime" was built into his overbooked schedule to allow him to rest up. In 1976, when he was stumping in New Hampshire for the nomination, reporters were taken with Reagan's resilience in part because he'd learned to pace himself.

The importance of the southern primaries was not lost on the field, especially George Bush and Connally, the two Texans, and Jim Baker, the son of the Volunteer State. In 1976 over one-third of Reagan's strength came from delegates below the Mason-Dixon Line, taking 388 to only 145 for Ford.[34] If they could slice Reagan in Iowa and New Hampshire, and then dice him in the South, it would be all over but the shouting, and Reagan would end his career as the new and ignoble Harold Stassen of the GOP, 0–3 at the national level and finished forever in elective politics.

David Keene, who had run part of the South for Reagan in 1976, was finally brought aboard the Bush campaign, in part because of his network of contacts and friends and his knowledge of southern politics, but also because of the brief Reagan imprimatur and his acceptable conservative credentials. He was also a world-class leaker, a schmoozer, and the national media liked him.[35] People who'd worked with Keene often refused to do so again, as his reputation for playing nice with others was poor, and his work habits spotty. It was not unusual for staff to quit because of Keene's unique ability always to let others take the blame while he took the credit. But he was always good to the press, always available for their phone calls. As a result, Al Hunt had the first real national story on the growing machinations inside the warring Reagan camp.

Keene also coldly assessed the landscape to the *New York Times*: "I was for Reagan last time but what is important is timing. Not only is he older, but the things he identified with, that he talked about in 1976, are not as timely."[36] Keene believed he'd been insulted and hurt by Sears and Reagan. Now he wanted to insult and hurt Reagan. He had already been feuding with Mike Deaver, but he'd also been hung out to dry by Sears as tribute to many inside the Reagan camp who did not want him back, and he didn't want to accept a lesser job in the 1980 Reagan campaign as a consolation.

The fact that Deaver was down on Keene was a sure sign that Keene had also lost favor with Nancy Reagan, with whom Deaver was very close. Indeed, years later, at a conservative dinner, the event planner had assembled the head table prior to going onstage. Tish Leonard, a former Agnew functionary, was the event organizer and told Mrs. Reagan she'd be seated between Keene and Pat Buchanan. Mrs. Reagan rolled her eyes and sarcastically said, "Well, that's just great!"[37]

Jim Baker offered Keene the number two position, and he took it—then rejected it and finally took it—even as he oddly told the *Washington Post* that Reagan was still his first choice.[38] Bush manfully stuck by Keene, but the seeds of doubt regarding Keene's true loyalties had been planted. Keene rationalized the choice of going with Bush by repeatedly telling people that Bush was a "conservative" even when Keene's fellow Bushies were telling people that Bush was a "moderate."[39] In fact, Keene was loyal only to himself.

Sears was also clashing with Dick Allen, a key foreign policy adviser to Reagan beginning in 1976. Allen was a hawkish conservative, and Sears believed Allen played to Reagan's worst instincts to bash Commies. Of course, Sears was right. Reagan and Allen were unreconstructed anti-Communists. Of the growing soap opera in the Reagan camp, Evans and Novak wrote ominously that "the campaign's power structure remains dangerously unresolved."[40]

Senator Jesse Helms, a Reaganite's Reaganite, and his powerful Congressional Club were remaining neutral, in part because Helms had also butted heads one too many times with Sears, going back to 1976. But also Jim Edwards, governor of South Carolina, one of only two governors to back Reagan in 1976, was withholding his support this time around, too. Reagan's support was slowly eroding across the South, but his campaign seemed unable or unwilling to stanch the bleeding. The defectors had their excuses, take your pick; "Reagan is too old and predictable"; or "Reagan's team is driving away people," as in the case of Helms; or Reagan was "not sufficiently wooing" people, as in the case of Edwards.[41]

The situation for Reagan became serious enough that a few journalists began to write defensively of their old plane mate. Others seemed as if they couldn't wait to see Reagan go down for a third and final time. One haughtily wrote of the moderates' plan for Reagan's undoing and the rise of Connally: "Reagan will, however, still retain enough strength to challenge Connally for the loyalty of the Republican conservatives, and this is the hope of the GOP moderates—particularly Senator Howard Baker and George Bush of Texas."[42] They were rooting for Connally to take out Reagan, and then for either Bush or Baker, in turn, to take out Connally, two candidates more to the liberal writers' liking.

———————

President Carter was not about to roll over and play dead for anyone, especially the hated Kennedys. He made his way back to New Hampshire, site of his breakthrough primary of 1976 and, just as juicy, Kennedy's own backyard. Though Carter, in his speaking style, had not improved much as a tub thumper, he still connected with voters better than many politicians in America. He joked to the crowd of fifteen hundred in Portsmouth "about how he was named one of the 10 best-dressed men of America in 1976 at a time when 'I had three blue suits—they cost $42 each.'"[43] He also wrote down the name of a little girl about the age of his own daughter and told the girl she would be invited to the White House where she could meet Amy in person.[44]

Carter had as good a touch for retail politickin' as anybody. He was down, but he certainly was not out. He also attacked his own Democratically controlled Congress, when he accused them of a "charade" in which they tried to "hoodwink" the American people by passing a windfall profits tax that would benefit the hated oil companies. Jimmy was going to stand in the doorway until the last and stop the Washington insiders from doing their dirty deeds.[45]

But what the American people wanted in 1979 was not more of the same. They'd had several years now of the down-home populist. Also, they were confused. How could Carter, as head of the national government for the last three years, all of a sudden proclaim he was the outsider . . . once again . . . who was going to clean up corruption in Washington. Where had he been for the last three years?

As always, the Carter White House was convinced that if it could negotiate the thicket of primaries against the Chappaquiddick-damaged but dynamic Ted Kennedy and the more recently damaged Jerry Brown, then it would mop the floor with that tired old man Reagan—if the Republicans were dumb enough to nominate him.

Brown looked down his nose at Reagan, Carter, and most everybody in politics, but it was he who was a granola-crunching California cliché. In 1977, Carter had worried that Brown, who had beaten him in five of the last six primaries of 1976, would be his main challenger for 1980, but Brown's helter-skelter lifestyle and Haight-Ashbury behavior caught up with him. Now Carter was focused on a couple of other guys who wanted his job. His campaign team thought Howard Baker "would be the toughest . . . their man, they believe, would have an easier time against a staunch conservative such as Ronald Reagan."[46]

On May 1, Ambassador George Bush joined the fight with a formal announcement of his candidacy for president.[47] There was just something about Bush, but for whatever reason, a goodly number of Republican voters could just not support him in the past. They thought he was some rich aristocrat, but in fact, Bush was only worth a bit over a million dollars, a staggering sum for most Americans in 1979, but nothing like what most of the scions of "old New England" had or even the nouveaux riches of the Houston Country Club.

Ironic also because some of them derided his manhood, but while Bush was getting his ass shot off flying in the Pacific for the navy, many of his conservative critics had never even seen the inside of a barracks, much less had a murderous Japanese pilot trying to shoot down them and their crew. Perhaps Bush might have benefited if he'd talked about it more, but he was of a culture and a generation of people who thought it impolite to boast, and did not publicly disclose their grievances or their problems. They simply moved on. Service to their country was expected, since their country had done so much for them. This Greatest Generation usually reacted to praise by digging their toes in the dirt, Tom Sawyer–like, and replying, "Aw shucks, anybody could have done it." When President Kennedy was once asked how he became a hero in World War II, he modestly replied, "They sank my boat." It was irresistible, and had Bush played the issue a bit more like JFK, he might have gotten more of a break from the conservatives and the media, who were attacking his "toughness." Bob Dole, another war hero, who was horribly wounded by a Nazi machine gun and, as the expression goes, lived to tell about it, didn't.

Bush thought he was playing to his strength. In addition, unlike Howard Baker, he had opposed the Panama Canal Treaties, so much of conservative angst was aimed at the Tennessean and not the Texan. He did tell the reporters that he was a "lifelong Republican," which neither Reagan nor Connally had been. He also took a shot at both by saying his campaign would be one "of substance, not symbols; of reason, not bombast; of frankness, not false promise."[48] Bush specifically disagreed with Reagan on a constitutional amendment to overturn abortion, and unlike Reagan, he favored a peacetime draft. He also opposed a substantial cut in personal income taxes, which Reagan favored.

Bush got good marks—better than a "gentleman's C," for his performance at his announcement press conference by the political reporters.

But they failed to understand that for the first time in years, Bush was not speaking for the party or for Ford or for Richard Nixon, but for himself, and in this he was freer and more comfortable. Bush was a long shot, an asterisk in the polls. The popular cliché among the media and in Reagan circles was that Bush had gone about as far as he could go the day he announced. Columnists Jack Germond and Jules Witcover wrote harshly about Bush and his chances with the best they could say of Bush's long-shot candidacy, "There is, of course, no suggestion that Bush is a hopeless case."[49]

Even though the press gave him decent marks for it, Bush's announcement had not gone as well as some had hoped. Bush ordinarily wore granny glasses, and his staff made him switch to horn rims and then wanted the farsighted Bush to give the speech without glasses at all. The type on his prepared text was extra large, and extra lighting was brought into the room. Even so, he stumbled a few times over the speech but did well in the question-and-answer session that followed. Still, Germond and Witcover, who clearly did not like Bush, compared him to George Romney, who had stumbled badly for the 1968 nomination against Nixon and quickly dropped out.[50]

Bush, the next day, tried to take President Carter to task on foreign policy, especially on his handling of Iran. Bush talked about U.S. efforts to "destabilize" the Iranian regime, but then backed off. A reporter asked him if he was "vacillating."[51] In another interview, he weakly cited, as evidence of his leadership capabilities, the fact that he had been captain of his baseball team at Yale.[52]

Some in the press clearly did not like Bush. At a press conference in New Hampshire, one reporter was overheard saying to another, "If we're going to maul him, let's sit up in the lights."[53]

———

Phil Crane's campaign quickly ran aground as top aides resigned en masse. Gone were his unusual pollster, Arthur Finkelstein; his campaign manager, Tony Palladino; and Mari Maseng, the campaign's press secretary.[54] Finkelstein had been the architect of Reagan's comeback win in North Carolina in 1976, as well as his win in Texas several weeks later. Finkelstein had hoped for a bigger role in the 1980 Reagan campaign, but with so many cooks and so few ladles, he went to greener pastures, or what he thought were greener pastures, with Crane. Maseng, a tall, attractive, and talented writer, had gotten high marks for her work as Strom Thurmond's press secretary in his difficult reelection campaign the year before.

While Crane's campaign was going down for the count, across the Atlantic, Margaret Thatcher won her championship bout with James Callaghan. Callaghan's leftist government had fallen after a vote of no confidence by Parliament. Thatcher was the first woman ever elected to lead a major Western power. She was a Tory, but not a defender of the status quo. She was much in the mold of Winston Churchill in that she challenged the nostrums and conventional wisdom of her era. Old policies that favored Labour and punished the entrepreneurial classes would be swept away by her free-market reforms. She was a staunch anti-Communist and believed that the West was dangerously close to slipping into oblivion. She and Reagan were already developing a friendship as ideological soul mates. This "Iron Lady" of Great Britain would prove formidable to those who foolishly underestimated her powers. She once said, "I am not a consensus politician or a pragmatic politician. I'm a conviction politician."[55]

Two weeks after Bush's announcement, Bob Dole became the seventh candidate for the Republican nomination.[56] Dole was renowned for his big temper, but also for having one of the biggest hearts in town. Without fanfare, he raised money and helped many children who were born with or had acquired physical handicaps. That was something with which he was familiar, having lost complete use of his right arm and most of his left in World War II. Dole's chest was a mass of scars, damaged in the war as well.

Dole also had hoped that some of Reagan's staff from 1976, including Charlie Black, John Sears, Paul Russo, Lyn Nofziger, and David Keene, would be with him. They all were good friends with Dole and liked the hell out of him. Alas for Dole, all (Black, Sears, Russo, and Nofziger) stayed with Reagan. Keene, after his clash with Sears, had moved over to Bush, but only after Dole upbraided him for doing so instead of coming with him.

Dole was not subtle or tactful about his opinion of another GOP candidate. "George Bush was born with a silver spoon up his asshole. He probably fucks with his socks on," he said acidulously of Bush.[57] "He was not happy about my going with Bush," Keene recalled.[58] "Not happy" is an understatement.

Dole started all his campaigns in Russell, Kansas, his birthplace. He visited Dawson's Drug Store, where he'd been a soda jerk before the war, visiting old friends and family, highlighting his roots. Dole was a compassionate conservative who believed in private compassion and public policy conservatism. In his statement, he talked of two men, one wealthy and one poor. Both were equal in the eyes of God and in a court of law. He said, "But it is arrant nonsense to suppose that because they may not be equal in ability and ambition, government should equalize their portion of the material advantages which flow from the unfettered exercise of ability and ambition."[59] The mellifluous statement was out of character for the shorthand-speaking Dole.

With the new polls showing Reagan and Ford tied, some in the party lamented a possible replay of the knock-down, drag-out fight between Ford and Reagan. "It would be awful . . . a pair of 69-year-old men re-fighting the battles of the past. Reagan would win, but who needs it," said Clarke Reed, the GOP's national committeeman from Mississippi.[60]

Another national committeeman considered introducing a resolution by the RNC asking both men to step aside for 1980.[61] Gerald Ford, through an aide, made his disdain for Reagan clear: the former president "will do whatever is necessary to keep Reagan from taking control, even if it means running again himself." Lyn Nofziger had a solution for avoiding any future unpleasantness: "Just keep Jerry out."[62]

As expected by everyone but him, Lowell Weicker took his ego and went home by mid-May. He dropped out of the GOP contest, having been beaten in the polls in his own home state of Connecticut by both Reagan and Gerald Ford.[63] He would not endorse anybody else. The GOP field was back to six candidates.

To make campaign strategies worse, in August 1979, Lyn Nofziger suddenly resigned from the Reagan operation. He was peeved that when the official campaign started, he was originally, once again, supposed to become Reagan's press secretary, but "Sears and some of the others had changed their minds."[64] He also said as much about Mike Deaver, with whom he'd had a good relationship, but now found himself facing a united "Sears-Deaver" front.

Reagan had been ambivalent about Sears running the campaign again. When he wrote to an old friend Harry Everingham in Arizona several months earlier with concerns about Sears, Reagan said that "[Sears] will be doing only those things that he can do well, and he will not be in charge of the campaign." He meant it.[65]

Yet, in the end, supported by Deaver—which effectively meant that Mrs. Reagan agreed—Sears became the de facto manager for the 1980 quest.[66] Reagan called Nofziger and asked him to stay, to no avail. He later found out at a meeting at the Reagans' home that most wanted him gone.[67]

Nofziger was another one of those who was popular with the media and who competed with Sears for the reporters' attention and affection. Nofziger had been put in charge of the campaign's fund-raising by Sears, which was akin to putting a sow's ear on a silk purse. Nofziger had zero patience or tolerance for the niceties of romancing high-dollar givers. He was born for the rough-and-tumble of political hand-to-hand combat and not as a handmaiden to the rich and famous. His most important assets, as a close confidant to so many conservative leaders and activists around the country and as someone who knew the national media and had their respect, had been put into a closet.

Mike Deaver was assigned to take over the fund-raising duties, and though smoother than Nofziger, he was equally ill-suited to the task. Some believed Deaver was being set up, just as Nofziger believed Sears had also set him up. Nofziger's animus toward Sears went back to 1975. He didn't like Sears. Period. It should have followed that as two garrulous men who enjoyed drinking and talking politics, they might have enjoyed each other's company, or at the very least found a way to get along. But Nofziger was also a "True Believer" who wanted Reagan to follow his conservative instincts. He believed that Sears was "moderating" Reagan's message and mistakenly holding Reagan back from being on the road.

Nofziger also knew that the more Reagan campaigned, the more he worked at it, the better he became. Having him to sit on the sidelines was like hitching a thoroughbred championship horse to a plow. The horse needs to stretch his legs, and so, too, did Reagan. It was the continuation of one long fight going back years, about Ronald Reagan and his ideology, his stamina, and how to "package him."

While in Washington in late June, Reagan got into a hypothetical conversation about the vice-presidential slot with a reporter he thought would keep his confidence. The individual surreptitiously tape-recorded the conversation and then gave it to the media. Embarrassingly, Reagan said on the tape, "[T]he press is telling the people of America every day that I'm so decrepit with age that I must appoint a boy in his teens if I'm the [presidential] nominee . . . to be ready to take over the day after the Inaugural. I don't think anybody would think that I'm able for the second spot if I'm not able for the first." Reagan was getting sick and tired of the damned "age issue." He composed himself, moved on, and then told the questioner, "My own view on the vice presidency is that it is described by an old rule of dog sledding—only the lead dog gets a change of scenery."[68]

In the years when he refused to fly, Reagan took trains, but unlike businessmen of the day, he did not hang around the club car tossing back drinks, chasing skirts, and swapping stories with traveling salesmen—although he was fond of jokes about the traveling salesman and the farmer's daughter. On those long trips from L.A. to Boston, or L.A. to Miami, or L.A. to Washington, he would get a private compartment; take along with him suitcases full of books, magazines, and newspaper articles; and read voraciously, alone, for days. A favorite book was *The Road to Serfdom*, by Friedrich Hayek, a treatise on the dangers of big government.

It was during those long train rides that Reagan really developed his understanding and commitment to free-market economics, an intellectual foundation far deeper and more grounded than any of his contemporary critics would ever fully comprehend or acknowledge. The press, pundits, and political pros might concede that former professional football player Jack Kemp could develop an economic "gravitas" from self-study, but in their minds a former movie star simply didn't have it in him.

They were wrong, of course, and anyone who took the time to read the elegant simplicity with which Reagan presented his ideas on freedom, capitalism, and liberty (beautifully written and superbly spoken) would have understood that even a former movie star can powerfully articulate the founding political philosophy of America. "Reagan was far from the uninformed dunce that many modern-day liberals considered him to be. He was thoughtful, well-informed, and knowledgeable about the core principles of free-market economics. Among his favorite economists were the nineteenth-century French classical liberal Frédéric Bastiat and twentieth-century 'Austrian' economists Ludwig von Mises and Henry Hazlitt," Richard Ebeling, former president of the Foundation for Economic Education and a professor at the Citadel, wrote.[69]

Reagan did for political speech what Henry Hazlitt (who wrote for the *Wall Street Journal*, the *New York Times*, and later *Newsweek*) did for economics when he wrote his 1946 classic *Economics in One Lesson*, a book Reagan read and reread during his long train trips.[70] Like Hazlitt, Reagan translated complex ideas into concepts that could be readily understood by the average person.

———

The American people had seen their government eradicate many childhood diseases such as polio and scarlet fever, struggle to right the wrongs of Jim Crow laws, and create the finest public education system in the world. The previous two decades saw a building distrust, however. After a plethora of failures, the government was additionally not solving the problems in the inner cities, and really, the last thing people could point to as an example of government succeeding was on July 20, 1969, when two American astronauts landed on the moon and beat the Soviets while fulfilling JFK's goal set eight years earlier.

But by the late 1970s, NASA, like much of America, had lost its way and its nerve. Under increasing pressure from Congress to cut flights to the moon, and in response to the public's lessening support for the space program, NASA, rather than making the case that reaching for the cosmos ennobled the human spirit and was an extension of Manifest Destiny and American exceptionalism, in an attempt to save itself, surrendered and changed its focus from the planets and the stars to earth. The agency changed its logo; significantly, the new one did not depict space or stars as the old one had, but was a presumably focus-grouped logo that could have been the sign for a new discothèque. It was nick-named the "worm" due to its stylized lettering of the NASA acronym.[71] Rather than looking outward, NASA, along with the rest of the Me Generation, would look inward.

Gone were the dreams of colonizing the moon or missions to Mars and the other planets. Now it was all about the earth—all about us. Now the focus would be on space stations and low-Earth-orbit Space Shuttles whose only purpose seemed to be grandstand PR stunts and not real science or exploration. How many times could you watch ants mate in space?

NASA made sure that the early Skylab, essentially a giant grain silo into which astronauts and cameras would be stuffed for weeks or months, sent back plenty of photos of the Earth, always pleasing on the front pages of the morning newspapers and the evening local news, but of dubious scientific value. One could look at cloud formations over Africa only so many times. Newspapers stopped printing them, and NASA sank into near oblivion, its budget cut so severely that it threatened no one. The capstone of NASA's public relations binge was a meaningless linkup in Earth's orbit between a tired, old Apollo spacecraft and an even more tired and older Soyuz in the summer of 1975.[72] The reason for the space program's existence had been reduced to a rationalization that without NASA, Tang orange drink, hand calculators, and Velcro might not have been invented. NASA in the 1970s was a metaphor for the country: riddled with excuses, failures, and incompletes, a cruelly diminished shadow of its former self.

For the first time that pollsters could remember, the American people were beginning to believe the future would be worse for their children than it had been for them. Alexis de Tocqueville in his landmark *Democracy in America*, written in the 1830s, saw America as "the country of tomorrow." Americans, including Reagan, from Philadelphia in 1776 to the dawn of the space age, had become deeply imbued with the ideas of Manifest Destiny, and that their country and her people were special in the world.

Reagan truly believed in American exceptionalism. He'd also been steeped in Horatio Alger novels as a young child growing up in Dixon and Tampico, Illinois. Alger wrote 135 books, all based on a similar theme: boys, always poor, who were noble and hardworking and who, in the end, were rewarded for their superior character.

The seat of American government was not dominated by superior character in the 1970s, however, just characters. Carter was in a battle with his party and the city it dominated. "People are beginning to wonder whether they are seeing the unmaking of a president or the unraveling of a presidency. He is at war with his own party and estranged from the country," wrote the personally gentle but feared columnist Mary McGrory.[73]

Up from Carterism

"Pouvez-vous dire un malaise?"

It was little consolation to Americans that as bad as their economy was, it was far, far worse for the Soviet people. The dirty little secret in the Soviet Union was that the Communist economy could not provide adequate basic foods, including produce, and "more than a third of all market fruits and vegetables come from private ground."[1] In the Soviet system, private ownership was officially banned. The "people" owned everything, but the chosen "people" who dictatorially ran things for the "people" were incompetent collectivists whose devotion to a failing system would eventually be their undoing. Everyone was equal, but some were more equal than others.

The masses lived grim lives of deprivation, while the privileged apparatchiks rode in limousines, shopped in special stores that sold coveted Western goods, and vacationed in comfortable dachas. It was hardly the "classless" society envisioned by Marx. American leftists may have pined for the "worker's paradise" of the Soviet Union, but they preferred their privileged and elitist status in the media and academia. Harvard may have taught the superiority of the Communist system and the sharing of scarcity, but make no mistake, they needed their Birkenstocks and tweed jackets from Brooks Brothers and their down vests from L.L.Bean. What Harvard preached, Harvard didn't practice. Harvard put the "H" in hypocrisy, and Reagan often made fun of the intellectually bankrupt Politburo on the Charles.

Each year, the Soviets announced with great fanfare that yet another five-year plan for the production of steel, wheat, or clothing had exceeded the quotas, and each year, the Soviet people ate less, wore less, and lived less long. The only things in ready supply were vodka, nuclear weapons, and the hegemony of the Kremlin. Inevitably, after yet another "record" wheat harvest, the Soviets would go, hat in hand, to the West, asking to buy some of its wheat—on credit. All their boastful pronouncements were greeted by Western anti-Communists such as Ronald Reagan as so much baloney, and they took to referring to the Soviets as a "third world country with nuclear arms." Spies were Russia's only worthy export, as George Will wryly noted.[2]

Two months earlier, former president Gerald Ford's autobiography was released. Entitled *A Time to Heal*, it wasn't deep, it wasn't philosophical, and it wasn't well reviewed, but it sold reasonably well. In the mostly ghostwritten book, Ford spends a considerable amount of time regretting: regretting that he didn't speak out against Joe McCarthy, regretting that he didn't speak out against Spiro Agnew, regretting that he didn't stand up for Nelson Rockefeller.

Ford, out on the hustings to promote the book, let Jimmy Carter have it—kinda. He said that all GOP candidates should concentrate their firepower on the "bad job" Carter had done as president on the economy, but avoided any criticism of Carter's foreign policy. It seemed like an odd prescription, but Ford was in basic agreement with Carter on foreign policy. Ford, the old internationalist, had supported the Panama Canal Treaties, had signed his own arms control agreement with the Soviets at Vladivostok, and was conferring with Carter about supporting SALT II. Most Republicans at this point, even RNC chairman Bill Brock, had declared bipartisanship on foreign policy as dead. But Ford persisted in his old notions.

Ford also dismissed Ronald Reagan's status as the front-runner for the GOP nomination, telling reporters that the race was wide open. "I think others . . . have made inroads. I don't think anybody has it locked up."[3] He also warned against the party going too far to the right, and this, along with certain passages in his book, pretty well summed up the animosity that continued between the two men. In many ways, the relationship had gotten worse. Though Ford was not actively campaigning for the nomination, he was making clear that he would be interested in accepting a draft or the nomination of a brokered convention, if only to stop Reagan. But he wanted the party to come to him, not the other way around.

If Ford wanted the GOP to shy away from criticizing Carter's foreign policy, Senator Henry Jackson from Washington State was under no such constraints. Jackson was one of a dwindling number of anti-Communist Democrats, and he blasted the president on concessions toward Moscow. Speaking to the Coalition for a Democratic Majority, Jackson compared Carter's policies to those at Munich, when British prime minister Neville Chamberlain knuckled under to Adolf Hitler. Chamberlain's weakness helped pave the way for World War II. Jackson, in fact, used the word *appeasement* when describing Carter's policies.[4] He condemned the SALT II treaty, including the waiving of requirements in Jackson-Vanik, a bill he created, that forced the Soviets to allow Jews to emigrate. The Soviets wanted the requirements waived in exchange for arms control agreements, and the Carter Administration was going along with this proposal. Jackson, utterly honest, ethical, and decent, was impossible to intimidate, and his words resonated across the political spectrum.

―――――――――

No one was happy in America, and people were getting downright angry. "Gasoline stations in the Washington area were the scene of panic buying, traffic jams and occasional fist fights . . . as motorists—apparently concerned about reports of a weekend shortage—rushed to fill their tanks."[5] Some stations closed early in the afternoon, when their supplies dwindled. Virginia's governor, John Dalton, issued an executive order instituting "odd-even" sales of gasoline. Some station attendants took to holstering guns to fend off crazed drivers. The Administration's only solution so far was to propose a standby rationing system, but Carter's own Democrats in the House eventually rejected this. The gas crisis dominated the news for weeks.[6]

Some states in addition to Virginia imposed an odd-even arrangement to get control of the gas lines, but if your license plate ended in an odd number, then those months that ended with a "31" were your extra friend. In the Washington metro area, they dealt with the problem by letting all buy gas on the thirty-first, if they could "find an open station."[7] Other states used Social Security cards, but anyone in a family could come up with cards that covered the odd-even restrictions. People were getting up early just to go find a gas station that would open in a few hours, hoping to be near the front of the line. Others more resourceful would park their cars the night before to hold a place in line. On some highways, gas purchases were limited to as little as three to four dollars. Stories of people waiting in gas lines for hours on end were common.

Meanwhile, Washington was in the midst of yet another sticky and oppressively hot summer. The only relief that would come would be the appearance of violent thunderstorms in the afternoon, which might cool things down for a bit, but as the rainwater evaporated, the town became even more uncomfortable, and the cycle of heat and humidity and rain and humidity would carry on unabated. Many Washingtonians fled each weekend to the Maryland shore or the Blue Ridge Mountains or farther, to West Virginia, to escape the onslaught. For those less fortunate, it was a matter of slowing down and pacing oneself. They didn't call these the dog days of summer for nothing. The swampy, malarial humidity of the Tidal Basin in summer was a reminder of why British diplomats, before the advent of air conditioning, had been given additional "hardship pay" for being stationed in DC.

The Potomac River offered no relief in the mid-1970s. No one in his or her right mind would even have gotten downwind of it, much less dared swim in the slow-moving cesspool. Indeed, the city government had had a law on the books for seven years that prohibited any physical contact with the toxic river. Some lunatics in the summer of 1978 tried to organize a "swim-in," to show that the river was safe. But the event was to take place far upstream from DC and the overflow from Potomac, Maryland. Even so, it was not reported if anybody ever actually swam, and if they did, if they came out alive.[8]

Grimier and grubbier than the Potomac was the revelation that members of Congress and favored staffers had no such difficulties getting gasoline, as they had their own secluded taxpayer-subsidized private gas station for their government limos and chauffeur-driven private cars, right on Capitol Hill. "There is a place where gas still sells for 67 cents a gallon," a *Washington Post* story read. "It is conveniently located. The service is good. And there are no lines. The only trouble is that the station has only two pumps. And they are reserved for a few bigwigs."[9] The rest of the populace in and around Washington was paying ninety cents per gallon, but actually the favored few never had to dig into their pockets, even to cough up sixty-seven cents. Every red cent was charged to the American taxpayer. The story also detailed how a newspaper photographer was harassed when he tried to take a picture of the pumps, and finally, a congressional bureaucrat looked down his nose and, in his best Marie Antoinette impersonation, sniffed, "That's not public territory."[10]

While all this was happening, gold was going through the roof and the dollar was crashing through the floor, as wealth fled dollars to commodities. In the span of only a few months, an ounce of gold went from $230 per ounce to $440, almost doubling.[11] People were going to jewelry stores to get estimates on the worth of their gold filigree pens or chains. On a daily basis, radio stations blared with updated market news: the stock market was tanking, and the economy was in a nosedive. The Congressional Budget Office forecast a severe recession for the rest of 1979 and well into 1980, "with inflation continuing at a double digit pace and the jobless rate rising to 7.5 percent."[12] Stagflation had arrived in America. The economic textbooks taught that it was impossible to have both stagnant growth and inflation. The Carter years proved the theorists wrong, handing hard-pressed consumers and businesses the worst of both economic worlds.

Jimmy Carter was under siege from all corners. Ted Kennedy was not overtly seeking the Democratic nomination at this point, but the siren song being sung in his ear by liberals in the party was starting to turn his head. Kennedy was a loyal Democrat, but his pals in Congress were exhorting him to do his duty and save the party. Besides, he didn't like Carter. "Draft Kennedy" operations were already under way in Florida, Wisconsin, Iowa, and New Hampshire. Five Democratic members of the House came out publicly for Kennedy and were attacked by their own party chairman, John White.[13]

Over the course of three years, Carter's depiction by the acerbic political cartoonist Pat Oliphant had continually shrunk until he appeared Tom Thumb size. That cartoon shrinkage mirrored the American people's opinion of Carter. His polling was plummeting, in the range of Harry Truman and Richard Nixon at their worst, in the low twenties. The senior staff gathered each afternoon to brainstorm and talk over new ideas, but the confab quickly became a joke and was labeled by some White House wags as the "five o'clock follies," an echo of the Pentagon press conferences in Vietnam that reporters such as David Halberstam derisively gave the same label.[14] The Carter White House was mired in its own quagmire. Nothing was working.

It seemed as if it were all spinning out of control. Even the Fourth of July fireworks on the Mall in Washington were canceled because of rain throughout the day. The Democratic Party was on the brink of a full-scale donnybrook. Given the party's history, at least that hadn't changed. Incredibly, the leader of one labor union, the left-wing Jerry Wurf, openly speculated that he might switch parties, and that his highly activist union could vote Republican in 1980.[15] Wurf had been a Young Turk in the labor and civil rights movements of the 1960s; now, like many liberals, he was so disgusted with what he perceived as Carter's betrayals of workingmen and -women that he actually threatened to vote for the GOP, a blasphemous act that many other true-blue Democrats were contemplating.

Reagan's campaign was stuck, and the candidate was bristling under the bridle his campaign manager John Sears had imposed on him. Sears in turn was getting nervous about his candidate. Over the span of a few days, Reagan had come out for decontrol of gas prices and speculated that fuel might go to $1.50 per gallon. Reagan tried to give the reporters a tutorial on the laws of supply and demand, telling them "a few years ago, pocket calculators cost hundreds of dollars. Today you can buy them for $20."[16] For this, Reagan was charged with shilling for the big oil companies in arguing for the free market. Several days later, he argued for giving arms to the pro-American dictator Anastasio Somoza in Nicaragua to fend off a Communist uprising.

Reagan had a history of making provocative statements. Both free-market and anti-Communist conservatives cheered Reagan, but Sears was determined to keep Reagan from being Reagan. Reagan's appeal over the years with conservatives was due in no small part to the fact that he would say things that other politicians did not have the courage to say.

Sears had Reagan's schedule and public appearances curtailed even more, and Reagan's speeches would be strictly according to script. Reagan began to fall in the polls. Meanwhile, Carter was telling the media that he considered Reagan unqualified to be president. Many in the media bought in, at least initially, to Sears's strategy to keep Reagan above the fray, referring to "the effectiveness of Reagan's low profile strategy."[17]

—————

In late May, the Iowa Republican Party decided to hold its first-ever presidential preference straw poll at the party's annual Lincoln Day Dinner. Party leaders touted it as an early precursor to the increasingly important Iowa presidential caucuses, which would be held in January 1980. New Hampshire still held the distinction as the first-in-the-nation presidential primary, but the Iowa caucuses, especially after the important role they played in Jimmy Carter's successful 1976 capture of the Democratic nomination, were now the first-in-the-nation presidential contest.

Reagan had still not announced his candidacy, but George Bush and his political team were all over the event. The *Des Moines Register* conducted the poll with 1,214 Iowa GOP faithful, each of whom ponied up fifty dollars for the privilege of the vote, along with the meal. When the straw poll results were announced, it came as a temporary shock to some that Bush was declared the winner, with 39 percent of the vote. Reagan came in a distant second, with 25 percent.[18]

The straw poll was meaningful for Bush for bragging rights and favorable publicity, and was seen by those who knew how the Iowa caucuses worked as a trial heat for "turning your people out." Bush's campaign had staked a lot on the straw poll, bought dinner tickets, and arranged for rides, and the gambit worked. Sears mistakenly comforted himself and the Reagan campaign with the most recent poll numbers out of Iowa, which had Reagan at 31 percent, Gerald Ford at 28 percent, the rest cluttered in the low double digits, and Bush pulling up the rear, just ahead of the self-destructing Phil Crane, at 3 percent.[19]

Conservatives were becoming increasingly dismayed with Reagan. Some had earlier gone over to Phil Crane, but his campaign was falling apart. John Connally just wasn't their cup of tea. Dole was too much of a Washington insider, and didn't go over with voters outside Kansas. Some conservatives were kicking the tires of Wall Street impresario Bill Simon, the conservative with the Midas touch who had made millions and had written *A Time for Truth*, which unexpectedly sold 250,000 copies. He was a veteran of the Ford cabinet, but had not been tarnished by the "moderate" brush and was tall, articulate, and handsome. Simon mulled for a time jumping into what was appearing to all to be a wide-open race, as Reagan seemed to be faltering.

In Mississippi, Clarke E. Reed emerged as his state's chairman for the tone-deaf Connally campaign. Reed had gained infamy among conservatives as the man who broke his promise to Ronald Reagan in 1976, at a critical time in the primaries and later again at the convention, costing

Reagan the nomination—or so the Reaganites had come to believe. It will never be definitively known if Reed's switch fully determined the outcome. Reed controlled 30 delegate votes and 30 alternates from the Magnolia State in Kansas City, and these flipped from Reagan to Ford, a swing of 60 votes. On an important rules decision and then, the next night, for the nomination, Ford won both, balloting over Reagan by a somewhat greater margin than those 30 votes.[20] Yet it remained the Reaganites' probably well-founded belief, nurtured by Reagan's high command, that those other state delegate votes that went to Ford had fled Reagan only when they learned of Reed and the Mississippi delegation's apostasy.

The yarn, fair or not, was born that Reed had betrayed Reagan, costing him the nomination, and Reed would be branded that way for the rest of his political life. Such are the vicissitudes of politics, but Reed did not back away from the political game. To his credit, he hung in there as Mississippi's Republican national committeeman and was working hard to "whittle away" Reagan's base in Mississippi and move those voters to Connally. Feelings had not healed in Mississippi in the three years since the fight over Ford and Reagan; indeed, they had hardened into downright loathing. Reagan's 1976 chairman and key supporter in 1980, Billy Mounger, said, "The Republican Party here is about 75% Reagan. Nothing happened to us then until Clarke Reed started selling us out."[21] Mounger and Reed had been lifetime friends until the 1976 convention, when Reed went with Ford, and Mounger never again spoke to Reed.

Another GOP operative in Mississippi who flipped from Reagan to Connally (although Reed actually went from Reagan to Ford to Connally) was Haley Barbour, the well-regarded former state party executive director. Some resentment over Barbour's move from the Gipper to Big John was inevitable, but Barbour was a smooth, charming, and polite operative who could toss back sour mash whiskey with the best of them. His man, Connally, was going right at Reagan's throat, telling audiences, "What has Ronald Reagan ever done? And he's how old—69 or 68? And that's all he's done?"—referring to Reagan's two terms in Sacramento and running for president twice. Connally took note of the fact that Reagan had come up short nationally and said, "In my world, second is fine and third is worse, but neither is acceptable."[22] Reagan's only supporter among the Magnolia State's party leadership was Charles Pickering, who was the former state GOP honcho and was running in 1979 for attorney general. Pickering had been a true-blue Reagan supporter for years.[23]

———

Ronald Reagan was slipping badly in the national polls. He'd gone from 40 percent among Republicans in January 1979 to only 28 percent by June, according to Gallup.[24] Gerald Ford continued to poll strongly, though he'd done nothing to help his chances. Reagan, while not engaged in a frenetic schedule due to Sears's "front-walker" strategy, was certainly out and about more than Ford. The *Washington Star* euphemistically referred to the Reagan slowdown as a "recent lull."[25] Reagan's slippage did not go to the other candidates, but instead into the undecided column, often fatal for any front-runner. Reagan had been more or less a national figure since his Hollywood days in the 1930s and '40s, a television figure in the '50s, a successful two-term governor, and the beloved national conservative leader. The Reagan sale should have been closed long ago with GOP voters.

More and more, however, the GOP register was showing "no sale" to Reagan. Adding more woe, at a time when California's economy was crashing and gas lines seemed to go from San Diego to San Francisco, Reagan was only tied with Carter in polls in his home state. The mood of the country was awful. Less than a quarter of Americans "felt things are going well" in the United States.[26]

Reagan's nascent campaign was stumbling badly in the South, as noted by columnists Evans and Novak. Not only were several state GOP operations turning to conventions as opposed to primaries, which was seen by all as a move not only to help Connally, the backroom wheeler-dealer, but also to undermine Reagan, whom they knew and feared was more popular with rank-and-file primary voters. The duo referred to the sickly "slow unraveling of Ronald Reagan's once impregnable southern base," and also noted a decided lack of enthusiasm for the Gipper in Arkansas. "Reagan's southern problem is no different from elsewhere in the country: the party faithful whispering that he is 'over the hill,' accompanied by complaints about shabby treatment from Reagan's campaign staff."[27]

One GOP presidential candidate who clearly had no momentum was liberal Illinois congressman John Anderson. His announcement in April barely caused a ripple in the fevered political reporting emanating from the banks of the Potomac. Little changed over the next several months. "Anderson was hard to figure out. He ran a campaign that did so many unusual things for a person pursing a presidential nomination that audiences had trouble understanding what he was doing and why he was doing it," Jim Mason wrote in his account of Anderson's quixotic campaign.[28]

As behind the eight ball as Anderson's campaign was, however, its plight was nothing compared to that of the president. New polling released by CBS News and the *New York Times* showed President Carter with only 30 percent approval overall (down again) and his numbers were even worse on his handling of the economy (an appallingly bad 20 percent approval) and the energy crisis (even worse, at 19 percent approval).[29]

Carter's fund-raising for a 1980 drive was also lagging, incredible for an incumbent president. Kennedy was looking omnipotent, as fully one-third of Republicans had a favorable opinion of the Bay Stater. Still, the vast majority of Americans saw Carter as a good and moral man, and more clear-thinking GOP pollsters were telling their clients that though Carter was politically wounded, he was still a tough political operator. Carter was not about to become yet another casualty of intraparty squabbling, at least not without a hell of a fight.

Reagan wasn't in much better shape than the president, and his age was becoming more of an issue with voters, if such a thing were possible. Sixty-two percent of those surveyed said they would be less inclined to vote for someone for president who would be over seventy years of age while in office. Moreover, Reagan's overall approval/disapproval was 44 percent to 36 percent. Okay, but certainly not great. The poll also showed a yawning gap in leadership in America. Americans were thirsting for it, wandering in the desert, going from mirage to mirage, but finding little in June 1979.[30]

Sixty-four people filed campaign committees with the Federal Election Commission, and surely among this group there was a leader. The group included a clairvoyant, Mrs. Nell Fiola, who promised to bring new knowledge to the world. She was also a moderate Republican, so the GOP's big tent would have to make room for her sideshow. Another candidate was only twenty-three years old, and when he held a one-dollar-per-plate fund-raiser for his campaign, only he and his campaign manager showed up. The fact that the young Mr. "Peter Titti" (whose name had to have been a joke) was constitutionally barred from serving because of his age, though, did not deter him. "If I can fight in a war, why can't I declare it?" Sound logic.[31]

Amid all these sideshows, "Teddy" was on everybody's lips. Kennedy held large leads over Carter among Democrats, and was handily besting Reagan by mid-June while also trouncing the rest of the GOP flock in national polls.[32] Clearly he was a tragic beneficiary of the assassinations of his brothers. Both JFK and RFK were seen by the American people as strong, articulate, and decisive. Of course, the circumstances of their deaths made them iconic, and people forgot that President Kennedy's approval rating had slipped to only 58 percent in mid-November 1963. JFK was in trouble in the South, and he was in Dallas to attempt to mend broken political fences within the Democratic Party in preparation for the 1964 campaign. An obvious indication of how much trust Americans had lost in their own government by 1979 was the fact that many believed their own government had conspired to assassinate President Kennedy.

RFK hagiographies to the contrary, it's not clear that Bobby Kennedy would have gained the 1968 nomination, even after winning the crucial California primary the night he was assassinated. President Johnson, though politically wounded in his lair at 1600 Pennsylvania Avenue, still controlled the lion's share of the delegates, and he hated Bobby Kennedy with as much passion as RFK hated LBJ. Johnson, the master arm twister, would have worked overtime to keep Kennedy from winning the nomination in Chicago. Even if Kennedy had won it, it's also not certain he would have beaten Richard Nixon, especially with George Wallace running as a third-party candidate sucking up Democratic voters in the South and leaving a divided Democratic convention in his wake.

Nonetheless, in the summer of 1979, the "last brother," Edward Moore Kennedy, forty-seven, damaged politically though he was, as well as unabashedly liberal, overweight, and a heavy drinker, was the hope of the nation.

———

As if things weren't bad enough for the country, "Mr. America," John Wayne, died of cancer at the age of seventy-two. In the course of his career in Hollywood, he'd made more than two hundred movies, many on the same theme: Wayne as the rugged, individualistic, patriotic American hero. He finally won an Oscar for playing his signature role, in *True Grit* in 1969. "The Duke" was a big fan of Reagan's, and the feeling was mutual. Reagan wasn't the boozer or A-list actor or hell-raiser Wayne was, but the Duke and the Gipper were good friends and often enjoyed each other's company.

Although they never made any movies together, they were admirers of each other's work, were products of the same culture and generation and region, and shared a conservative political viewpoint. Their only major policy disagreement was over the Panama Canal Treaties, but since Wayne was a fishing and drinking buddy of Omar Torrijos of Panama, his support was understandable.

Wayne was "iconic" before the term came into popular overuse. Congress had struck a medal for him the year before, the same as cast for the Wright Brothers; thousands of young boys had been named after Wayne over the years. Wayne was an anti-Communist as far back as the 1940s, when he'd been accused by political enemies of supplying names to the House Un-American Activities Committee, but denied it. Reagan wrote a touching tribute to his old friend for *Reader's Digest*. In character, Senator Barry Goldwater said of Wayne's passing, "John Wayne was just one hell of a guy, a hell of a man."[33] The medal created by Congress to honor the Duke was simply inscribed "John Wayne, American." Before Wayne passed away, Carter kindly paid a visit to him at his hospital, telling him that "he had the love, affection and prayers" of the American people and people around the world.[34] On his deathbed, Wayne converted to Catholicism.

It's telling, however, that during the 1970s a new kind of protagonist was emerging in Hollywood, with roots in the moody Method acting of the 1950s. As a counterpoint to Wayne, existential antiheroes such as Dustin Hoffman, Al Pacino, and Jack Nicholson were becoming role models for the nation's impressionable youth. In 1970 the gritty movie *Midnight Cowboy*, rated X at the time and starring Hoffman and Jon Voight as two drifters in a morally bankrupt New York City, did much to loosen cinematic morals in Hollywood and reflected the widening cultural divide in America. Indeed, a different kind of cowboy was emerging: ironic, detached, and openly scornful of the traditional values represented by the "square" Wayne in his classic John Ford Westerns or, for that matter, by Reagan in his Westerns. Hollywood was winning the race to the bottom.

Reagan's campaign had not completely stalled. In New Hampshire, the GOP state chairman, Gerald P. Carmen, resigned after a successful tour there and signed up with Reagan, after John Connally had put an unsuccessful full-court press on Carmen to garner his support. Carmen was a conservative through and through, and the diminutive businessman loved Ronald Reagan almost as much as he loved political combat.

Relishing a good fight, Carmen was known for getting his way and not suffering fools gladly.

The field in New Hampshire was divided into two groups according to the commentariat. In the first group were Reagan, Connally, and Howard Baker as the anointed front-runners. In the second grouping were Bob Dole, Phil Crane, John Anderson, and George Bush. Indeed, a mail-in poll conducted by New Hampshire congressman Jim Cleveland had Reagan in front of both Gerald Ford and Jim Baker. The surprise was dark horse Bush, ahead of Connally.[35]

Republicans gathered for their annual summer meeting in Minneapolis tingling with excitement at their newfound respect. Jimmy Carter was mired deep in the polls, pulling the Democratic Party down with him. The Grand Old Party was more unified than it had been in years, at least on the surface. Money was flowing in, and party elders believed that many governorships, statehouses, and possibly Congress itself were within reach in 1980. They also assumed that once the intramurals of the presidential primaries were over, their nominee would knock off the incumbent Carter and make the Georgian the first incumbent to lose re-election since FDR beat Herbert Hoover in 1932. Many of the members of the national committee were secretly pulling for Baker, believing, as the Carter White House did, that he would be the toughest candidate against the Georgian in November.

Bill Brock, the RNC's chairman, had committed resources to the rebuilding of the party as never before. In 1978 the RNC state and local division handed out $1.7 million to local office seekers and won 282 state legislative seats and control of 10 new chambers.[36] They also sent out skilled political operatives, who worked with these often-naïve candidates to help them get elected—everything from fund-raising to media to organizing voter lists. Brock's gamble had paid handsome returns.

———

The *Detroit News* released a survey of the fifty state-GOP chairmen, and two-thirds said Reagan was the favorite in their state.[37] This did not mean, however, that they actually supported the Gipper. Reagan had received no help from the party establishment in 1976 and expected to get little from them in 1980. His campaign would have to make it based on the conservative grass roots and not the establishment Republican grass tops. Ed Blakely, who had produced the House Republicans' effective campaign commercials for years, and no conservative himself, said what was really going on inside the party committees. "They hated Reagan," he bluntly stated years later.[38]

Reagan was keeping up with his correspondence, his commentaries, but for a man who had one last chance at the brass ring, he was incredibly underbusy, staying close to his homes in California. He was ducking Republican invitations, conservative invitations, and journalists' requests for interviews. More and more and wider and wider, the stories spread that Reagan had become old and brittle and that Sears was hiding him.

Reporters, when they were granted access to Reagan, invariably complained that his answers were not specific enough, or when he was specific, they were outlandish. They complained that he was too nice, but too aloof, that his home was like a movie set. To a man, their "assumption has been that if Reagan is the Republican nominee the election of a Democrat is certain."[39]

Reagan was also fighting a complaint made to the Federal Communications Commission against him and radio stations that were carrying his daily commentaries. A former FCC commissioner, Nicholas Johnson, filed the complaint, charging that radio stations were in violation of the "equal time" provisions and that they must stop broadcasting Reagan or give the same amount of time to all the other candidates—in other words, that the stations in effect were paying Reagan for the commentaries. Johnson and his front group, the National Citizens Committee for Broadcasting, failed on a 4–2 vote. A ruling against Reagan would have cost him around $500,000.[40]

Johnson was a Democrat with deep ties in Texas politics. LBJ had appointed him to several federal posts, including the FEC. John Connally obsessed every day that he needed to catch up to Reagan in terms of national support. Nicholas Johnson knew Connally well, and filing the complaint against Reagan helped his fellow Texan.

———————

On the broader world stage, significant developments were afoot. Eight months before, the Vatican had selected the first non-Italian pope in five centuries to lead Catholics worldwide. Puffs of white smoke were spotted over Rome when the College of Cardinals chose Karol Wojtyla, a Pole, fifty-nine, the archbishop in Krakow, as the new pope.

John Paul II was unlike previous popes. He was sophisticated in modern media, used technology to get his message across, and was a dedicated anti-Communist, by virtue of having seen up close what the Communists had done to his native country. In early June, he made a triumphant return to Poland, where millions greeted the "local boy who made good." The *Washington Post* called the handsome pontiff a "media star" and wrote that "the faithful respond to the well-built, athletic, spiritual leader who encourages contact with the masses instead of giving the impression of fleeing from them as did his predecessors."[41]

In nine days of touring his homeland, John Paul II gave hope to millions of his oppressed compatriots and fearlessly took on the Soviets. The paper also noted that John Paul II made liberals, Communists, and the Soviets "uneasy."[42] Of course, it was the Soviets who only two years later tried to murder him. The *Post*'s lead editorial on the papal tour, and the remarkable man, noted, "It is that the power of faith is more than equal to the power that tries to deny it."[43] The Soviets took a different view and issued a paper calling for a strict enforcement of Soviet atheism.

———————

There was little consensus among the presidential hopefuls over what should be done to alleviate the gas shortage. President Carter's plan was for controlled deregulation of the price of oil with a huge tax on the oil companies, which free-marketers saw as a disincentive for those companies to produce more. Ronald Reagan's was a straight free-market approach: immediate decontrol, no new taxes so that the companies would be free to produce more, and with that, supplies would grow and prices would fall. Reagan, for once, banged Carter hard, saying the president shouldn't "be so willing to agree with the OPEC nations that we are greedy."[44]

Carter was stymied on how to alleviate the calamity, and Democrats fretted that the whole party was on the decline. Senator Daniel P. Moynihan, a nondoctrinaire intellectual, told the *New York Times*, "If they just go on giving us explanations about why nothing can be done, they're explaining this administration out of office."[45] Carter was in Asia and was planning on stopping in Hawaii for a vacation on the way back, but this was scrapped. The imagery of a president frolicking on a sandy beach while Americans could not get away for the Fourth of July because they couldn't get any gasoline was all wrong, so he headed back to Washington.

Peter Hart, a well-respected Democratic pollster, wrote that "voters are looking for relief from the present chaos." Meaning, not Carter.[46] It wasn't just gasoline. Inflation was surging at over 13 percent for the year, and the economy was in the toilet, contracting 2.4 percent in the second quarter of 1979. Because Social Security benefits were tied to inflation, the jump triggered a 9.9 percent increase, putting even more strain on the creaking federal retirement system.[47]

Teddy White, in his important book, *America in Search of Itself,* named one of his chapters "The Great Inflation," and detailed how insidious inflation wreaked the Sung dynasty in China a thousand years earlier, devastated Revolutionary France, the Confederacy, and post–World War I Germany, and how, by 1979, currency inflation was destroying America. In a fascinating aside, White documented how only twice in Western history had inflation been successfully defeated. In post–World War II Germany, all the Western Allies were printing money to pay their war debts and Germany was using all four currencies. General Lucius Clay defied the Allied Powers and created a new, independent currency for the German people, thus saving their economy. General Charles de

Gaulle also did much the same in the Fourth French Republic. Simply put, in 1979, it was far more likely that inflation would destroy America than the other way around.[48]

Carter was now losing even to the mostly silent Reagan in national polls. A poll taken by Lou Harris for ABC News showed Reagan ahead of the incumbent president, a first time in this campaign, by a 51–44 margin.[49] However, the poll was more about the American people losing faith in Carter than coming to believe that Reagan had the answers. Across the spectrum, most everybody doubted that Reagan would either be the GOP nominee or win the general election.

Like Carter's poll numbers and the American mood, NASA's Skylab was also falling—literally. The decrepit space station was coming down through a decaying orbit. NASA's rocket scientists could not keep it in space, and the seventy-seven-ton abandoned satellite could come crashing down anywhere in the world.[50] Astronauts had not visited the station in years; indeed, no one had been in space for the United States in years. NASA had no idea where its white elephant might land, or on whom. Skylab stood in stark contrast to the can-do spirit of the Apollo moon program; the doomed space station had become a macabre metaphor for the current state of the nation, an object lesson in a downward motion.

―――――――

President Carter departed Washington for Vienna, and his first summit with Soviet leader Leonid Brezhnev, for the signing of SALT II. Brezhnev was in poor health, walked slowly, and slurred slightly. With a flourish, the two leaders signed the pact for the benefit of the media and then stood up and bussed each other on the cheek. Carter would rue the day this moment was recorded on film. The treaty, of course, would have to be ratified by the Senate. In an ABC poll, 69 percent of Americans supported SALT II, but a majority said they knew little or nothing about the details of the agreement. A plurality believed the Soviets were ahead militarily.[51]

CERTIFICATE OF ATTENDANCE

AWARDED TO

Name : _____

Frank Islam Athenaeum Symposia Event

Author Craig Shirley
Reagan Rising: The Decisive Years

Tuesday, February 27 @ 7 PM

Multicultural
Diversity Training

Reagan and John Connally had taken a pass on going to the RNC annual meeting, but they both made appearances at the Virginia GOP's annual bash. Both were hugely popular with the Republicans gathered in Roanoke, but Connally upstaged Reagan, giving the better speech. The take by observers was that Connally was making inroads into Virginia. Reagan had carried the bulk of the Old Dominion's delegates in 1976, 36 to Reagan and 15 for Gerald Ford. One embarrassing story said that Reagan was little more than a "sentimental favorite," that his support was "soft" and "eroding," that he looked "tired," and that he did not perform well at a press conference. Of course, the article also harped on his age.[52]

Reagan's barbecue did sell out, however, and miffed Reaganites were offering one hundred dollars for the twenty-dollar ticket to eat pork with the Gipper. Former governor Mills Godwin was supporting Connally. He, too, had once been a Democrat. Reagan was being supported by Helen Obenshain, widow of the former state chairman and Senate nominee Dick Obenshain, who had been killed in a plane crash while campaigning in 1978.[53]

Reagan's populist message at the gathering was a brushback pitch thrown at the hard-charging Connally and Bush. "Let us reveal to the people that we are not a stodgy, fraternal organization beholden to big business and the country club set. We're the party of Main Street, the small town, the farmer, the city neighborhood where the working people live. Our strength comes from the shopkeeper, the craftsman, the farmer, the cop on the beat, the fireman, the blue collar and the white collar worker."[54]

Some revelers complained that Reagan spoke too long. They wanted to get back to the barbecue and the bourbon, but others liked how mellifluous he sounded. A book called *Talking Fast Between the Lines* came out at the time, and the authors, Julius and Barbara Fast, wrote this of Reagan: "His tempo is well moderated, he has good timbre and a rich tone. His volume is low. He has proper inflection and good enunciation . . . pleasant, easy to listen to, confident and assured." Of Carter, they were harsh in their assessment of his speaking style, using terms such as *uneven* and *flat*.

"There is a complete absence of melody . . . His pauses are not logical . . . It is one of uncertainty, of depression. There is a strong sense that he is unsure of himself."[55] The problem was John Sears not allowing Reagan to get out and give voice to his vision of American conservatism.

───────────

"WHAT ARE YOU UP TO MR. PRESIDENT?" the *New York Post* understatedly screamed.[56] And *Newsweek* nervously wondered about "the nearly total disappearance of the government."[57]

Again, cocktail chatter in Washington's salons speculated on Carter's mental condition, as it had periodically over the past several years. Political and economic hacks were shuttled to Camp David to meet with President Carter, but when Carter went AWOL, "his aides and the guests were having a drink and chatting about the 'breakdown' rumor." The White House aides told the guests at Camp David that the nation's governors had met with Carter the Friday before and "reported he looked fine."[58] A politician's ability to judge mental competence was not questioned in 1979. Shake-ups in the White House staff and the Cabinet were in the wind. It was the summer of 1979, and Carter had canceled yet another national address to the nation on energy, now convinced that Americans weren't listening to him. Something radical was needed to rattle things, and Carter had retreated to Camp David for what was charitably called a "Domestic Summit." The summit took days, and Americans were quite frankly wondering if their president was indeed suffering from a nervous breakdown. Dozens, including the young governor Bill Clinton of Arkansas, came traipsing to the sheltered Maryland mountaintop, summoned by Carter. City counselors from Los Angeles and Connecticut also came to meet with Carter, as did the mayor of Mayorsville, Mississippi. Overall, Carter met with nearly 150 individuals. Potential rivals Ted Kennedy and Jerry Brown were not among those summoned to the summit.

The next day, he choppered off to West Virginia to visit a retired Marine Corps major and some of his neighbors, who gathered in his living room to meet with the president and Mrs. Carter.[59] The Carters came back from their visits to "ordinary Americans" full of what they thought was astounding commonsense insight, which they bubbled about to their guests at Camp David. However, the White House staffers who had accompanied the Carters thought all they heard were the typical banalities.

An Atlanta business executive wrote in the *Wall Street Journal*, "People have said a lot of bad things about Nixon. They called him a liar and a cheat and a crook. But they never called him ineffectual. Ineffectual is just about the worst thing you can say about a President, and it's what people are saying about Carter."[60] Leading Democrats were openly criticizing Carter, jumping over to Kennedy or withholding their support. Of the thirty-two Democratic governors, only twenty agreed to sign a letter endorsing Carter's reelection. Only two-thirds of House Democrats, according to a survey by the *New York Times*, were ready to support him. These defections were nothing less than shocking for an incumbent president who as yet had no active opposition for his renomination. Even fewer, about one-half of the 276 Democratic members of the House, thought Carter would win reelection.

Gallows humor took over at the unattended White House. "It's like working for [Nicaraguan dictator] Somoza," one told Fred Barnes. "It's been like a bunker here," lamented another.[61] Carter also trundled in much of the national press corps, including Walter Cronkite, David Broder, and Jack Germond, for off-the-record briefings at Camp David. In seclusion, Carter buried himself in two anti-American screeds, *The Culture of Narcissism* and *The Fall of Public Man*, both of which conclude that because of the free market, Americans were no good, lacking spirituality. If they lacked spirituality, it was because the Supreme Court had kicked God out of public schools and the public commons years before. Still, the books' conclusions perfectly matched up with Carter's philosophical and political outlook.[62]

———

Finally, after days on end, Carter came down from the Catoctin Mountains, and when he arrived back at the White House, photographers were barred from taking pictures. The White House asked NBC, ABC, and CBS to broadcast a national address that night by the president—July 15, 1979. Carter was finally to speak on television, giving a speech that, it was hoped, would make him "a thoroughly chastened leader who accepts the evidence of the opinion polls that he has lost . . . most of his credibility with the voters."[63] Hendrik Hertzberg, a White House speechwriter, and Carter pollster Pat Caddell were working hard with the president on the new speech. The problem was that the American people looked to be deliberately tuning out Carter. Eighty million of his fellow citizens watched him give his first speech to the nation in 1977 (in the much-maligned cardigan sweater), and just a few months earlier, for his speech in April 1979, that figure was down to thirty million.[64]

Some who hadn't yet reached the brink of complete cynicism in America wanted to tune in that night to watch their president. In homes and bars, in air-conditioned clubs and hot military bases, in firehouses and police precincts, the American people in the depths of their despair, losing a Cold War, losing the economy, were hungry for leadership. Advanced copies of Carter's address were not released to the media, breaking a long-standing custom. Carter was aiming for maximum effect and suspense.

———————

In a memo to the president about the national mood, Caddell had used the word *malaise*, and it subsequently leaked out. *Malaise* was not a manly word for Middle America. It was a word for the hoity-toity sophisticates of the Ivy League or the elites in Manhattan, sipping white wine in a private club, when describing their polo ponies' performance or how well their trust funds were doing. It was not a word for the hoi polloi of the American Legion or the Knights of Columbus halls that dotted America and served cold draft beer. Period. It was French—unmanly, feminine. Corrupt elites used such words, but not the honest and ethical men and women of America.

Caddell was known for his excruciatingly long memos—whole forests quaked at the sight of him—and the one he sent Carter was no exception. Weighing in at 107 pages, it was referred to as the "Apocalypse Now" document by White House insiders.[65] In it, he recommended that the president trundle in as many people as possible and listen to their plights, and then deliver an address to the nation. Caddell had been urging for a "crisis of confidence" speech since the beginning of the year, and now the twenty-nine-year-old was getting his way.

The president took to the airwaves at 10:00 p.m. Eastern Standard Time, Sunday. The speech was scheduled for the late hour, so viewers on the West Coast could see him as they sat down in front of their television sets for the evening. The weather across the country was hot and humid, with violent thunderstorms racing across the Midwest, causing some flash flooding and several deaths. The nation's capital was hazy and still, with nary a breeze. In what came to be seen as yet another symbol of his vacillation and lack of conviction, Carter had oddly changed the part in his hair from the right side of his head to the left.

Addressing the nation, Carter used words such as *losing, disrespect, tragedy, murders, suffer, paralysis, stagnation,* and *drift.*[66] He never used the word *malaise.* Officially, the address was dubbed the "Crisis of Confidence" speech, but it felt more like a sermon delivered by an old-time fire-and-brimstone preacher excoriating his congregation for their evil ways. Fittingly, his televised address delayed an airing of *Moses the Lawgiver* on CBS.[67]

"The symptoms," Carter said, "of this crisis of the American spirit are all around us. For the first time in the history of our country, a majority of our people believe that the next five years will be worse than the past five years. As you know, there is a growing disrespect for government and for churches and for schools, the news media, and other institutions. This is not a message of happiness or reassurance, but it is the truth and it is a warning."[68] Carter then related the quotes of more than twenty Americans he'd met over the previous ten days. Two or three quotes might have been fine, but Carter lacked the discipline to leave anybody out. It was the same old problem with him: indecision. More troubling was that most of the quotes from "average Americans" were banal nothingnesses.

Carter gesticulated, clenching his fists, appearing to want to pound the table, so frustrated and furious was he. Sixty million of his fellow citizens watched the speech. At first, Americans were happy for some talk, some action, something at last—but the upward movement in the polls did not last.

———————

Despite revisionist history, most reviewers in both parties initially gave Carter good marks for the so-called malaise speech. It was only after several days that opinion began to shift. One initial criticism from Republicans was that Carter should have given the speech earlier. Reagan himself, later that same month, said that "there really isn't any crisis in the country. There's just a crisis in the White House."[69] The focus on energy independence was where most people's attention was, and not on the lecture Carter gave Americans about their "loss of confidence."

Indeed, the next day, Carter went to Kansas City to speak to a convention of county officials and was met with cheers and praise. There, he went even further, telling them that America had come to "worship self-indulgence and consumption."[70] The long story from the *Washington Star* examining both speeches also said ominously, for the first time in print, "The result of the current malaise . . ."[71]

The irrepressible—some would say overbearing—Republican senator from Alaska, Ted Stevens, growled, "As I go around the country I don't find people licking their wounds. The only one I find licking his wounds is the president."[72] Stevens also questioned Carter's mental stability, and George Will wrote testily, "America is not as sick as you might gather from what the president said about it last Sunday."[73] He also nailed the inconsistency in the first part and the second part of the speech. Carter assailed the emptiness in American life and said that materialism had led to this state. But in the second half of the speech, he proposed ways in which to create more energy to keep the economy working so it could produce more material goods and, presumably, more emptiness.

Evans and Novak wrote publicly what Vice President Mondale had expressed privately to Carter: "The president's critical new view of his fellow Americans contains dangerous potential for rationalizing and minimizing his administration's shortcomings."[74] The lively columnists also startled their readers when they reported that while Carter was at Camp David meeting with lawmakers in an effort to solve the nation's problems, he spent much of his time obsessing about the immoral media coverage contained in *People* magazine.[75] Mondale had sharp political antennae, and had lobbied hard against the speech; he actually contemplated resigning his office, he was so opposed to it.[76]

Reagan's pollster, Dick Wirthlin, deeply probed the grass roots in the summer of 1979, just as Pat Caddell had, though the two came up with wildly different conclusions. Caddell's polling showed that America's "malaise" predated Carter, and that his was just one in an unbroken line of failed or incomplete presidencies.

Wirthlin queried 250 Americans for up to six hours each and found that the vast majority of them believed the country was on the wrong track; they also believed that happiness came from within and not from material possessions. However, they also felt that, with the right leadership, the problems of America could be solved, and that those were not with the people but with their elected officials. Over 75 percent thought it was extremely important who was elected president in 1980. Reagan saw Carter's "malaise" speech as an "abdication" of his responsibilities as president. The problem was he wasn't telling anyone.[77]

―――――――

A few days after the "Crisis of Confidence" speech, Carter fired those members of his Cabinet who were getting on his nerves, including Michael Blumenthal at Treasury and Joseph Califano at Health, Education, and Welfare. Both drove Carter bonkers. In all, five Cabinet secretaries departed, including energy secretary James Schlesinger, making him one of the few men to be fired from the Cabinets of successive presidents.[78]

Carter had in fact asked for the resignations of the entire Cabinet and the senior staff at the White House, and then decided which to accept.

Hill Democrats denounced the mass firings, and Ted Kennedy hinted darkly that tobacco interests had forced out the smoker turned anti-smoking zealot Califano. The Georgia Mafia, however, stayed intact, and indeed, White House aide Hamilton Jordan was promoted to chief of staff, a position that Carter in 1976 had vowed Jordan would never have.[79] Jordan promptly fired a dozen staffers, including some in the Office of the First Lady. Carter did call in the remaining staff, to give them not a pep talk but a dressing-down. Morale, already low, disappeared into the subsoil.

This Slow Motion Massacre—it took three days to fire everybody—drove Carter right back down in the polls. Nothing he did could alleviate his malaise or the nation's problems. Carter's biblical belief in "commandments" and philosophical belief in governmental mandates couldn't fix the plight of the country or his presidency. His next move, strangely, was to take a cruise down the Mississippi River on the *Delta Queen* to promote his newest energy proposals.[80] The White House staff was excited about the visuals of the president on the river, but had still not gotten the message from the American people that what they wanted from their president was action, not pretty pictures. The capital was going through a record heat wave of temperatures over ninety, day after day, but it did not come close to how hot the American people were about the mess in Washington.

═══════════

In other depressing news, NASA's Skylab space station finally came crashing back to earth, giving a memorable fireworks show to the people of Australia. Twenty-six tons of what was left of the flying albatross went into the Indian Ocean. Ten years earlier, NASA had landed a man on the moon, but now it couldn't even say where Skylab would hit. Walter Cronkite interrupted his annual two-month vacation sailing off Martha's Vineyard to anchor a special on the death of Skylab. Bumper stickers appeared in California: "Chicken Little Was Right."[81]

And in Iraq, President Ahmed Hassan al-Bakr unexpectedly retired because of poor health. He was replaced by the vice chairman of the Revolutionary Command Council, General Saddam Hussein, whom many regarded as a pro-Western improvement over Bakr.[82]

———

The initial burst of support for Carter's malaise speech began to fade within a matter of days. William Winpisinger, head of the International Associations of Machinists and Aerospace Workers, blasted the speech four days later by saying that "blaming . . . the American people is a cop-out."[83] The economy was continuing to tank, and it was the worst GNP report since the end of the 1975 recession.[84] Inflation continued, up a full 1 percent in the month of June alone. In 1979, it took $217 to purchase what $100 had bought in 1967. Gross earnings of the average worker declined by nearly a full 1 percent in June as well. Productivity also tanked.[85] In a memo prepared for Reagan, economic adviser Marty Anderson wrote, "By a wide margin, the most important issue in the minds of voters today is inflation." The nine-page memo spelled out what must be done to curb inflation, including increase productivity, decrease the federal tax rates, and slow the growth of federal spending.[86] It wasn't called "supply-side economics" yet, but that's what Anderson wrote.

In case anybody missed the bad economic news of the past several years, Carter's new treasury secretary-designate, William Miller, told Congress that the economy was in a recession, with little hope in sight. He had few answers for the current dilemma, beyond job training for youth, but he told Congress that belt tightening on the part of the government and cutting income taxes was a mistake.[87] Carter's Cabinet and new appointees were still unsettled, two weeks after his malaise speech. It was clear that when Carter asked for his Cabinet members' resignation, a list had not been drawn up of suitable replacements.

More Democrats were moving away from Carter. Jill Buckley, a Democratic consultant, put it best, saying, "He's the first president who told us that our children's lives won't be better than our own."[88]

Unbelievably, Carter's job approval rating sagged even lower than before his speech. According to a new poll conducted by Harris for ABC News, it was now down to 26 percent approval; and a staggering 71 percent disapproved of his presidency.[89] On energy policy the numbers were even worse.[90] It seemed that it couldn't get worse for the White House—but it did, when the owner of Studio 54 alleged publicly that Ham Jordan had used cocaine at the establishment. Carter said the allegations against his aide were "lies." The Carter White House was completely defensive.[91]

Collectivism in the West was dying, especially in Great Britain and the United States. The unfulfilled promise of the New Deal and the command economy it tried to create were not meeting the needs of the American people. Carter simply did not understand this, and his presidency had begun just as the old order was crashing. Government activism in the marketplace was being rejected in Great Britain, Canada, and the United States. The Carter Administration was mystified as to how to solve inflation, and knew only that "too many workers are working too much, business is doing entirely too much business, producers are producing too much and consumers are consuming too much," as Congressman Jack Kemp wrote, describing the Administration's beliefs. Therefore, according to Kemp, it was perfectly logical for the Carterites to tell everybody to slow down what they were doing and inflation might just go away.[92]

Columnist George Will totted up the problem: "Egalitarian economic policies that were supposed to foster feelings of social fraternity have not done so. Instead, they have intensified the politics of envy; they have fostered irritable social comparisons; they have invested too much social energy in redistributing rather than producing wealth."[93] The changing makeup of Congress reflected this in the 1979 voting analysis produced by the American Conservative Union. It showed that the ninety-six new representatives and senators in 1978 had moved the Congress as a whole to the right. Conservatism was on the rise.[94]

One young college professor turned congressman understood this: Newt Gingrich of Georgia. He had observed the tactics of the Tory Party of England when it swept to power several weeks earlier. The Tories stitched together a unifying, national message and talked about not programs for the government but goals and the kind of future they wanted for the British people. The GOP heeded Gingrich's advice and the example of its ideological cousins in England and called its new economic proposal the "budget of hope."[95] Gingrich had been in Congress only one year, but the new Republican Party was becoming more pluralistic, less interested in rank, and more interested in new ideas—and hence, it didn't matter whose file the ideas were coming from.[96]

What was indeed curious was that a role reversal was taking place. The Republicans had historically been opposed to tax cuts while the Democrats had supported them, especially as a means to stimulate the economy. Now the GOP was all about individual tax cuts and the Democrats opposed them and talked up budget austerity. Reagan had worked his will on the GOP, and Carter had worked his will on the Democrats. Most traditional economists were counseling against any tax cuts. The old switcheroo of the two parties was a significant development.

═══════════

As yet another symbol of the crumbling of the old order, Herbert Marcuse, a well-known international Marxist, died at the age of eighty-one. He had toiled mostly in obscurity until the 1960s, when the antiwar movement took note of his writings and he became a celebrated critic of America. He was especially popular on the campuses of Brandeis, Harvard, and other Ivy League and California schools, earning himself the moniker "Father of the New Left."[97] His death earned a front-page obituary in the *New York Times*. Reagan had criticized Marcuse more than once when the German American political theorist taught at the University of California at San Diego, while Reagan was governor. Some considered him nothing more than a tired and eventually old cliché. According to the *New York Times*, "he believed a new coalition of student radicals, small numbers of intellectuals, urban blacks and people from underdeveloped nations could overthrow forces that he saw as keeping workers from an awareness of their oppression."[98] It was pure radical chic. The only thing missing was the white wine.

In a meeting with reporters, Carter told the assembled that he believed Ted Kennedy would not be a candidate for the 1980 nomination, but then he dropped a small bomb that a recent private meeting between the two had been "strained." He also predicted that the Republicans would choose Reagan as their nominee, but said he personally preferred to run against Howard Baker.[99]

Carter had nothing but contempt for Reagan. He also astonished the reporters when he told them that the crisis of confidence he'd spoken to the nation about was due to the fact that America had never admitted defeat in Vietnam and that the United States was "immoral" for having conducted that war.[100] A year later, Reagan referred to the Vietnam War as a "noble cause."[101] Nothing better summed up the difference between the two men than their elucidation of the Vietnam War.

Yet Reagan was still holding back on actively joining the fight. He was earning an excellent income, much of it from public speaking. In the previous eighteen months, he had raked in more than $900,000.[102] By the summer of 1979, it wasn't clear that he was hell-bent for leather to run one more time for president. If he didn't, the lecture circuit was always available, but he was watching the national issues, including the contentious stories. He sent a letter to Congressman Henry Hyde expressing his support for a pro-life amendment, but indicated his preference for the issue to be returned to the states to be sorted out. "I would hope that Congress itself would propose the amendment and send it to the states for ratification." He also had no problem with a constitutional convention to address the issue.[103]

The Californian ducked yet another "cattle show," this one in his birth state of Illinois, where more than a thousand rabid Republicans jammed a Holiday Inn close to O'Hare airport to hear John Connally, George Bush, and John Anderson, among others. Reagan was on vacation and sent his regrets. His political director, Charlie Black, haughtily said, "We've never done one where we've been on the platform with another guy."[104] It was just one more in a long line of strikes against Reagan in 1979.

Howard Baker was trying to cut through his own fog, or, as it was called in Evans and Novak, "the Panama Canal blot."[105] Senator Baker needed his own signature issue to trump the field, including Reagan. He calculated that he could support the Canal Treaties early, to appease the establishment, oppose SALT II late, and thus still make his bones with conservatives in the party. He miscalculated—the conservative movement had a long memory, and they were not about to let Baker off the hook for his apostasy.[106] Conservatives were holding him close.

In desperation, he finally threw down the gauntlet in late June, calling the SALT treaties "fatally flawed," as it would leave the Soviets in an obviously superior position with regard to missile capability. Undeterred by the 1974 nuclear treaty that banned the detonation of atomic bombs of more than 150 kilotons, the Soviets detonated one that U.S. officials estimated to be closer to 200 kilotons. The Soviets were flexing their nuclear muscles once again, but the Carter Administration gave them an out when a U.S. official told the media that maybe the Russians had done so because the 1974 Test Ban Treaty was unclear about underground explosions.[107]

The Soviets seemed offended that the American Constitution provided for the Senate to "advise and consent" on all treaties negotiated by the president, and Ambassador Andrei Gromyko said as much publicly. Virtually every senator got his back up at this untoward interference, and Barry Goldwater spoke for his colleagues when he said, "Tell that Gromyko to go to hell."[108]

The Joint Chiefs of Staff supported SALT II, but it was widely assumed they'd been strong-armed into doing so. Walter Mondale traversed the country to campaign for the treaty. Two years earlier, the Carter Administration had been caught with its pants down over the

Panama Canal Treaties and was forced to scramble to shore up support. Its intention now was to not let that happen again.

———

While Reagan was on vacation at his ranch, the Federal Election Commission released his personal financial filings, which were required by law for all federal candidates if they had registered a campaign committee with the FEC. They showed that Reagan had received nearly $73,000 in speaking fees for the previous year and a half, but had grossed overall $817,082 in that period for speeches, his syndicated column, his radio commentaries, and dividend interests from multiple investments.[109] However, over the course of a year and a half, he'd spoken to only twenty-two Republican audiences. The rest of the time, he'd spent talking to trade associations, colleges, and business groups, including the University of Southern California, the Pet Food Institute, and the Distilled Spirits Institute.[110]

———

The gathering of the Young Americans for Freedom in Washington in the late summer of 1979 was one of the group's most successful. Yet again, Reagan did not speak, although he almost always did for them. Instead, he stayed at his ranch in Santa Barbara the entire month of August. At a time when the other candidates were madly scrambling for the nomination, Reagan was not meeting with his most fervent supporters. He was isolated, saddling horses and clearing brush. The YAF convention was one of the few he'd ever missed. A taped speech by him addressing his young fans was played for the disappointed crowd. The heavy lifting on his behalf at the convention was being handled by the Fund for a Conservative Majority, which was rounding up voters and volunteers for the coming Reagan campaign—if it indeed was coming.

YAF, almost twenty years old, had purged the racists, the anti-

Semites, the isolationists, and most of the libertines by 1979, and had developed a mature and sound philosophy, based on minimal government and maximal freedom, just like their hero Reagan. A straw vote was taken, and Reagan, as expected, won 78 percent of the kids' votes.[111] Crane came in a pathetic second, with 18 percent.[112]

Frank Donatelli, the former executive director of both YAF and its umbrella organization, the Young America's Foundation, had signed up earlier in the year as one of Reagan's field men, handling states in the Middle Atlantic from New Jersey to Ohio. But he felt the tug of his old friends at YAF and attended, where he perceptively told the *New York Times*, "We've outlived Woodstock, we've outlived Students for a Democratic Society."[113] YAF was dead center in the conservative movement, and a long and deep affection existed between YAF and Reagan. This did not stop Jules Witcover of the *Washington Star* from comparing the convention to both *Animal House* and *Romper Room*.[114]

An older admirer of Reagan's was Raymond Massey, the famed actor nominated for an Academy Award for his portrayal of Abraham Lincoln in the 1940 film of the president's life, *Abe Lincoln in Illinois*. When Massey came out with his autobiography, *A Hundred Different Lives*, he wrote of Reagan that "the screen's loss was the nation's gain."[115] Millions of Americans weren't convinced.

Reagan's leisurely political pace through the spring and summer of 1979 led more and more Republicans and more and more reporters to keep harping on the "age issue." They were convinced that Reagan's measured tempo was because of his advancing years and not because Sears was holding him back until the fall kickoff of the campaign. The controversy also boiled over because many in the media had heard Reagan give one variation or another—or so they thought—of "The Speech" going back to the 1950s. The issue, which had started mostly as a whispering campaign at the beginning of the year, was now being shouted from the rooftops.

Fervent supporters such as Congressman Bob Walker of Pennsylvania suggested to Reagan that he pledge to serve only one term, but both Reagan and Sears rejected this advice, as it would only have invited further speculation about Reagan's age-related health. At the rare events he did attend, GOP officialdom and the rank and file looked to see if he lost a step, or if his hair had grayed a bit, or if he'd developed any new wrinkles. In fact, Reagan was still the best on the stump, with only John Connally a competitor. "Nor has the smooth-as-silk technique of the greatest Republican orator of his day diminished,"[116] wrote Evans and Novak.

Reagan, however, did sign up an important northeastern moderate Republican, Drew Lewis of Pennsylvania. In 1976, part of the reason Reagan took Senator Richard Schweiker of Pennsylvania as his running mate was to shake loose some uncommitted delegates from the Keystone State—and the key to the state GOP was Lewis. He and Schweiker had grown up together, and their fathers had been close, so John Sears and company had hoped that Lewis would follow his old friend, who had been feuding with the Ford White House, into the Reagan camp. Lewis, however, stuck to his guns, and to Ford, and held that handful of delegates in line for Ford, helping to prevent Reagan from getting the nomination in 1976. This time around, he was a welcomed addition, especially since Schweiker was now Reagan's northeast chairman.[117]

President Carter had also gained, albeit only a bit, as his approval rating had crept up from a comatose 23 percent to a sickly 32 percent, according to Gallup.[118] Nothing had really changed, except the gas lines were slowly disappearing as drivers stopped panicking each time the needle on their gas gauge went below seven-eighths of a tank. Delivery of supplies sped up also, as even the heavily regulated free market responded to the problem. In a trial heat among Democratic voters, Kennedy was mopping the floor with Carter, 52–18.[119] Among the reasons cited by the few Democrats who backed Carter as to why they supported the president over the senator was the fear that Kennedy, if elected president, could be assassinated.

In the end, whatever Jimmy Carter may have hoped for in restarting his presidency and recasting himself as an outsider was lost in the ether because he'd taken so long to give his malaise speech, to reorganize his Cabinet, and to reorganize his priorities. Had it all been done in one quick and decisive stroke, he might have benefited. Even after the Slow Motion Massacre, he then issued evaluation forms to his Cabinet and White House staff on their subordinates, including what time they showed up for work. Accountability is always important, but one would hope the president of the United States would have bigger things on his mind than what time the woman from the steno pool arrived. He finally junked the forms as some foreign regimes, not understanding the nature of Washington, seriously believed the American government had collapsed.

Yet even with all his problems, Carter, in a *Time* magazine poll, was losing to the nonexistent Reagan by only 42–38 percent. Ted Kennedy was ahead of all by a wide margin.[120] However, one ray of sunshine for Reagan was that while one year earlier 45 percent of those polled said he was unacceptable to be president, now 62 percent said he was acceptable. A third still said he was "too conservative."[121]

The FEC released the filings of all the major candidates, and John Connally was leading the pack by a wide margin, with over $2.2 million raised. He also had a healthy $750,000 on hand. Phil Crane had raised over $1.7 million, but had little on hand, as 85 percent of what he'd raised had gone to pay his direct-mail debts to various companies owned by Richard Viguerie. What no one could ever understand was that direct mail was a matter of patience. It took time to build a "house file" of reliable donors before net money could be used for campaign purposes. Viguerie's overhead included postage, list maintenance, printing, and staff, but friction was growing between Viguerie and the Crane operation. Bush also turned in an impressive report, and had almost $500,000 cash on hand.[122] Dole came in a poor last, with only $250,000 raised and $33,000 on hand.[123]

Viguerie was another of those talented conservatives who eventually walked away from the Crane for President operation because of the bloodletting that went on each day inside the operation. He signed on with Connally's campaign, and thus gave the Texan another entrée to the conservative movement. Some conservatives were not happy, including Terry Dolan of the National Conservative Political Action Committee, who said he believed that Reagan had the far greater claim on conservative support than did Connally.

In the late summer of 1979, Senator Howard Baker suggested that Carter rule out running for reelection to save what was left of his presidency. Normally, reporters would have taken such a clearly partisan proposal for what it was and put down their pens, but this one was taken seriously, and reported upon at length. Baker had that much credibility and Carter that little with the national media. Baker was also trying to hit the trail more than just on the weekends. His duties as minority leader left him far behind the other GOP candidates, some of whom, like Connally, Bush, and Reagan, were not burdened with anything as trivial as actually having a job. They were essentially unemployed men, now seeking to be employed by the federal government, like several million other Americans. The prize, though, would be the best public housing in America. As Bob Dole once quipped, "It's indoor work with no heavy lifting."[124]

Before anyone would hand any of these men the key to the front door of 1600 Pennsylvania Avenue, a minefield had to be traversed. That minefield was not just the caucuses and primaries and the innumerable early-morning coffees and the endless cocktail parties and small talk and big speeches and strange cities and the mind-numbing exhaustion that came from campaigning day in and day out, week in and week out.

On top of all the fund-raising, interviews, strangers, and indignities, even the best campaigns had internal strife, and the worst had open warfare, sometimes staff against staff, sometimes staff against the media, sometimes staff against the candidate or the candidate's wife. Campaigns, especially presidential ones, are filled with fear, terror, giant egos, and small men. Small victories are cherished, and losses crushing. To most, it was not small beer. Republicans or Democrats, liberals or conservatives, most worked because they had a philosophical commitment and a certain view of the world. These underpaid and overworked staffers often went without paychecks, slumming with friends of the candidates. One GOP staffer was both too busy and too broke to buy deodorant, and so used the Lysol in the campaign headquarters bathroom.

Tears were just as common as laughter. Most staff became emotionally involved and took everything to heart. It was hard not to be personally invested when you worked, lived, drank, and sometimes slept with the same people for hours, days, weeks on end, with all pulling hopefully in the same direction, trying to get your man those keys to the White House. You read his speeches—after someone next to you wrote them—and the press releases and the brochures, and you watched him on the Sunday talk shows and eagerly read the newspaper each morning and afternoon, searching for favorable coverage of your man and unfavorable coverage of his opponents.

And, of course, along the way, if your guy won, you got your own just deserts. It was all ludicrously glorious.

———————

The notion was beginning to take hold in the national media that Reagan could not go the distance for the GOP nomination. Now virtually everybody said he was too old. His campaign was a mess, "riven with feuds." Sears "wanted, and made no bones about it . . . total control."[125] While Carter was telling the American people about their crisis of confidence, another type of crisis, of morale and direction and paralysis, was metastasizing within the embryonic Reagan for President campaign.

"Big John" Versus "Poppy"

"There was a haziness about exactly where he stood politically."

I n the three years since John Connally was skipped over by Gerald Ford to be his vice-presidential running mate, little had changed as to how the political world viewed "Big John." No one in American politics was without an opinion about the outsize Texan. *Opportunist, smart, cunning, crass, mean, tough,* and most especially *wheeler-dealer*—all these words had been applied to Connally over the course of his political career.

A partial survey of the 1976 GOP delegates in Kansas City had shown an enormous preference for the Texan for veep: at 224 to 97 for Ronald Reagan and to 93 for Senator Howard Baker.[1] It was only a small group of the more than 2,000 delegates to the convention, but it was a sign of Reagan's weakness and of resistance on the part of the GOP to the Californian. Connally was eventually passed over in favor of Bob Dole, and the humiliation was more than the proud Texan could bear. Ford had ruled out Connally and all that he brought to the table, including a federal indictment (for which he was found innocent), a larger-than-life ego, the Lyndon Johnson association, and a real fear that he might simply overshadow Ford.

Connally could throw mud with the best of them. In 1960, at the Democratic National Convention, while helping his friend Lyndon

Johnson try to wreck the near nomination of John Kennedy, Big John spread rumors about JFK's physical infirmities, "suggesting that John Kennedy suffered from Addison's disease."[2]

That there was intense jockeying among the growing number of prospective would-be GOP presidential suitors was not surprising. By the fall of 1979, most everybody had bought into the well-founded notion that Gerald Ford was perfectly happy raking in the big bucks from the many corporate boards he served on, his lectures, and the endless stream of invitations to play golf at the best and most exclusive clubs with no waiting for a tee time. A rationale for a Connally candidacy was emerging.

Gerald Ford's fires of anger at Reagan over 1976 had been banked only a tad by 1979, and he often carped about the primary challenge.[3] Where it was once assumed he would head a "Stop Reagan" effort, practicality and cooler heads prevailed, as Ford realized that if he were to accept the nomination of a brokered convention in 1980, he would need Reagan's support.

As far as Reagan was concerned, he was the nominal front-runner, but for all of 1979, neither hide nor hair could be found of him, and more and more were convinced that he had a glass jaw that would shatter. Among the feisty competition in the field the one in second place would benefit from Reagan's demise. "It was assumed that he was too old, too blunder-prone, simply too improbable," wrote Elizabeth Drew. "Another assumption [is] that if Reagan is the Republican nominee the election of a Democrat is certain."[4] Drew was a well-respected political observer, and her writings and pronouncements spoke for many and were read by all.

Ford often ribbed Reagan in public. At a speech at the National Press Club in late September, Ford took a big swipe at Reagan, Jimmy Carter, and Ted Kennedy, saying, "Nostalgia cannot bring back the military superiority we had in the early sixties . . . [I]t's gone. Nostalgia cannot bring back the leader of the early sixties . . . [H]e's gone. Neither government by nostalgia nor government by ideological reflex will meet America's needs any better than government by inexperience has."[5]

═══════

In the meantime, Reagan was of course staying up with his correspondence. Each week, Helene von Damm, his personal assistant, would hand him a bundle of letters (as many as twenty) that she thought were "representative." They were not all necessarily positive, but "Reagan enjoyed also answering this mail himself." It was all he could do to answer twenty per week, as he was getting "hundreds every week." The mail, especially the letters from young people, was important to Reagan because it was his window into America.[6] Von Damm remembered one persistent letter writer, a young man from Pennsylvania, Marc Holtzman, who later quit college so he could work on the campaign.[7] The Lehigh University undergraduate would end up doing more than just putting stamps on envelopes for the Reagan effort.

═══════

John Connally was feared and admired, but not beloved. On the other hand, many in the GOP, even those who thought he somehow lacked "toughness," liked George Bush. Bush did, however, use the word *golly* quite often. Bush decided "what the hell," and embarked on his own effort to win the nomination.

Bush had friends and family, it seemed, in every nook and cranny of the country, people on whom he could count and who could count on him if he were ever to seek the presidency. The rap on Bush in the late seventies was that he was not a polished public speaker. Certainly, when compared to Ronald Reagan and John Connally, he came up short, but Bush could be an exceptional and fiery speaker when so motivated. The *New York Times* took note, in a story by Adam Clymer, that "Mr. Bush is one of the hottest properties on the Republican dinner circuit."[8] In 1978, though there was no official Bush campaign operation to speak of yet, and though he had not announced his candidacy, he was determined to run for the 1980 nomination. He enjoyed campaigning; didn't mind or complain about the travel, the bad food, or the cramped hotels with beds too short for his tall frame, and he genuinely liked meeting people.

Bush was kinda, somewhat conservative . . . moderately, although he eschewed labels. His beliefs came more to him as a result of his faith, his upbringing, his culture, his war experiences, his business know-how, and from being a husband and father. He wasn't a fan of outright conservatism. It was the Eureka College–educated movie star from humble origins, not the Yale-educated blue blood, who was more well read and thoughtful on political and economic philosophy. Bush's 1966 congressional media man, Harry Treleaven, wrote a report for Congress on Bush's successful campaign. Of Bush's ideology, he wrote, "[T]here was a haziness about exactly where he stood politically."[9] Bush was attracting followers, though, many "based mostly on his personality."[10]

Bush's PAC was called the Fund for Limited Government, and Connally's was the John Connally Citizens' Forum. Bush traveled one hundred thousand miles to forty-two states in 1978, stumping for GOP candidates big and small. Everywhere, business cards and names were obtained and handwritten thank-you notes sent.[11] The discipline required to keep up such a volume of personalized correspondence was enormous, but for Bush it was just another element of a political candidate's duty. He had his own reasons for wanting to take on Carter. He told a reporter with *Time* magazine, "Carter just has no class, and I don't mean in a social sense. It's a shame for the presidency to have that little guy in there."[12]

The Bush family may have been considered dysfunctional by upside-down political standards, because they were so close and preferred the company of one another to that of friends. Ambassador Bush always said that his proudest accomplishment was "that my children still come home."[13] The family members were asked their opinion of their father's running for president, and all enthusiastically supported him. All eventually went to work for the campaign except for the eldest, George, who was getting his oil business going in Midland. The family cutup, Marvin, moved to Iowa and was a hit on the College Republicans party circuit. He got an Iowa vanity plate for his car that read "BUSH 80."[14] There was an Osmond Family–Brady Bunch goofiness mixed with a PG-13 rating to the Bush family.

Ambassador Bush jogged nearly every day, rising at 6:00 a.m. to do so. Wherever he was on the road, this physically active man could not sit still. He seemed to be always in motion. Bush was spending an inordinate amount of time in Iowa and New Hampshire, hoping to replicate Carter's success in 1976. Retail politics: one-on-one stuff, small basements, town meetings, village libraries—that is what worked with the voters in these two mostly rural states.

The early-morning jogging by the fifty-five-year-old Bush also provided a nice, energetic contrast to the sixty-eight-year-old Reagan. "I'd like to see Reagan doing this. We couldn't even get him up this early," Hugh Gregg, Bush's state chairman, nastily said.[15] Gregg had headed Reagan's Granite State operations in 1976, but he flipped to Bush in preparation for the 1980 primary. Bush stuck to his résumé on the stump. "I make no apology for having a good education. I believe in excellence. One of Carter's many failures has been a lack of excellent people around him."[16]

———

Of the three, John Connally and Reagan had grown up dirt poor, while Bush grew up having a driver and limousine take him to grade school each day in Connecticut. He was to the manor born. Connally and Reagan both became wealthy, but Reagan never forgot, unlike Connally, what it was like to go without.

Bush was also animated to take on Reagan because Reagan had supported the primary opponent of his son, George W. Bush, in Texas for a House race there in 1978. Of Bush, *New York Times* columnist William Safire wrote that he "is out to throw a body-block into Reagan whenever he can."[17] A wound had been opened between the two men over the hotly contested primary. Reagan had given $3,000 to Jim Reese, and

Nicholas Kristof, at the *New York Times*, wrote that "the Congressional primary became a proxy battle between two national Republican giants."[18] The younger Bush won the primary but lost the general election to the Democratic candidate.[19]

By 1979, Bush officials had become more pragmatic and had planted stories that the GOP would find a Reagan-Bush ticket acceptable, as it would unite both sides of the party. The cagey Bush officials did this for three good reasons. First, to suck up to Reagan's conservatives. Second, it made some sense, and they wanted the Reagan high command thinking this was a possibility, especially because, if Reagan lost the 1980 general election, it would set up Bush nicely as the 1984 nominee. Third, they needed to gloss over the residual anger the Bush family had for Reagan after he'd campaigned for young George's primary opponent in Lubbock. One close family friend said that the elder Bush would have "killed" Reagan, he was so angry about the primary.[20] Perhaps not an exaggeration, either.

———————————

John Connally was a mystery to most Republicans, who viewed Big John as an unwanted gift from Richard Nixon to the GOP. Connally had been Lyndon Johnson's protégé and a lifelong Democrat, switching parties only in 1973, at the behest of Nixon, so Nixon could nominate Connally, with whom he was utterly fascinated, to be his vice president. Spiro Agnew was slowly twisting in the wind, and eventually resigned. Nixon, always the game player, also wanted to grease the skids for Connally to be his successor as the GOP nominee in 1976.

Connally lacked a fundamental understanding of the party and its growing base of conservative supporters, attacking them as "uncompromising and unyielding in their devotion . . . these people do not represent the majority."[21] Big John had been the Democratic governor of Texas when Bush was trying to build the GOP in Houston. Bush gamely sought the U.S. Senate in 1964, but was slaughtered, 56 percent to 44.[22] Meanwhile, Connally that year won over 70 percent of the vote running for reelection.[23] It was the worst state and worst year to run as a Republican.

Even so, Connally had surreptitiously helped Bush in 1964 because he hated incumbent Democratic senator "Smilin' Ralph" Yarborough with a passion. Yarborough, a holdover from the New Deal, was a courageous liberal of utter integrity. Unlike his colleague Lyndon Johnson, he refused to sign the "Southern Manifesto," which opposed racial integration. Yarborough voted for civil rights legislation, along with Barry Goldwater, in the 1950s while Johnson was opposing it or scuttling it as Senate majority leader. Only after he became president did Johnson become interested in a "Great Society."

Bush had better luck in 1966, winning a House race in Houston, which had become a Republican enclave. The year 1966 was also a comeback year for Republicans. Newly elected GOP officials included Howard Baker as a senator from Tennessee and Ronald Reagan as governor of California. Bush was easily reelected in 1968, as Connally was stepping down as governor and was widely thought to be considering a run for the presidency. At this point, Bush and Connally had not yet tangled with each other—until 1970, that is, when Nixon talked Bush into running once again for the U.S. Senate. Everybody expected this to be a rematch for Bush with Yarborough, and a left-right race in Texas would favor the candidate on the right. By 1979, Texas had become more hospitable to the GOP.

Bush expected some sort of help from Connally, since the chasm between Big John and Smilin' Ralph Yarborough had only deepened. But Lloyd Bentsen unexpectedly upset Yarborough in the Democratic primary, a conservative more to Connally's liking, one whom Connally had talked into running in the first place.[24] Connally liked keeping his options open, and Bentsen beat Yarborough by portraying him as a big-spending liberal who had opposed the war in Vietnam.

Connally worked hard for Bentsen to defeat Bush—too hard, Bush and his family thought, and the animosity between the two men deepened. As a consolation prize, Nixon appointed Bush to be ambassador to the United Nations. Bush was expecting to carry the Texas franchise to Washington and New York for the Nixon Administration, but he was trumped only months later when Nixon appointed Connally to be secretary of the treasury. They fortunately didn't have to deal often with each other, and that suited each just fine. But sometimes they were required to be in the same room, as at Cabinet meetings, where they eyed each other with mutual suspicion. Connally had the upper hand over Bush in the early '70s. Secretary of treasury outranked ambassador to the United Nations by a wide margin. But the tables were turned when Connally was caught up in the collateral damage of the Watergate scandal. He was indicted by the Watergate Special Prosecution Force, which claimed he "took $10,000 in bribes while he was Secretary of the Treasury under President Nixon."[25] Jake Jacobsen, a Texas attorney, was the main witness.[26]

The trial came down to who was telling the truth, Connally or his accuser Jacobsen. The strategy of Connally's attorney, famed trial lawyer Edward Bennett Williams, was to bring in dozens of high-profile political leaders to testify to Connally's character as a man of high integrity who would never even consider taking a bribe. The list of Connally's potential character witnesses was long, and included former First Ladies Jacqueline Kennedy Onassis and Lady Bird Johnson, Democratic congresswoman Barbara Jordan from Texas, former secretary of state Dean Rusk, evangelist Billy Graham, and George Bush. All the others testified on Connally's behalf, but Bush declined the invitation. Still, the star-studded list did the trick, and Connally was exonerated. The jury believed Connally's character witnesses, and he was found not guilty after a mere six hours of deliberation. Connally had been sticking it to Bush for years and now had the audacity to ask Bush to say things he didn't possibly believe about Connally. Ironically, Bush may have done Connally a favor, though, because, forced to tell the truth under oath, he might have told the courtroom what he really thought of Big John. "George wasn't going to testify to the character of a guy he thought was a slimeball," journalist Vic Gold said. Gold found it telling that Connally and Richard Nixon admired each other so much.[27]

Bush had been a Republican since birth and viewed Connally as an interloper. Though Bush was uncomfortable with the more conservative elements in the party, he at least understood them, or what their hot-button issues were, no matter how much or little he thought of them or they of him.

Connally was smooth and wore expensive tailored suits. Bush was a bit rumpled and sometimes looked as if he'd just arrived after a quick game of handball at the club. Bush was "preppy," as George Will de-scribed him: "Nothing wrong with that. Many people educated at An-dover and Yale have nevertheless lived useful lives." Bush sent Will a teasing note in response. "I am . . . going to the shinier polyester double-vented, slash-pocket, pleated-front look. My friends at Mory's won't know me."[28] Of course, he did no such thing. His charge account at Brooks Brothers was safe.

This animus between Bush and Connally would ironically be the cause of their mutual undoing as credible presidential candidates in 1980 as they "mouse-trapped" themselves in Florida and South Carolina. Nei-ther could stand the idea that the other might just win the nomination, and each lost all reason when the other was a part of any calculation. Truth be told, Connally had almost as much contempt for Reagan as he did for Bush, saying, "They just put speeches in front of Reagan to read," and "Bush just sat on his butt in those appointed jobs."[29]

As the fall of 1979 approached, GOP officials in Florida hit upon the quote unquote bright idea of holding a state convention that would feature a large straw poll to determine Sunshine State support for the various can-didates. The convention and poll would not offend Republicans in New Hampshire and Iowa, who held the "we go first come hell or high water" breeding papers, but would still bring attention to their state. No delegates for the national convention would be selected via the Florida straw poll.[30]

These shadowboxing games were a pain in the neck for front-runners such as Reagan. They were a "no win" in that Reagan would be expected to win, and expected to exceed expectations or at least meet them. Anything short of that would send the media into a frenzy about what was wrong with the Reagan campaign. It meant committing precious time, as well as personnel and financial resources, for a meaningless exercise. The Reagan folks hated the whole idea, but they had to play. After all, this was a southern state where Reagan had only narrowly lost to Ford in 1976. Reagan was occasionally on the stump, making good money for his speeches and his commentaries, but once he looked like an actual candidate, the paid speeches, columns, and radio broadcasts would cease, and going to Florida to compete might just make him a formal candidate whether he liked it or not. As far as the FEC was concerned, the determining factor was not announcing a candidacy but simply behaving like someone actively seeking one's party's nomination.

For someone struggling for credibility like Bush, a small investment of time and financial effort in Florida could pay huge dividends. He wasn't required to win, only to "beat expectations," which were exceedingly low. Bush's campaign was also doing an excellent job managing those expectations. Keene's deputy, David Sparks, said to Bob Shogan of the *Los Angeles Times*, "It's not very important to us."[31]

The process began with mini-caucuses in the counties from which delegates would be selected to attend the state convention in Orlando. There, around 1,450 state delegates would be wined, dined, courted, and counted by all the candidates, but the selection process was a joke. Of the county conclaves, a "special committee" would choose 20 percent, and the other 80 percent would be chosen by the luck of the draw. This meant each campaign might put in hundreds of man-hours and thousands of dollars to get its candidates' people to these county meetings, and then pray for good luck. The whole process was puerile and pointless, but if one candidate went in, they all had to go in—or so they felt. Bush's political director, David Keene, summed up what everybody else thought about this Orange Crush: "You'd like to say this is just crazy and ignore it. But nobody will."[32] Eddie Mahe called the process "phony," and a Reagan supporter complained bitterly at one county conclave that Reagan hadn't done so well because "our people arrived early and the cards with their names on them were at the bottom of the box. The box should have been shaken harder."[33]

One who argued for avoiding the whole mess was Mahe, Connally's campaign manager. He was a shrewd operator who was attempting to overcome the mossy Texas gang of political hangers-on who were always around Connally. As Louisiana governor Earl Long said in the movie *Blaze*, he had "the finest bunch of 'yes' men ever assembled."[34] Mahe would have disputed that: he went to the other GOP contenders and suggested a unified front, with all campaigns boycotting the whole nonsense. It would have worked except for one small problem: nobody trusted anyone else to keep his word and stay out. The word *sandbagged* was thrown around a lot.[35] The process had gotten out of control.

———

Back in Washington, Carter's problems continued. It was looking more and more as if Ted Kennedy would challenge the president in the primaries. Carter's ambassador to the United Nations, Andrew Young, was forced to resign after it was revealed he'd had surreptitious, authorized meetings with the PLO and then lied about them. He was another of the Georgia Mafia who had become an embarrassment to Carter. Over the previous three years, Young had called Gerald Ford a racist, had asserted that there were "thousands of political prisoners" in the United States, and had called the Ayatollah Khomeini a saint.[36]

Another political operator whom no one ever called a saint, but whom many regarded as savvy was John T. "Terry" Dolan of the National Conservative Political Action Committee. Dolan was young, and impatient with the niceties of party process and politics. He'd been raised in a "Reagan Democrat" household in Connecticut, though the term had yet to be coined. His friends knew or suspected that Dolan was gay, but it didn't matter. He was charismatic, creative, quotable, and well read, with a marvelous sense of humor. Dolan's friends and supporters were legion.

In August 1979 his organization was growing exponentially and announced a massive independent expenditure against incumbent liberal Democrats in South Dakota, Idaho, Indiana, and California. All were running for reelection in 1980, and all were from states thought to be far more conservative than their voting records reflected.[37] Dolan also knew that based on the voting patterns in Senate races in 1976 and 1978, it was entirely possible that the GOP might pick up as many as eight seats in 1980. Others saw this possibility as farfetched.

Newspaper editorialists and liberal interest groups howled in protest, but there was little they could do. The Supreme Court, in its historic *Buckley v. Valeo* decision in 1976, upheld the FEC's one-thousand-dollar limit on contributions to federal candidates and PACs, but struck down, as a violation of the First Amendment, the thousand-dollar limitation on independent groups banding together to affect the outcome of an election.[38] Dolan and NCPAC and other conservative groups, as well as liberal groups, were free to raise as much money as they liked (with limitations on the size of the gifts from each individual and no corporate money) and spend it as they liked to help or hurt a candidate. NCPAC also frequently had to file mountains of disclosure statements with the FEC.

═══════

As its county caucuses began, Florida's GOP was also confronting a rearguard action by presidential candidates protesting its outlandish straw poll and convention. At each county meeting, delegates to the state convention were selected by drawing names out of a hat. So the trick for each campaign was to drive as many of their grassroots supporters to each of the county meetings to increase the odds that their people would be randomly selected. John Connally's campaign sent buses to retirement homes and filled them with senior citizens to attend the caucus meetings. After this first round of caucuses in several dozen counties, Reagan was ahead, albeit only slightly over Connally. Bush was mired with Phil Crane near the bottom.[39]

The whole nonsensical Florida process received top political coverage from the national media, even though all knew the exercise was a farce. It dragged on from August to November. Bill Brock dismissively referred to the whole exercise as "the great Florida crap shoot."[40] The process was so ridiculous that Connally's campaign, which had done a good job getting its people out to the county meetings and witnessing names drawn out of a drum by a blindfolded GOP official, saw the group arbitrarily add thirty votes to Reagan's total.

Eddie Mahe had argued for months with Governor Connally to skip the nonsense in Florida, but Connally wouldn't hear of it. His old enemy George Bush was contesting Florida, and Connally just couldn't resist the opportunity to show him up. By mid-September, Connally had already spent $200,000 in Florida and would shell out a lot more before it was done. Bush's team, though, was exhibiting restraint, unlike Connally. While they were working Florida just as hard as Connally, they were also keeping "a low profile" through a "disciplined" operation.[41]

The only direct confrontation between Ronald Reagan and George Bush in Florida was a goofy one. Reagan was headed into a hotel in Orlando, looking spiffy in a suit and tie, waving and smiling to all. Out of nowhere, a sweaty, jogging Bush appeared in shorts and a T-shirt and ran right up to Reagan and said, "Hi, Governor," to which Reagan replied, "Hi, George," all in front of the media. In fact, Bush had been watching Reagan through a window, and when Reagan got out of the car, he darted out to presumably show Reagan and the media how healthy he was. It was contrived, it was silly, and it was odd.[42]

The real story of Florida was that while Connally went in because of Bush, the Reagan forces wanted Connally in to sap Bush's resources and time and, they hoped, to embarrass Connally with a big loss. Of all the challengers to Reagan, Connally was the candidate they feared the most, and if Big John wanted to waste untold hundreds of thousands of dollars on a meaningless exercise, that was just ducky with the Gipper's folks. Connally's campaign also let expectations get foolishly out of control. "We've got to have a breakthrough," Mahe said.[43]

But heading into the Florida state convention, Connally gave a policy address in which he excoriated Israel and said they should withdraw immediately from the occupied lands of the West Bank, the Gaza Strip, and the Golan Heights. He also called for direct talks with the PLO, which Bush had called "an international Ku Klux Klan."[44] In his address, on Palestinian statehood, Connally said that America should pursue a policy in the Middle East of "self-determination" rather than support Israel.[45] Even for Big John, it was a mouthful. No one called him an out-and-out anti-Semite, but the thought was on many people's minds, including Republicans in Florida, where there was a large Jewish population. Gentile Republicans in Florida were sympathetic to their cousins from the other side of the Jordan River, as Connally's comments would eventually come back to bite him in the Sunshine State. The Republican Party had become more and more pro-Israel over the years, and Israel's most reliable supporters in America came from the ranks of the Christian Right, who were becoming big players in Republican politics.

Reagan had been a famous supporter of Israel since day one. Only a few weeks before, he gave a speech in which he said, "Only by full appreciation of the critical role the State of Israel plays in our strategic calculus can we build the foundation for thwarting Moscow's designs on territories and resources vital to our security and national well-being."[46]

Connally, after his foreign policy speech, stepped all over his "line of the day" when a reporter asked him about the milk fund scandal in which he was tried but acquitted. Connally regarded most reporters as swarming gnats and responded furiously, "Milk fund scandal? Well, what about it? I was tried and found not guilty. I never drowned anybody. I was not kicked out of college for cheating."[47] Connally was of course referring to Ted Kennedy and the scandals of Chappaquiddick and Kennedy's having been expelled from Harvard when a friend took his Spanish test for him. Connally was staking out his own position on Israel. Even Ted Kennedy was a hard-liner not only in supporting Israel but also in opposing talks with the PLO.[48]

Bush's FEC reports for the year showed he'd raised a healthy $2.4 million, impressive for someone who was generally referred to as an "asterisk" in the race.[49] He was working Florida assiduously, but in a CBS poll of Florida Republicans, Reagan was first, with 31 percent; Connally was second, with 21 percent; and Phil Crane third, at 9 percent. No one else, including Bush, was yet on the radar screen.[50] Bush hit Connally hard over his Israel speech. Over the years, Bush had been accused of being a closet "Arabist," but he saw the opening and took it, in a speech before the National Conference of Christians and Jews, calling for continued support of Israel.

In mid-October 1979, Bush bounded into New York City to raise money, give speeches, and meet the press. After much pestering by his staff, he'd finally gotten rid of the "granny glasses," which made him seem prissy. His confidence was soaring in the wake of the winning results of the second straw poll in Iowa in early October in Ames. It was the only time there would be two straw polls in Iowa during the year immediately preceding the presidential caucuses, and Bush won them both.

The second time around in Iowa, around twenty-five hundred Republicans, with nothing better to do, paid fifty dollars apiece to attend the annual dinner, which featured all the GOP candidates save—you guessed it—Reagan.[51] Rich Bond organized hard, and beat the Reaganites again. Bush got nearly 36 percent, Connally received 15 percent, and Reagan received a paltry 11 percent.[52] Bush was, as it turned out, having his cake and eating it, too. He was scoring important psychological victories over the rest of the field, especially Reagan. Yet, in poll after poll, Reagan was the clear front-runner. A *Des Moines Register* poll had him with a two-to-one lead over Bush.[53] Indeed, the Gipper had actually gone up by 10 percent since the newspaper's last statewide poll.[54] John Sears, who, it was believed, knew all about politics, was making a rookie mistake by putting his faith in the polls rather than listening to Reagan's people on the ground. Bush's feverish on-the-ground up-and-at-'em work included claiming to meet, at this point, some 5,500 individual Republicans in the state, and no one doubted Bush for a moment.[55] Setting aside their personal preferences, the Republicans attending the dinner were asked who they thought would be the nominee—and again, Bush thumped the field with 32 percent to only 13 percent for Reagan. There were a number of local straw polls in the state, and Bush won them all as well.[56]

Bush was focused on the next rung of the ladder, but it was Connally who was now becoming more obsessed with Reagan. He was going for a knockout punch against the Gipper. He launched a five-minute national television ad campaign on CBS that left skilled operatives scratching their heads. Why would a campaign run national advertising when it would be far more prudent to go to the homes and hearths of GOP voters in Iowa and New Hampshire? The five-minute commercial repeatedly used the word *leadership* and was essentially a "bio" spot. Produced by Roger Ailes, it talked about Connally's service to four presidents, though it did not mention any by name. Still, a photo of Connally and JFK was shown, curiously, and the narrative said that Connally was "gravely wounded in the same car in which President Kennedy died." Connally did not forget his audience entirely, though, as he talked about the "forgotten American" and paid tribute to the people who paid their taxes and went to church.[57]

Connally also went to California and proudly bragged about raising a million dollars in Reagan's home state. Instead of diligently building his campaign house by house, block by block, the way Bush was doing, the retail politickin' of long hours and hard work, Connally wanted to go wholesale, using aerial bombardment with no ground troops. His campaign was now openly bragging about taking out Reagan. Connally's aide Julian Read said that Connally's early success in Florida "proves Reagan is not invincible."[58]

———————

Yet one more candidate was to enter the fray for president, but for neither the GOP nor Democratic nomination. Forming his own Constitution Party, former governor of New Hampshire Meldrim Thomson jumped in because none of the Republican candidates was conservative enough for him. Thomson had been a huge fan of Reagan's for years and had been a key, though controversial, supporter in the 1976 primary.[59] If he had stayed on board or kept his mouth shut, he might have ended up with a minor appointment in a new Reagan Administration, but Thomson was unsteady, to say the least, and had become an embarrassment to the state and its Republicans. After all, he once proposed issuing nuclear weapons to the National Guard. Blissfully, he suggested only low-yield bombs. His campaign chairman was somebody named Rufus Shackelford.[60]

Thomson held a press conference in Washington to announce that he was leaving the Republican Party for good, and the overwhelming reaction in the party was "Don't let the door hit you in the ass, Meldrim." To Thomson's credit, he abolished many taxes during his three terms as governor, including a sales tax on liquor. When Massachusetts tax collectors were seen in New Hampshire writing down the license plate numbers of Bay Staters who were buying liquor there, Thomson had the agents arrested and was toasted across the Granite State for his actions.

The madness in Orlando finally came to an end, and for the GOP presidential candidates, not a moment too soon. Reagan called the entire exercise "absolutely meaningless," and he was right. His campaign had devoted minimal time and resources to the GOP sideshow across the street from Disney World. A new CBS poll of Florida had Reagan in first place, Connally a close second, Phil Crane in third, and coming in fourth but climbing was Bush, at 9 percent.[61]

The speakers in Orlando had gone in an odd order. Reagan got the best position, going first. Connally was second. Following Reagan was not something most men were capable of doing—but Connally did well. Bush was the last speaker, and many thought it was the worst position, but Bush made the most of it and gave a barn burner of a speech ripping Carter, one that had the delegates standing on their chairs, whooping, and cheering for Bush. Dean Burch, a longtime apparatchik in the GOP, had introduced Bush by telling the crowd that Bush was "a man of universal acceptance and unsullied reputation, a man in his physical prime, and for the 80s." Burch of course was referring to Connally and Reagan in his remarks.[62] Bush's campaign, on the verge of dispatching Bush's old enemy Connally from the field, was now turning its guns on the sixty-eight-year-old Reagan.

Bush got the lion's share of the favorable media coverage, while Connally got the bum's rush. Out of 1,326 delegates casting ballots, Bush received only 74 votes fewer than his fellow Texan. Connally had spent over three hundred thousand to beat Ronald Reagan, had bragged that he would do so, but in the eyes of the political establishment, Big John was now the big "loser."[63] In his life, he'd been called tough, arrogant, self-assured, a "wheeler-dealer," successful, and brash, but the "loser" moniker had never been hung on him before. His Florida escapade was shaping up as just one big expensive dry hole. As they said in Texas, it was a bust.

Once again, the venerable David Broder of the *Washington Post* read the tea leaves correctly. "George Bush has made better use of 1979 than any of his rivals."[64] The Reagan campaign heaved a sigh of relief that it had dodged a bullet in Florida with no apparent damage, yet it did not see the underlying problems with the campaign. Reagan's Florida director, former congressman Pat Hillings, had been fired when Charlie Black became nervous about the outcome. Former Ford aide turned Reagan field man Paul Manafort replaced him. The Florida straw poll came at the end of Reagan's twelve-day announcement tour, one that the *Los Angeles Times* called "unimpressive."[65] Reagan escaped Florida unscathed only because of the Connally-Bush rhubarb.

Connally and his campaign had foolishly told reporters that they expected to win the Florida straw poll. Even worse, they had eschewed any retail campaigning whatsoever. Big John was running a "national campaign," which would have been wise, except he was not the nominee of a national party. His Texas Mafia had raided the campaign and was running an operation based on Connally-size hubris served up with a large dose of know-it-allism.

Connally was slapdashing across the country in his private jet, giving big speeches to big audiences. The problem was that not one of those people in attendance was a precinct worker in Waterloo, Iowa, or Nashua, New Hampshire. The big Texan was bypassing Iowa and would make only a token effort in New Hampshire. His revised plan was to lie in wait for Reagan in some southern states. Connally conveniently forgot that Reagan had barely lost to Ford in Florida in 1976, and squashed Ford in Georgia and Alabama.

Reagan was more popular in the South than Elvis.

Georgia Versus Georgetown

"Yeah, we hate Carter."

The Democratic Party was to make even more headlines. And not in a good way, either.

A bearish, Irish American backslapper in the ward-heeler mode, Ted Kennedy liked to party, he liked women, and he liked a good joke. As was once said, all Irishmen are not saloonkeepers, but all saloonkeepers are Irishmen. Ted Kennedy was a born saloonkeeper.

Given all that the youngest Kennedy had been through, all the pain and suffering, all the agonizing, paralyzing heartbreak over brothers, sisters, and friends killed horribly in war and monstrously in peace, the terrible travails of his mentally disabled sister Rosemary, and the cancer that cost his young son Teddy Jr. his leg, it was small wonder that Kennedy turned to the bottle.

As the *Washington Post* wrote, "[T]he Kennedy family saga has become an incredible, ongoing, heart-wrenching novel-for-television."[1] *Newsweek* saw the romantic reverie in another Kennedy mission: "the dream endures."[2] The national media saw the GOP front-runner as the most threatened in the Republican field as they concluded quickly that Kennedy would be running on a clean slate, unlike Jimmy Carter; that youth would like him, unlike Ronald Reagan; that moderates would support him, again unlike Reagan. All that showed why newspaper reporters make poor campaign managers.

The ghosts of JFK and RFK had haunted America, and particularly those who wrote about or participated in politics barely a decade after their murders. Edward Moore Kennedy was also haunted by the memories of his brothers, and the heavy mantle of national leadership he was expected to inherit from them.[3] The ghost of dead Kennedys also haunted Carter, like his predecessors. One Carter insider said, "No matter how he succeeds as president he is never going to be admired like a Kennedy. The envy and unfairness of it all gets to Carter the way it got to Johnson and Nixon."[4]

In 1976, a special effort had to be made in Massachusetts late in the campaign to show the world that the Democrats were "united." Massachusetts was going to go heavily Democratic. Carter needed to be campaigning in more critical states, but there he was in Boston, wasting precious time, beaming for the cameras and the cheering crowds with Kennedy.[5]

"Yeah, we hate Carter. After a while, even a dog knows the difference between being tripped over and being kicked," said a Democrat in Wyoming. "It requires a profile in courage for a Democrat to mention the President," wrote a columnist in New York.[6]

More "Draft Kennedy" campaigns sprang up around the country in the summer of 1979; and in New Hampshire one was run by a woman with the improbable name of Dudley W. Dudley. She was a Reagan hater through and through. "Ronald Reagan must not become president of the United States," she stormed. Dudley cited the polls that showed Reagan beating Carter but not Kennedy, and used this as her rallying cry for New Hampshire's Democrats to support Kennedy.[7] John Persinos, a liberal journalist who worked on the 1980 Kennedy campaign in New Hampshire, recollected, "Sure, we were frightened of Reagan. We perceived the rise of Reagan as the coming of fascism. But we reserved our very special hatred for Carter, whom we despised as a betrayer of Democratic Party principles. For us, Teddy running for president was akin to the Second Coming."[8]

Kennedy had for months used a carefully crafted line that he "expected to support Carter"[9] and that he "expected the President to be renominated,"[10] but now told old hands to "stay in touch" and "hang loose." That was political in-speak for "I'm running and I want you to support me." Frantic Carterites were calling Kennedy the "enemy."[11] The Kennedy family held a patent on revenge, and these insults only fueled the already rising flames.

By 1980 the Democrats expected to hold thirty-six state primaries in which over 70 percent of their national delegates would be chosen. Each primary was looking to shape up as a street brawl between the Carter country boys and Kennedy's Boston street toughs.

Yet Carter, meanwhile, was falling, and he was falling hard. He bottomed out in the national polling when he hit 19 percent approval according to a new NBC/Associated Press tabulation. It was the lowest in the history of the poll.[12]

Carter did not help his cause when he collapsed in a 6.2-mile race while the other joggers ran around their president's prostrate body. It was his first race as a runner, and he fell after 4 miles. By the time Secret Service agents arrived to help him, he was "very, very pale."[13] His subsequent failure to finish the race was seen as an awful metaphor for his presidency. The American people wanted their president to be a hero, and heroes win the race, any race, or at least finish it. The photographs of a collapsing Carter, with his mouth agape in the agony of defeat, were splayed across the front pages of every newspaper the next day.[14]

Americans now pitied their president. It was devastating.

After moving closer to a conclusion over the summer of 1979, Kennedy now dropped the charade. Though he had not made a final decision, protesting that he was not thinking about it was ridiculous at this point. "I have not ruled out the possibility of a candidacy," Kennedy told the *Boston Globe*. Polling conducted by ABC News/Harris showed that 70 percent of the American people believed that Carter would not be re-elected.[15] When asked if the president inspired confidence, only 20 percent said yes, while 76 percent said no.[16]

Carter continued to have his troubles with the Soviets, as they tested him once again when it broke that the Kremlin had secretly emplaced a brigade of some two to three thousand troops in Cuba, had stationed ground-attack aircraft there, and was sending nuclear submarines there for "tending." The subs of course carried missiles armed with nuclear warheads. Sub-launched nuclear missiles situated that close could hit the continental United States in seven minutes. A cry went up from even the most liberal members of Congress that Carter needed to confront the Soviets, Jack Kennedy style, and force them to withdraw their provocative forces. Carter weighed how much he could stand up to the Soviets without their walking away from his cherished SALT II negotiations.

The *New York Times*, whose editorial pages had been an often reliable supporter of Carter's, took him to task: "A weakened America will be led for the next 14 months by an alarmingly weak President. Congress knows it. The Russians know it. The Israelis and Arabs know it. The oil companies and labor unions know it. Does Jimmy Carter understand that fact and know how to deal with it? The signs are not encouraging."[17] The paper was moving toward supporting Kennedy, but it first had to denigrate its relationship with Carter, to ease the pain of separation.

The Soviets' actions in Cuba went against the Monroe Doctrine of 1823, which posits that foreign powers are no longer to colonize or interfere with the affairs of the independent nations of the Americas. The USSR was also reneging on its post-missile crisis deal with JFK. But the Soviets cared little for such inconvenient things as treaties and agreements, unless they could be hung over the West, and especially the United States, to the benefit of the Kremlin. In their house organ, *Pravda*, the Soviets said American fears about their new military presence in Communist Cuba were "totally groundless."[18] Americans were not comforted.

Reagan was less worried about the collapse of SALT II or the niceties of diplomacy. He said bluntly that the United States "should not have any further communications with the Soviet Union" until the combat forces left. "Détente must be a two-way street or there is no détente."[19] After that outburst, he strangely went silent.

The Administration and some of its allies on the Hill argued that there should be no linkage between the Soviet military presence just ninety miles from Florida and arms negotiations, but no one explained why there should be none. (Reagan and the conservatives, on the other hand, thought linkage was just dandy.) President Carter went even further, saying there should be no linkage between the treaty and anything the Soviets did.

All the GOP candidates predictably blasted Carter, except for Ronald Reagan, who said nothing for two weeks after his initial outburst about the Soviets. No statements, no speeches, no interviews—nothing. Anticommunism, especially anti-Soviet communism, had been his bread-and-butter issue for nearly thirty years, and yet, unbelievably, his campaign was handing it over to others, just at the moment when Reagan was being proved right once and for all about Soviet intentions. Indeed, the *Washington Post* solicited responses to Carter's actions from all the GOP candidates, and all replied except Reagan.[20]

Someone who did have an opinion—about Reagan, that is—was Fidel Castro himself, who told Barbara Walters in an exclusive interview for ABC News that Reagan was the one candidate he would not like to see elected president. Castro was ambivalent about Carter and expressed his preference for Kennedy.[21]

———————

Reagan was closely, and oddly, watching the developments on the Democratic side. In a letter to Winfield Schuster, he wrote, "I must say that Senator Kennedy does remain an enigma in the coming race. I find myself believing that he truly would like to stay out of this race, waiting for 1984, but may be pushed into it if things continue to worsen for the president . . . Whatever happens, it's going to be an interesting year coming up."[22]

At this point, there was little linkage between Jimmy Carter and Ted Kennedy. Carter wanted to sound out Kennedy's intentions. Kennedy told Carter that he was considering challenging the president in the Democratic primaries, and Carter "replied that he respected Mr. Kennedy's right to do so but would fight him at the convention for every last delegate."[23] Carter was then humiliated when he was forced to deny a rumor that Kennedy had asked him to drop out, as reported the previous day in the *Atlanta Journal and Constitution*.[24]

Carter was also fighting a rear-guard action in his own backyard. "Draft Kennedy" organizations had taken hold in five states in the Deep South, including his own Georgia. There were also people in the South, conservative Democrats, who could not tolerate the liberal Kennedy but who had been disappointed by Carter. They were shopping for a candidate, and many of these populist conservatives were potential Reagan supporters, if he got the chance to ask them for their vote in a general election. "If Ronald Reagan is the Republican candidate, I might have to reassess myself," said one member of Carter's famed "Peanut Brigade," the original one hundred supporters for his long-shot presidential bid beginning in 1974.[25] With Kennedy to the left of him and Reagan to the right, Carter was being run down, and a "pickle" is never a good position to be in, either in baseball or in politics.

How bad off was Carter? One of the more respected Democratic political consultants at the time, Joe Rothstein, struggled and then came up with a slogan for Carter's reelection: "No matter what you think, he's been a good President." Carter's people held their noses and searched for something else.[26]

If the general consensus was that Carter did not have a prayer for re-election, hundreds of thousands in the nation's capital gathered for real prayers. Pope John Paul II came to America for a six-day tour and celebrated Mass on the Washington Mall, which many of the city's young professionals treated as a social event, spreading out blankets and picnic baskets, sipping wine, and munching on gourmet bread as the pontiff spoke. Vendors sold buttons and pennants of the smiling pope. Atheist Madalyn Murray O'Hair filed suit in federal court to prohibit the Mass from taking place on federal property.[27] That didn't stop this pope, whose charisma and moral authority would do much in time to stop Soviet Communist expansionism.

Democrats in Florida had organized their own straw poll and statewide convention, much like the Republicans, at which not a single delegate would be selected for the national convention in New York. Carter won with a plurality in the first round, though of course with the White House working for him. Kennedy had no formal campaign organization, as he had not yet announced and only a ragtag group had assembled a draft organization in Florida. Jody Powell, Carter's White House press secretary, was pointedly on message when he said, "I think we've shown that if you thought the Democratic nomination was Teddy's for the asking, you'd better think again."[28]

Both pushed on, and Kennedy and Carter journeyed to Chicago for a fund-raiser, Democratic Party Windy City–style. It included no fewer than twelve thousand people. Jane Byrne had become "Herzzoner" after the machine left over by Richard Daley collapsed with his death in December 1976, only weeks after he helped Carter win the presidency. Kennedy elicited sly giggles in tribute to Byrne, telling the audience, "I admire the leadership you have provided . . . and I just hope you remember who has loved you in Chicago longer."[29] Byrne, whose stewardship of Chicago was marred by screwups, was suffering under an unwanted moniker the Chicago media had bestowed upon her: "Calamity Jane."

After the Chicago speech, with indications from Mayor Byrne that she would bolt Carter and back Kennedy as soon as the latter made it official, all signs pointed to go, despite the organizational disarray. "So far, he has had to rely on informal and uncoordinated draft organizations," the *Detroit Free Press* said.[30]

Carter did not inspire confidence, though, when he told the thousands of Democrats that "I am the fourth president who had tried to deal with inflation and so far we have not been successful."[31] He then proceeded to tell them that if the cost of energy were set aside, the galloping inflation of 1979 would be much like what it was in 1978 and 1977. He did not explain how anyone in America could set aside energy.

Florida Democrats moved on to the next stage and selected delegates for their November state convention. President Carter won again, handily, over Kennedy's draft forces, 508–291. Jody Powell, remembering his Mark Twain, told the media that "reports of the political demise of the President are very greatly exaggerated."[32]

Carter held a press conference and, even with the win in Florida, displayed a public insecurity that the media had not seen previously. He knew the outcome in the Sunshine State was the result of his all-star team against the Kennedy scrubs. Kennedy was now getting ready to put his best lineup in the game. The *Washington Star* said of Carter, "Yesterday, there were a few limp gestures and his voice seemed almost as subdued as his body language."[33] Kennedy's confident forces shrugged off Florida as an aberration.

The Florida loss was a mere bump in the road, a hiccup for Kennedy. Still, Kennedys did not lose elections. Indeed, one of the leaders of the Florida effort for Kennedy told Jules Witcover of the *Washington Star*, "There has been no battle."[34] But now the professionals would take over and gently nudge the amateurs off to one side. Kennedy would never lack for volunteers.

Kennedy was flowing on a course he could not alter. He said his brother Jack had taught him how to "sail against the wind."[35] As an old sailor, however, he knew that while one could maneuver through prevailing winds, it was easier to go with them. This favoring wind looked like a freshening breeze across the beam, perfect, it seemed, for Captain Ted and his *Odyssey* to reclaim the lost mantle of president and restore the

Camelot dynasty to its rightful place. Kennedy would learn only later that, in politics as in sailing, there is an "apparent" wind and a "true" wind. Right now, though, it was looking to all like a Homeric tragedy in which the young hero, full of wanderlust, returns to avenge his brothers' deaths and complete their unfinished work.

However, Kennedy was already coming back to earth from his unrealistically high numbers. A new poll by *Time* had him ahead of Carter at a more reasonable 10 percent. He was still walloping Reagan, however. Portentously for Kennedy, the heretofore-whispered issue of Chappaquiddick was now being loudly talked about among his fellow Democrats. Forty-four percent of Democrats were bothered by the issue.[36] Carter naturally rated higher on the "home and hearth" issues such as being a good family man, while Kennedy rated higher on the leadership and strength issues. As Kennedy's campaign continued to meander into late October, Stephen Smith and a number of Kennedy family advisers from the JFK glory days pushed Teddy to accelerate his announcement.

It wasn't only Carter who was wringing his hands about the Kennedy challenge: Republicans were right there with the president, as Kennedy was swamping everybody on both sides in the polls. Robert Teeter, Bush's pollster, fretted, "Kennedy will just run over everything."[37] Unlike other GOP operatives, however, who were clearly distraught about Kennedy, Dave Keene of the Bush campaign had a far more insightful view: "When was the last time he had a tough race? I think he can be rattled. Once you force him out of his set piece, he's in trouble."[38] One happy development for the Carter campaign was that Bob Strauss, wiseman of the party, had agreed to head the Carter effort. Strauss was widely known, and was respected in the media and, indeed, in both parties.

John Connally was arguing that he could match Kennedy charisma for charisma, but others snickered that he could also match Kennedy scandal for scandal and ego for ego. Still, no one could miss, yet no one would write or talk about, the awful irony of a possible Kennedy-Connally

1980 pairing. Kennedy could be running against John Connally, through whose body the bullet that killed JFK had passed; Connally, who was JFK's secretary of the navy before being elected governor of Texas in 1962, and whose wife uttered the last words John Kennedy ever heard: "You can't say Dallas doesn't love you, Mr. President."[39]

Carter had already ordered Secret Service protection for Senator Kennedy in late September. There had been no reported new threats to Kennedy's life, but given the family history, there was a great deal of apprehension that one more lone nut would want to write himself into history by taking out the last Kennedy brother. Ted Kennedy had been afforded protection by the Secret Service twice before: once, after the assassination of Bobby Kennedy in 1968, and the other time, after the attempt on George Wallace's life in 1972.

Federal law had been that once someone was the nominee of his party, he would get a protective Secret Service detail, but it took the murder of Bobby Kennedy to change this, and now all "major candidates" would henceforth receive protection. Ted Kennedy was not a candidate yet, but "Carter ordered the protection for Kennedy not under the terms of the candidate law, but on his own authority after receiving an assessment of Kennedy's security situation from the Secret Service."[40]

The way the Service now determined if a candidate was a "major candidate" was simple. If he raised a million dollars and qualified for matching funds, he received protection. In October, five Republicans were each offered their own detail: John Connally, Ronald Reagan, George Bush, Phil Crane, and Howard Baker. Only Bush refused to accept the offered protection.[41]

The economy continued to corkscrew downward. Detroit reported that its overall sales in October were down 20 percent, and Chrysler's were down a stunning 56 percent.[42] More layoffs were announced in Motor City. Interest rates in late September continued to rise, and their upward spiral was continuing with no end in sight. Unemployment also was quickly moving up.[43]

A new poll was released, conducted by the political science department at the University of New Hampshire, and Kennedy was mopping the floor with an eye-popping three-to-one lead. Three-quarters of New Hampshire Democrats thought Carter was doing a bad job as president. Kennedy increased his advantage over Carter in all categories (among liberal, moderate, and conservative Democrats) compared to the survey taken in the spring of 1979, when he'd led Carter 48–23.[44]

It wasn't only the president who was taking a beating. Mrs. Carter was getting her share of lumps as well, because congressional investigations revealed that her staff of twenty-one was the largest in the history of the Office of the First Lady. In fact, it was almost as large as Vice President Mondale's. "We get a lot of mail . . . from people who say, 'Who is running the country, the president or his wife?'" a Hill staffer complained.[45]

In Buffalo, Kennedy hit Carter hard on the windfall profit tax to be imposed on energy companies. He also poked the president on his plan to slowly deregulate the cost of gasoline. An enthusiastic crowd of five thousand greeted Kennedy. The mayor of Buffalo, Joe Crangle, introduced Kennedy effusively, though he'd already endorsed Carter's reelection. Crangle's slither was seen as just another example of Carter's waning power. One of the first announcements out of the Kennedy for President Committee was to install Massachusetts lieutenant governor Thomas P. O'Neill III, son of House speaker Tip O'Neill, as head of the New England operations, meaning Tommy would be in charge of the all-important New Hampshire primary.[46] In his own backyard, Kennedy was expected to win the Granite State. Yet another respected Democratic operative, Tim Russert, who worked for New York senator Patrick

Moynihan, declined Kennedy's attempt to sign him up. Moynihan also was dubious about the Kennedy insurgency. He had little respect for Carter, but Kennedy's operatives didn't seem to know New York politics, and they were recruiting some of Moynihan's mortal enemies into Teddy's effort.

New polling by CBS and the *Times* showed that an amazing 81 percent of Democrats and 79 percent of all voters saw Kennedy as a "strong leader." Reagan also got good numbers, but they were still dramatically lower than Kennedy's, with 66 percent of Republicans and 58 percent of all voters seeing him in the same light. The poll also highlighted the general altruism values of the American people of the era: most wanted strong leadership, many thought honesty was the most important quality, and barely anyone thought the most important factor in a good president was supporting policies the respondent agreed with.[47]

Carter's numbers were appallingly bad. His job approval was still on life support, and in the head-to-head matchup, Kennedy was besting Carter everywhere, even in Carter's native South and even among self-described conservatives. To the American people, Carter was looking like a one-term president, and all polling showed they did not want him to seek another term.

Hence, the worry at the White House was not the GOP nominee. The focus was on the Democratic ticket. Carter's people thought the Republicans would choose either John Connally or Ronald Reagan, both of whom they thought they could deal with. It was Kennedy who was the problem. Especially with the expected slobbering special they presumed CBS would air the first Sunday of November. Indeed, Carter went out of his way to attack the media, always a fool's errand. At a small rally of about five hundred supporters, he gave a speech in which he plaintively pined for the day when "the news media, for a change, will accurately assess what we have fought for and achieved."[48] But Kennedy was about to experience his own problems with the media, largely self-inflicted.

On the evening of November 4, CBS ran a one-hour special they called simply *Teddy*, an in-depth profile of the last of the Kennedy brothers. CBS produced many superior specials, such as several years earlier when it had done one entitled *What About Ronald Reagan?*, a fine and fair study of the Gipper.[49]

Veteran CBS broadcaster Roger Mudd had interviewed Kennedy on two occasions, and film from each interview was edited into the final broadcast. The first took place at Squaw Island, at the Kennedy Compound in Hyannis Port, on September 29, and covered the family and Kennedy's personal life. The second interview, on October 12, was held in his Senate office and was on Kennedy's view of politics.

In the second interview, Mudd posed his famous question, "Why do you want to be president?" It was a simple question, one that every presidential candidate had been asked and answered routinely that year— except one. Kennedy's response broadcast that night was embarrassingly incoherent. "Well, I'm . . . uhh . . . were I to make the announcement . . . to run . . . the reasons that I would run is because I have a great belief in this country that is, it has more natural resources than any nation of the world," he began.[50] It got worse as he continued to ramble. It was a shattering image for his supporters, who found out that night that heroes can indeed have clay feet. The irony was that Mudd was a friend of Kennedy's, and everyone had expected the interview to be a cakewalk for the senator.

Kennedy's people, rather than moving on, foolishly kept attention on the disaster by attacking CBS. The supposed professionals running the Kennedy operation claimed that he'd been set up and "sandbagged." CBS executives sagaciously said, "Whatever was seriously damaging to Kennedy's image was brought on by himself. It comes from what he did."[51] The back-and-forth between CBS and Kennedy continued unabated for days, with Carter watching it all from the sidelines.

Within a matter of a day or two, however, conversations spread across newsrooms, in the halls of Congress, in the drinking salons of Washington, the meeting rooms of organized labor, and around the watercooler at the Democratic National Committee, and then spread across the country, especially among Democrats. A consensus was beginning to form and harden: Teddy had blown it, big time. CBS officials said that no broadcast before had generated so much interest, editorials, or controversy. Kennedy's suit of armor had fallen off, and everybody in the party now saw him as bewildered, befuddled, and ungainly. Because he was so high in the polls, expectations for his performance in the show had been lofty; it would have taken a miracle for him to meet those expectations.

The CBS special's competition that night from the other two networks would be nothing less than monstrous. ABC was broadcasting, for the first time on network television, the megahit movie *Jaws*, and NBC, no slouch, served up the most popular movie of all time, *Gone with the Wind*. Still, the Kennedy special received a 15 percent share, meaning 15 percent of all televisions were tuned in to it. Presumably, every television belonging to a Democrat was tuned in to the Roger and Teddy Show.[52]

There was another dynamic at work, this time in the national media, which also would not serve Kennedy's purposes. For years, some in the media had been supplicants to America's Royal Family, coddled, bought, rented, plied with alcohol, bestowed with cherished invitations, leased, fêted, and frozen out when necessary by all in the family. The Kennedys were their own best press agents. They were photogenic and charming (when they wanted to be), and as far as they were concerned, ordinary rules didn't apply to them. Images of the photogenic Kennedys had been plastered on *Time*, *Life*, and *Look* magazine covers. The toothy, smiling, forelock-tugging, athletic Kennedys seemed to be everywhere, playing touch football, sailing, tussling with their children, and jogging on Cape Cod beaches with their dogs. These images of the Kennedy clan became part of the country's political DNA.

Yet, beginning with Vietnam and Watergate, a new crop of cynical journalists was coming out of "J school" who were not captivated by the Kennedy magic. Stories and books stripping the bark off the family began to appear, and Camelot started to lose its luster and take on a decrepit, sad, even sordid image. The tabloids each week detailed some arrogant or excessive behavior by one or more of the Kennedy family, especially the children, some of whom not only were spoiled by easy wealth but also had lost their fathers to violence. Meanwhile, historians were unearthing links between the Kennedys and the mob, and JFK's compulsive womanizing was finally coming to light.

In this last Kennedy campaign, Teddy would not enjoy the laudatory media coverage the rest of his family had received, or what his brothers had received in 1960 and 1968. Teddy himself had gotten the kid glove treatment from the media in 1969, when the tragedy of Chappaquiddick took place. It was only one year after the assassination of Bobby Kennedy, and many in the media surmised that Teddy had not completely healed. The widespread view was that he had been prematurely and unexpectedly vaulted to the head of the family, suddenly anointed the father figure for his brothers' suffering children, and the repository for the country's unrealized hopes and dreams. What mere mortal was ready for such a burden? The fact that a young woman had died, possibly in an act of malfeasance on Kennedy's part, was conveniently glossed over. In 1980, though, once reliably pro-Kennedy papers would cover his campaign critically, warts and all.

Kennedy had disappointed his fans, and there is nothing sadder than a hero who falls far short of expectations and hopes, even if those hopes were out of reach. The *Post*'s Tom Shales reflectively wrote, "The greatest danger of the Kennedy candidacy may lie in its potential to reawaken in a discouraged nation unrealistic yearnings for a super-hero to lead us into Utopia."[53]

Kennedy's imminent entry into the presidential campaign led to many television specials and newspaper features, reexamining the events of ten years earlier at Chappaquiddick. Attempts to interview the parents of Mary Jo Kopechne were fruitless, and the rumor around Washington for years had been that the Kennedy family paid them a large settlement in exchange for their silence about their only child. All the old unanswered questions were raised again; what was a married man doing with an attractive, single young woman late that night? Why didn't he report the accident right away? Why did he use lawyer tricks to stop an inquest? How much money did the family spread around? Why was he caught in so many lies? He'd been taken to the party by his chauffeur. Why hadn't the chauffeur driven Kennedy and Kopechne to their respective homes? The ticking time bomb exploded.

Edward Moore "Ted" Kennedy formally entered the race for president three days after the disastrous Roger Mudd interview aired.[54] He eschewed announcing his candidacy in the Caucus Room on the third floor of the Russell Senate Office Building, as his brothers had done, opting for picturesque Faneuil Hall in Boston instead. The Caucus Room had seen much over the years besides their announcements. The hearings on the sinking of the *Titanic*, Teapot Dome, McCarthy-Army, and Watergate had all taken place there.

In Boston, he nailed Carter on the now-infamous malaise speech: "Before the last election, we were told Americans were honest, loving, good, decent and compassionate. Now the people are blamed for every national ill and scolded as greedy, wasteful and mired in malaise."[55] Kennedy had phoned Vice President Walter Mondale the day before, out of friendship and courtesy. Mondale told him, "Ted, I am very sorry about this. I am sorry for Carter . . . and I am sorry for you and I'm sorry for the Democrats because this is going to get mean and nasty no matter what we say. We intend to win this thing," he recalled telling Kennedy.[56]

The fallen standard had finally been picked up, a standard that had been ripped away in 1963 and again in 1968, and that Teddy had refused to pick up in 1968, 1972, and 1976. When Bobby Kennedy was assassinated in 1968, many Democrats urged Teddy to run in his place, but he refused. Still, it would be hard to imagine the Democrats denying Teddy the nomination in 1968, given the circumstances, and it is also quite easy to imagine he would have won the presidency over Richard Nixon and George Wallace. With a Ted Kennedy presidency then, in 1969, there would have been no paralyzing Watergate, and Mary Jo Kopechne probably would have lived to be an old woman.

Just a few hours before Kennedy's ill-fated interview with Roger Mudd was broadcast, an even more important event took place on the other side of the world. A large group of young militant Islamists, followers of the new ruler, the Ayatollah Khomeini, overran the American embassy in Tehran, Iran, and took hostage dozens of Americans working there.[57] They would hold them for 444 days.

Khomeini had lived in exile in France for years, returning to Iran only when the Shah of Iran was weakened. He was seventy-eight years old when he returned to seize power. A civilian government had begun during the demise of the Shah, but everybody knew who the real authority in Iran was. The civilian government was dissolved subsequently, and shipments of Iranian oil to America began to dwindle. Renewed fears of another bout of long gas lines shook the American consumer. Several leading Capitol Hill Democrats predicted mandatory rationing by the end of 1979.

Some Americans responded with their own form of protest. The ladies of the Cat's Meow escort service in Denver announced they would no longer accept "dates" with Iranian men. Other Americans took to the streets, carrying signs that read "Punch Persian Punks" and "Nuke Iran." Iranian students in America attempted their own pro-Khomeini protests, but were often met with slugs to the jaw by otherwise civilized American college students. Not all American college students supported their country, though, and some sided with the anti-American Iranians. It got deadly when four Iranian students were charged with conspiring to kidnap the governor of Minnesota, Albert Quie.[58]

President Carter deliberated about the illegal Iranians in the country and the Immigration and Naturalization Service said that tracking them down and expelling them was their "number one priority."[59] But beyond this and Carter's ban on oil from Iran, Americans would start to feel impotent and outraged. Three American women living in Iran, married to Iranians, held a press conference and said the hostages should be tried as spies. Pro-Iranians living in the United States held protests in all the major cities, including Washington. Pro-American counterprotesters were mostly spontaneous, but it was not unusual to see an old lady walking on the streets of Washington hurl a rotten egg at the Iranians and then go on about her business.

Originally, sixty-six hostages had been taken.[60] The number that remained throughout the crisis was ultimately fifty-two. The Ayatollah and his mobs also never harassed the Western media or the camera crews, and Khomeini also got away with something that no Western politician would dare: Mike Wallace of CBS wanted an interview, and the Ayatollah responded by demanding the questions in advance and then approving only those he would answer. Wallace dutifully did what he was commanded to do. The Ayatollah received better "press" from CBS than Teddy had. Especially since he released the American blacks and women held hostage, showing an understanding of the racial- and gender-obsessed American media.[61]

Ronald Reagan was also getting his fair share of bad press. In a devastating piece for *Human Events*, his old friend Stan Evans, longtime leader and writer in the conservative movement, authored a long piece, eviscerating Reagan's campaign and especially John Sears. Reagan was forced to write a new round of letters to angry conservatives, defending his campaign and explaining Sears's role in it. Reagan was so alarmed that he called Evans personally to reassure him.[62] Evans was not reassured, and the strife in Reagan's campaign continued.

Kennedy's inept performance on the road and in interviews with the media only served to spread the chatter about his ability as a candidate. His campaign itself was disorganized, and no one seemed to know from moment to moment where the candidate was or where he was going. Reagan's manager John Sears, often the brightest analyst in the room, nailed the problem: "He . . . moved attention away from Mr. Carter's job performance to the race . . . He's managed to create a race when there wasn't one before."[63] Carter and his folks were taken aback when some members of the Administration resigned and then endorsed Kennedy, including Carter's ambassador to Mexico, Pat Lucey, the former governor of Wisconsin. Lucey was an old Kennedy consort and loved the family. He'd been a key player in JFK's big win in Wisconsin in 1960 over Hubert Humphrey. Lucey immediately became deputy campaign manager, but as battle-tested as he was, he was surprised that Chappaquiddick "loomed as a bigger issue" than anyone in the campaign had assumed.[64]

Carter could gloat, though, when he won the final, nonsensical, meaningless, nonbinding Florida straw poll, crushing Kennedy, 1,114–351.[65] He beat Kennedy like a rented mule. It was the first head-to-head contest between Carter's political pros and Team Kennedy and the bohemians, and he beat the Bostonians without even breaking a sweat. This wasn't Kennedy's first team on the field. It was a bunch of scrubs, but expectations were so high for him that the crushing loss stunned many people.

In that time from early November, when Kennedy was looking as if he could defeat all comers—until only twenty days later, he dropped precipitously in the polls against both Carter and Reagan. He was now leading Carter 46–30, and was leading 51–37 over Reagan. Carter was also leading Reagan 47–42 in the *Washington Post* poll.[66] Many in both parties had failed to understand how much Teddy's performance had taken the focus off Carter and put it on him, and how much the developing Iranian hostage crisis would benefit Carter's political fortunes, at least in the Democratic primary.

―――――――

Kennedy thought he had an idea that could turn his plummeting polls around. President Carter had reluctantly admitted the Shah of Iran into the United States in late October 1979 so that the Iranian exile could receive medical treatment for cancer in New York City. The insult this act posed to the new regime in Iran was the pretense used by the young Islamic militants who stormed the American embassy and took the Americans working there hostage.

Kennedy announced in early December that the United States should send the ailing Shah back to Iran. Asked by a reporter in San Francisco what he thought of Ronald Reagan's view that the United States should instead grant the Shah asylum, Kennedy delivered another self-inflicted wound. The Shah of Iran, he said, "ran one of the most violent regimes in the history of mankind, in the form of terrorism and the basic and fundamental violations of human rights, in the most cruel circumstances, to his own people."[67] The Shah, while a benevolent authoritarian, had also installed many reforms in his country, unlike so many other Middle Eastern regimes. Regardless, most Americans thought Kennedy was running up the white flag, something his brothers never would have done.

Kennedy then stumbled even more badly, posing this rhetorical question: "How can we justify, in the United States, on the one hand accepting that individual because he would like to come here and stay here with his umpteen billions of dollars that he's stolen from Iran and, at the same time, say to Hispanics who are here legally that they have to wait nine years to bring their wife and children to this country?"[68]

Americans, historically slow to war, strongly supported military action if the hostages were harmed or put on trial. And by a three-to-one margin, they opposed sending the Shah back to Iran. Kennedy had blundered badly on this, listening to the elitists instead of the populists.[69] Even the *Washington Post* skewered him for his proposal, saying, "Unbeknownst to Sen. Edward Kennedy but symbolic of his burdened presidential candidacy, the United States had virtually completed arrangements for the shah of Iran to go to Argentina when Kennedy's blast at the toppled shah as one of history's worst tyrants killed the deal."[70] The piece continued: "[T]he failure of Kennedy and his entourage to comprehend the damage of his attack on the shah is by far the most dangerous setback. Some politicians now are asking each other: is it possible that Ted Kennedy really is a stumble-bum?"[71] For many Democratic voters, the answer seemed to be, increasingly, yes.

———

In Carter's native South, the saying went that "the sun don't shine on one dog's ass all the time," and this was also true in politics. Carter had been, by and large, unloved and unwanted by the elites. They considered him and his band of yahoos déclassé. He'd disappointed the American people. It looked to all as if he would never get off the mat and that Kennedy would waltz to the nomination. The irony, therefore, of their rapid role reversal was right out of O. Henry.

Along with the hostage crisis, Carter was benefiting from handing out federal goodies in Florida and other important states. He had railed against pork-barrel politicking four years earlier, and indeed vowed never to use it to his own advantage. That stance was now over the side. Millions of dollars also were going to New Hampshire in the form of highway improvements and a "$1.5 million loan guarantee to American Skate Factory in Berlin."[72]

Administration officials had flooded the state of Florida, handing out federal largesse. Carter had also spread around a lot of perks in Florida, including making sure that many of the state's Democrats got good seats at the White House reception for John Paul II.

Carter was now following the old Washington maxim: "Where you stand depends on where you sit."

Adrift

"Ronald Reagan . . . would be an easy Republican to beat."

The above-the-fray strategy to keep Ronald Reagan muzzled had been campaign manager John Sears's idea. "The race usually can't start until the frontrunner gets in."[1] In 1978, Reagan had given dozens of speeches and conducted hundreds of interviews and had crisscrossed the country many times. By 1979 the total was at best a handful. Sears was also directing a campaign that was aggressively pushing away Reagan supporters of long standing, going out of his way to make enemies out of onetime friends.

All were disappointed. But one man in particular, Ernie Angelo, was more disappointed than all the others. Angelo was a tall, soft-spoken, and fiercely competitive Texan. He'd been one of the architects of the big Reagan win in the Lone Star State in 1976 and was raring to go for 1980 as well. Sears was opting for more establishment figures this time around. Nevertheless, Angelo set aside his disappointment and for a time took a backseat to the new governor of Texas and Reagan latecomer Bill Clements.[2] The irony was Reagan didn't even like Clements.[3]

Reagan did pick up the endorsement of Pete Wilson, the mayor of San Diego. Wilson had campaigned aggressively—too aggressively for some Reaganites—in 1976 against the Gipper in New Hampshire. It cost Wilson in 1978, when he was considering running for governor, but was coolly received by conservatives in the Golden State and came in dead last in a five-man field, receiving only 9 percent of the primary vote.[4] This time, Wilson wisely saw not a Ford but a Reagan in his future and signed up for the ride. The move held Wilson in good stead for the rest of his career, as he eventually went on to become the state's governor and U.S. senator.

The Carter people were gunning for Reagan, and *New York Times* reporter Adam Clymer took note of this: "Ronald Reagan . . . would be an easy Republican to beat." Clymer's byline seemed to appear each day with fresh stories about the Republican Party, the conservative movement, and all the GOP candidates. He had emerged as one of the best and most important political journalists in America.[5]

Sears was putting the heavy "woo" on major Republican figures in the Northeast to sign up for Reagan. Fred Biebel, the Connecticut GOP chairman and a moderate who supported Ford in 1976, was brought aboard for Reagan. Sears also went to the Republican retreat in Mackinac Island, Michigan, to court Governor Bill Milliken, another moderate who was popular with the big-business Republicans and especially anti-Reagan Republicans.

Reagan had never been a fan of big business. They, in turn, were not fans of Reagan, as he had often complained that small businessmen were just as endangered by big business as they were by big government. John Connally and George Bush were the candidates of big business, but not Reagan, the populist. Bush and Connally, as Texans, were getting the bulk of their money from big oil. Still, Reagan had to defend himself against the charges that he was "moderating" his positions, saying, "I am what I've always been, and intend to remain that way. What I say is what I believe."[6]

One of Reagan's few trips in 1979 was to speak to a group of Republican women. Recognizing the political realities of the day, Reagan told the women, "We are deeply indebted to the one man who has single-handedly rejuvenated the Republican Party—Jimmy Carter."[7] With all the scandals and failures that occurred the previous years, it wasn't too far off the mark. He received a generous round of laughter and applause.

A positive development was that Bill Simon of New Jersey, after considering a presidential bid of his own, decided instead to endorse Reagan.[8] Suddenly Reagan, who hadn't even contested the New Jersey GOP primary in 1976, was looking somewhat better and better in the Garden State. With crooners Frank Sinatra and Dean Martin helping Reagan raise money, New Jersey appeared more palatable and less toxic, at least politically, for Ronald Reagan.[9] Still, many conservatives who were staying put with Reagan fretted that he was allowing Sears too much of a free hand in the message of the campaign. Without one primary vote yet counted, Sears was already talking about "the attention a front runner will get" to "begin to address ourselves to a broader constituency."[10] John Sears's strategy of keeping front-runner Reagan "on ice" yielded a bitter October harvest for the campaign in Iowa, when the GOP faithful conducted the state's second straw poll at another fifty-dollar-a-plate fundraiser, this time held at a field house in Ames, Iowa.

Before this second straw poll in mid-October, Reagan's political director, Charlie Black, oddly told the *New York Times*, "I would think Baker and Connally would be a little leery" of attending the dinner.[11] Senator Baker was offering free beer to anyone who turned out at the straw poll for him. A newly released statewide poll curiously had Reagan up since the spring. It confirmed to Sears that he was right about keeping Reagan out of Iowa.[12]

═══════════

In the Reagan camp, for his announcement in New York City, Helene von Damm had been sweating bullets, trying to raise money under the supervision of Charlie and Mary Jane Wick, close friends of the Reagans.[13] Wick named the early effort the "Ground Floor Committee."[14] Ultimately, sixteen hundred supporters, some paying five hundred dollars apiece, attended, even though only days before, the event looked like it would be a bust.[15]

Von Damm's personal story was in and of itself a testament to Reagan. An émigrée from Austria, she heard Reagan give a speech in the

1960s in Detroit and came away mesmerized. In 1966, when she heard he was running for governor of California, she quit her job and headed west with no work and not knowing a soul. Thrillingly for von Damm, she ended up as Reagan's private secretary in Sacramento and later worked for him via the firm of Deaver and Hannaford. When the 1980 campaign rolled around, von Damm wanted a battle-line job on the campaign and became the finance director for the Northeast.[16] Attorney William Casey, who had previously been chairman of the Securities and Exchange Commission under Nixon, introduced her to prominent Wall Street types and helped von Damm in fund-raising. One, whom they approached to help Reagan but who turned them down cold, was a Wall Street banker by the name of Don Regan.[17]

Sears was obsessed with the Northeast, but it was also a practical decision. In New York City, Reagan could once again count on the support of George Clark, a reliable conservative leader from Brooklyn, and of Mike Long, an influential conservative activist.[18] Mike Deaver, senior Reagan aide, said, "We haven't done a very good selling job in the Northeastern states."[19]

Reagan had a very effective political operation on the ground in the Northeast that included Roger Stone, a twenty-seven-year-old who had been dusted by Watergate but who came back to win election as head of the Young Republicans and helped found NCPAC. Stone handled a handful of states for Reagan in the Northeast, including New York and Connecticut. Reagan needed moderate Republicans on board, and Stone went after his prey relentlessly. In Connecticut alone, he brought in more than three hundred elected GOP officials, mostly moderate to liberal, to support the Gipper's eventual bid.[20] Reagan's Northeast team also included Clark, who was an understated but highly effective conservative.

John Connally, meanwhile, was still attempting to engage the Gipper. The Texan made it personal with an increasing number of backroom remarks about Reagan's age, his energy level, and his ideas. Reagan was biting his tongue, which frustrated the Texan and his troops even more. George Will took it upon himself to defend Reagan: "Connally's veiled references to Reagan's age . . . are nasty. Perhaps nasty people deserve a candidate." Will also spied something that eluded others: "Reagan has never been as dogmatically conservative as some careless or cynical detractors say."[21]

At one point, Connally suggested that all the GOP candidates meet in one room and have a "gang-bang" style debate; that proposal went over like a lead balloon, the sexual innuendo aside. He was more successful, however, in raising money from those, including oilmen, who did business with the Arab states. In fact, he was so successful that he was considering not taking "matching funds" from the FEC. This would free Connally from having to abide by the state-by-state spending limits imposed on candidates who did accept matching funds.[22]

The University of New Hampshire poll showed Reagan with a narrowing lead, 40 percent to 12 percent over Howard Baker. He'd been up and down in New Hampshire over the previous several years, and by this measurement Reagan was down again from four years earlier, when he'd gotten almost 50 percent of the vote against an incumbent president. This poll, therefore, was not good news for the Gipper. Bush had improved the most over the past several months, coming in a strong third with 10 percent of the survey. Connally was right on his heels, though, at 7 percent.[23] Bob Dole, John Anderson, and Phil Crane were nowhere. Bush reaped the reward for plain old hard work and his zest for frenetic campaigning.

Now, however, Reagan was making preparations for his third and final announcement of candidacy for the GOP nomination. Central to his economic plan would be the 33 percent tax cut touted by Jack Kemp and Senator Bill Roth of Delaware. It would reduce individual tax rates and index the tax code to prevent "bracket creep," along with a reduction in capital gains taxes to encourage savings and investment. In 1979 the top rate for the capital gains tax was 70 percent.

Reagan also proposed the immediate decontrol of oil and gas prices and called for the government to make way for more domestic exploration. Kemp had signed on to the Reagan campaign after briefly flirting with the notion of running for president himself. His friends Jude Wanniski, Art Laffer, and others, as well as his staff, had encouraged this, whispering in his ear that Jack French Kemp could be the next JFK. Indeed, Kemp wrote a book in 1979, *An American Renaissance*, that expounded on his economic theories, conservatism, and the GOP, but did not once mention Ronald Reagan. By the fall of 1979, however, Kemp was touted as a potential running mate for Reagan, though a few close to Reagan saw Kemp as uncontrollable. Plus, there were rumors, some of Kemp's being gay and others of his being a Lothario.

Aside from the quarreling presidential campaigns, the state of the GOP one year before the 1980 election was one of superficial harmony. Most of the old issues that divided the party were gone or at least masked over well by the widespread contempt for Carter. No one, save for John Anderson, was prescribing government activism as a means of solving the problems of America. Virtually all the major candidates called for the reduction of taxes, most were calling for the restriction of the federal government, and all were calling for junking détente—Reagan held the hardest line on all three. And with the notable exception of Connally, they all supported Israel.

But if one were to list ten economic, cultural, or political issues, most Republicans would agree with seven or maybe eight out of the ten. The party had discarded the advice of establishment Republicans to broaden the appeal. In fact, the party was broadening the base by narrowing the appeal. Instead of trying to be all things to all people, the GOP, with Reagan's gutsy leadership, was becoming one thing to all people. The party was becoming commodious by standing on principle and not power. It was also becoming less and less like the party of Nixon, which was a "power flows downward" Tory party, and instead becoming an American-styled conservative, populist party in which power flowed upward. Nixon's Republicans defended the status quo. Reagan's conservatives challenged it.

The imagery of the party was changing dramatically as well. Although it still had a goodly number of "country club" Republicans, the picture was becoming more decidedly middle class, suburban, southern, Western, Catholic, and entrepreneurial. Reagan had brought the party to him, through the force of his will, and was leading millions of conservative Democrats out of their historic home and into the once-hated Republican Party.

In Maine, an abbreviated caucus was held, and once again the Reagan campaign did not participate. Ambassador Bush won the vote; Howard Baker was not far behind. Baker didn't help himself when he told reporters, "Almost winning was gratifying indeed."[24] He exacerbated the problem the next morning when he went on ABC's Sunday show *Issues and Answers* and tried to explain to the befuddled interviewer that, although he had not won, he hadn't lost, either. Bush, not missing a trick, made ten thousand copies of the news stories on his win in Maine and mailed them out to supporters.[25]

Bush was undoubtedly helped by a political backroom deal made between Dave Keene of his staff and Charlie Black of the Reagan campaign. Both saw Baker as their more immediate threat, and politics is the art of convenience. Keene proposed to Black that since Reagan was not competing, Black should tell the Reagan forces in Maine to vote for Bush to stop Baker. Black initially turned down Keene's suggestion, but John Sears, smiling, later refused to deny there had not been a deal to count out Baker.[26]

A new national Gallup poll showed Reagan had improved his position somewhat as the choice of Republican primary voters, now getting 37 percent, up from the 28 percent he'd sunk to over the summer. His favorable/unfavorable numbers, however, had narrowed among Republicans. Connally was second among Republicans, and Baker was third. Bush was last.[27]

The problem for Reagan was that as the weaker candidates fell by the wayside, the remaining, not Reagan, could pick up their support. Pollsters knew the old line that "a front-runner gets what a front-runner gets." That meant that Reagan's 37 percent was about all he could expect to count on, and if the race got into a one-on-one contest, and if the status quo held, then Reagan would not be the nominee of the Republican Party. This dynamic, combined with the Reagan campaign's inability to squirrel away cash for a rainy primary day, was giving hope to the others.

———

The off-year elections of 1979 were disappointing for the GOP. The party had fielded strong candidates in the gubernatorial races in Kentucky and Mississippi, but conservative Democratic voters there were still not ready to cast their lot with the Republicans—not yet anyway. Republicans failed to make any real significant gains in the legislative seats in New Jersey or the many mayoral races, so the day counted as a setback for the grassroots efforts of Bill Brock's RNC. True, many Republican candidates fell just short, as opposed to previous years, but the party had developed a taste for winning, not just coming close.

One impressive pickup for the GOP was in Cleveland, where George Voinovich won. A Republican in this heavily Democratic labor city, he defeated the irritating little mayor, Dennis "The Menace" Kucinich. Kucinich was, by most accounts, overwhelmed as the city's bond rating sank as if into the polluted waters of Lake Erie. Kucinich, who appeared to be around five foot nothing, was humiliated when photographed at a press conference standing behind a lectern on a stack of phone books. Kucinich's stewardship had guided the once-proud city to becoming a national joke and picking up its new moniker, "the mistake on the lake."[28] Just a year before, Kucinich had barely survived a recall election, keeping his position with fewer than 250 votes more than his opposition, of the more than 120,000 cast.[29]

═══════════

As Reagan was on the brink of his declaration, he was still on cruise control per the edict set forth by John Sears: minimal campaigning and maximal seclusion. Staff tensions were beginning to boil over. The no-longer-whispering-but-shouting "on background" message from his opponents was "that Reagan is really a fading Class-B TV movie hero whose one-liners will last during the winter theater season but won't bloom in the spring."[30] He was, in fact, as the saying went, "tanned, rested, and ready." Before Reagan spoke to an audience, he liked to have a cup of black coffee with a dollop of saccharine. Waiting in the wings with a reporter once before going on, he joshed that "Jimmy Cagney told me that each morning before his father left the house, he'd check his fly, tip his derby and cross himself."[31] Reagan also had his doubts, and betrayed them one night to the *Los Angeles Times*, saying, "I always wonder, is the feeling for me out there? Do they still want me?"[32]

On the eve of his third try for the presidency, the *Washington Post* ran a long and mostly kind profile of Reagan. "He is 68 years old. The face is outdoorsy, ruddy, like an apple just shined. Up close, you see the lines and creases and crow's feet; leather finely cracked, like a saddle or the seat of an old Jaguar. These lines seem indigenous. They don't detract."[33] As Reagan moved closer to the long-awaited date, the FCC alerted television stations around the nation no longer to show any more reruns of Reagan's old movies and television shows, unless they were prepared to offer "equal time" to all the other candidates—which of course they were not.[34] Reagan had gone through the same nonsense in 1976, when he challenged Gerald Ford in the Republican primaries. Mindful of some of the low-quality movies he'd made early in his career, Reagan joked at the time, "Somebody must have goofed, because I've made some movies that—if they put them on television—I'd demand equal time."[35]

The political establishment had always held Reagan in minimal high regard, and even as his issues over the past several years had led the GOP out of the wilderness, those who did respect Reagan considered him a Moses-like figure; he led them to the Promised Land, but he could not go there himself. It was up to younger leaders such as Joshua—or, this time, Howard Baker and George Bush—to bring the party into Canaan. There were many, though, who had never respected Ronald Reagan and never would.

Yet there was one group, growing in numbers, who did think quite a lot of Mr. Reagan: young voters. Between Reagan and many young Americans was a special bond. Reagan of course loved speaking to and meeting with young people, such as those with the Young Americans for Freedom and the Intercollegiate Studies Institute. Reagan spoke at most of their national conventions in the 1960s and '70s.

The *New York Times* reported on what may have been the most unusual young supporter Reagan was to have in 1980. Anthony J. Nania of Connecticut had gone to Chicago to support Senator Eugene McCarthy in 1968 and four years later was working on the campaign of George McGovern. He started out in politics working for the very liberal congressman Allard Lowenstein of New York. By 1979, however, Nania was a locally elected Republican official in Hartford working hard for Ronald Reagan. An incredulous reporter pressed him on how he could be supporting Reagan, and he replied, "Now the common theme is probably a distrust of government in general."[36] Nania also talked about how there was more room in the GOP for young people than in the Democratic Party.

Reagan's Republican Party had been evolving into an anti-establishment, anti-status quo movement. Carter and the Democrats represented the existing order and centralized power. Reagan represented a challenge to that existing order and believed that power flowed upward rather than downward. He preached about a future of opportunity. It was appealing to young people.

Reagan also had a strong libertarian streak. He opposed the draft, he opposed restrictions on gay teachers, he made many speeches about the dignity of the individual, and he favored tax cuts as a means of enhancing the hope and entrepreneurial skills of Americans. All this further served to bring young people to his movement and eventually his party.

The Democrats had been widely assumed to be the party of younger voters from the time of the New Deal right up until the 1970s, but young Americans were not about to accept at face value the arguments of Carter that their future was now of paucity and despair. And among the Republicans, Reagan was the only one talking about a hopeful future.

———————

It was go time.

This was to be his third time running for president of the United States. He knew and understood that this was his last chance. His 1968 run was ill conceived and late starting, and was ultimately unremembered. His 1976 run, on the other hand, put him in the national spotlight, having him lose to incumbent president Ford by the narrowest of margins. The run in 1976 was everything the run in 1968 had not been.

Rumors had been swelling the previous years on the Gipper's running for the '80 race. If anything, the polls seemed to agree he was an even more popular candidate. Even then, it was a risk to enter. Reagan understood that.

He knew what he was doing this time around. Or at least believed what he had to do, provided he listened to his own instincts and not ever-opinionated consultants.

Ronald Reagan delivered the speech announcing his candidacy for the Republican presidential nomination in 1980 twice on the same day. The first time was on the morning of November 13, 1979, in a half-hour television address taped in a New York City studio that was produced and placed with independent stations across the country for broadcast that evening at a cost of about $400,000.[37]

He looked tanned and handsome. He began the speech behind a desk, but then he got up, moved gracefully to a globe of the world, sat a cheek on the corner of the desk, and continued his remarks. He paid tribute to American ingenuity, saying that "there remains the greatness of our people, our capacity for dreaming up fantastic deeds and bringing them off to the surprise of an unbelieving world."[38]

America, he noted, was a wholly unique country. It was exceptional to the entire world, a belief that many doubted in the late '70s and that ran completely counter to Carter's beliefs. He stated that "our country is a living, breathing presence, unimpressed by what others say is impossible, proud of its own success, generous . . . never mean and always impatient to provide a better life for its people in a framework of a basic fairness and freedom."[39] It was a thoughtful and somewhat well-received speech.

At his live speech that same night in New York, Reagan was interrupted by supporters' applause thirty-four times.[40] Though he was taking a worldlier approach on issues that concerned the Western Hemisphere, less imaginative conservatives failed to recognize that closer trade and military relations with Canada and Mexico would bring those countries into the anti-Communist fold and strengthen the Monroe Doctrine. On domestic affairs, Reagan believed in federalism and called for a "planned orderly transfer" of federal programs that handed welfare and education to the states, along with the funding to pay for those programs.[41] Reagan had previously gotten in some trouble in 1976 about redirecting authority to the states from the national government, but in four years the country had changed, even as Reagan had not.

This time, he identified not only the problem, but the solution as well. He also called for a massive across-the-board reduction in personal income taxes. Following his speech at the Hilton in New York, Reagan's family joined him onstage, including Nancy, Maureen, and Patti Davis. Reagan quipped that "none of them are looking for jobs in the government."[42] Echoing from nearly half a decade earlier, he said that the world depends on the United States and the American ideal, the same goal as in Colonial America. He declared that "a troubled and afflicted mankind looks to us, pleading for us to keep our rendezvous with destiny; that we will uphold the principles of self reliance, self-discipline, morality and, above all, responsible liberty for every individual."[43]

Of the struggles ahead for America, Reagan concluded that his fellow citizens could "expect to be tested in ways calculated to try our patience, to confound our resolve and to erode our belief in ourselves."[44] His taped address eventually brought in around $500,000.[45] At one tender point, Reagan choked up and was close to tears when he spoke about how his father had received a pink slip on Christmas Eve during the depths of the Great Depression. "I cannot and will not stand by while inflation and joblessness destroy the dignity of our people."[46]

Many of Reagan's aides heaved a collective sigh of relief. What could have been a half-empty room quickly filled up as late telegrams and invitations were sent for his announcement dinner, with a free plus-one, provided he or she was less than twenty-five years of age. The dinner didn't raise much, but the event was packed with media and influential GOP officials and everyone declared the event a success.[47]

Helene Von Damm was mildly disappointed that Mrs. Reagan would not let her call any of her wealthy friends in New York to help with the event.[48] But Nancy proved vital in other areas. Dan Terra, who eventually became national finance chairman, only reluctantly signed on after Mrs. Reagan charmed him into doing so.

Reaction to Reagan's announcement and subsequent campaign outings were muted, at best. Some argued that it was the candidate who was muted, hence the reaction. Reporters took note of Reagan's "shaky start." Also, his taped announcement was "bland."[49] There was one notable exception, however. Following the taped speech, actor Michael Landon of *Bonanza* and *Little House on the Prairie* fame came before the cameras to make an appeal for contributions to the Reagan presidential campaign, telling the audience, "I know him and I like him."[50]

The whole mess surrounding the campaign launch ended up in the Evans and Novak column, though John Sears had sworn Jack Kemp to secrecy; Kemp in turn told Jude Wanniski, swearing him to secrecy; and Wanniski in turn told Bob Novak, whose job was to pry people loose from their secrets. Sears called Kemp and "screamed" at him, hurling obscenities.[51]

The "age issue" was still festering. Reagan, in an interview with NBC's Tom Brokaw, pointed out that if he were elected, he would be younger than any world leader with whom he would be dealing except Margaret Thatcher.[52] Still, Reagan was comfortable in his own skin, win or lose. In a feature story, the *Washington Post* described him as "a man unembittered by past defeats, a man who might actually enjoy himself and other people if let loose by his staff for five minutes. Some people think Ronald Reagan is a prisoner of his staff, that a strange cocoon has built up almost insidiously over the past dozen years."[53]

The media covering Reagan joked about his leisurely style of campaigning. Indeed, Reagan had been out only for the last year, fewer than fifteen days out of every month. The plan since he'd announced was to have him on the stump four to five days alternating with three days off, but his admirers would be kept at arm's length. At one Boston appearance, an overzealous advanceman was heard saying he wanted nine men to surround Reagan at a fund-raiser. "We don't want any ordinary people coming up and touching him."[54] National reporters were spun by Reagan's "handlers" and wrote that they have "repackaged him with exquisite caution, wrapping him in cotton batting."[55] Nixon had insulated himself from the press and the public beginning in 1968. Sears had observed how this worked for a time for Nixon, but many thought he should have known better than to try the same tactics with the very public Reagan.

From New York, Reagan headed to Washington, Boston, Manchester, Philadelphia, Chicago, Milwaukee, Grand Rapids, and Orlando for tightly controlled events. He was getting his licks in against Carter, as when he told a crowd in Boston that "a graduate of Annapolis is at the helm of the ship of state, but that ship had lost its rudder."[56] He fumbled a question in New York about the size and terms of the federal bailout of the city, and the media harped on this.

However, Reagan was finally in the race, much to the relief of his friends and supporters. It was out on the stump, talking, meeting, and answering questions, where he shone. His early forays were getting good reviews, but he had to respond to the "age issue" repeatedly. To one woman, he proposed challenging his opponents to arm-wrestle. To a reporter, he said, "the whole issue of my age will be resolved when the people of this country see whether I can go the distance."[57] He picked up the pace and campaigned for five days at a rate that would have taxed his younger opponents.

He put in an appearance in Florida, but he had been to Orlando only briefly, and Floridians took note. "Absence and silence have not made the Florida heart grow fonder or silenced nagging doubts here about Reagan's age," wrote Evans and Novak.[58] A joke among reporters was that if you wound up a Reagan doll, it would go for an hour before it had to take a nap.

Others saw Reagan's being an outsider to Washington as yet another problem. He didn't know the levers of power. After all, hadn't Carter run as an outsider? And look at the mess he'd created. Two conservative columnists, Bill Buckley and James J. Kilpatrick, battled over it. They were friends, fellow Irishmen, and devoted Catholics. Kilpatrick saw Reagan's lack of Washington experience as a weakness, and Buckley saw it as strength.[59]

Iowa Agonistes

"Reagan is a political hemophiliac."

Ronald Reagan was fighting more and more headwinds in his own party over the renewed/old charges that "Reagan is too conservative to win a general election." Yet age had "displaced ideology as his most serious negative."[1] The liberal opinionist Richard Reeves of *Esquire* also wrote harshly about Reagan, calling him the "backwards" candidate, which pretty much summed up the cultural elite's opinion of the California conservative.[2] Power was shifting in politics from the candidates to the consultants, as the columnists and reporters almost never wrote harshly about the consulting classes, but often did about the candidates.

Many knew this, and understood how it would work. Staff turnover was high for both candidates and elected officials. Meanwhile, the candidates themselves meant little to the political civilization except as a medium to make money, gain publicity, and have fun. As Michael Corleone said to Sonny, "It isn't personal . . . it's strictly business."

The Soviets invaded Afghanistan, but there was internal debate in the Carter Administration over whether it was even worth raising the issue with the Soviets, since Afghanistan was not part of the defense perimeter of the West. The internal policy debate leaked out. Carter fecklessly announced that American athletes would boycott the 1980 Olympics to be held in Moscow, in response to the Soviets' invasion of Afghanistan. He called on other nations to join the boycott, but few did.[3] Reagan himself first supported then opposed the boycott, seeing the young athletes as taking the brunt of Carter's policies. He also came up with the interesting idea of holding the Olympics in Greece every four years, thereby bypassing the billions of dollars countries and corporations paid out in competition to host the games. Besides, the Olympic Committee was notorious for ignoring human rights violations.

To prevent any dangerous ideas such as personal freedom from spreading among schoolchildren in Russia during the Olympic Games, Communist teachers there were telling kids to beware of the Americans because they would give them chewing gum that was filled with infectious diseases. The Soviets also put on trial in Czechoslovakia six human rights activists accused of sedition.[4] NBC was looking at a devastating cash loss if the Olympics were canceled or curtailed. They had forked over more than $80 million of obscene profits for the right to broadcast the games to the anticapitalist purists who ran the Soviet Union.[5]

———

In 1976, those who had the courage to take on an incumbent president by working in a challenger's campaign were not, by and large, considered to be the cream of GOP operatives, though they would eventually make the entire Republican establishment look foolish. John Sears, who managed the '76 campaign, took on board the best campaign staff he could. At least, that's how the insiders saw it.

Sears had been the architect of Richard Nixon's successful delegate-hunting strategy of 1968. He was in the White House, however, for only a short time before Attorney General John Mitchell, paranoid about the relationship between Sears and Nixon, forced his ouster, but not before ordering the FBI to bug Sears's office. Mitchell had also kept the young attorney from gaining greater responsibilities than he himself had in the early days of the Nixon Administration. Hence the paranoia Sears exhibited. By 1980, Sears had channeled both Nixon and Mitchell, and woe to anyone who crossed his path.

Sears was now running the campaign of the front-runner, and he knew, better than almost anyone else did, how close Reagan truly was to becoming the fortieth president of the United States. The country had moved rightward. The Democratic Party had started to fall, while the Republican Party had started to rise. Sears saw that President Carter was weak and flaccid. Everything was falling into place. Sears, however, had been too cautious and too tone-deaf for his own good, or for Reagan's, in trying to protect Reagan's advantages.

———

Someone who was unquestionably focused on the task at hand was Reagan. It was rumored throughout his political life that Reagan was some sort of pawn of Sears and the moderates. But Reagan was always his own man, even though Sears played a significant role in the selection of the campaign personnel. Of course, Reagan had strong personal ties with friends on his campaign with whom he'd become comfortable over the years, such as Pete Hannaford, Mike Deaver, Marty Anderson, Dick Allen, Lyn Nofziger, and Ed Meese.

Sears was wary of those on the campaign who had close personal ties with Reagan. He'd been in politics since the early 1960s, he knew how it worked, and he knew how to throw elbows when he needed to, but in the 1980 Reagan campaign, doing so was unnecessary.

It all came to a head just a few weeks after Reagan formally announced. Elisabeth Bumiller of the *Washington Post* recalled a story: "His staff was at war. But that afternoon Reagan and his wife served as mediators. Sears, flanked by aides Jim Lake and Charles Black, complained that Deaver wasn't handling the campaign's money well. Deaver maintained this was just an excuse; the real problem, he said, was that the imperial Sears wanted sole power."

"You're going to have to make a choice," Nancy Reagan told her husband hours into the discussion. But Deaver would have none of it. "No,

Governor, you don't have to make a choice. I'll resign," he told Reagan and the others in the room. After Reagan walked Deaver to the door, he angrily shouted to the remaining aides, "Goddamn you guys, the biggest man in this room just left."[6] After Deaver's departure, Ed Meese soon came aboard the campaign full-time. His positive presence would prove to be a number of important things to a number of important people, especially one Ronald Reagan. Less so for Sears.

As George Bush's campaign was gaining more and more respect, a previously skeptical media now turned their attention to him. Who was this guy? He talked about how his fund-raising was close to Reagan's, about his political and personal life, and so on and so forth. He talked about how he would assist Iowa's ninety-nine counties, how he was prepared to go the long distance. What Bush did not understand was that he was benefiting from Reagan's foolish inattention to Iowa, and if its voters were miffed at Reagan and gave Bush the chance to speak, it had to be about them and not him.

Iowa should have been a cakewalk for Reagan, given his strong roots there. He'd been a local celebrity on the radio in Des Moines in the 1930s, and many people still remembered fondly listening to Dutch Reagan on WOC and WHO radio. Iowans did not like to be taken for granted, and Reagan, it appeared, was taking them for granted. They were proud, independent, and patriotic. But they also did not like to be treated like hicks or second-class citizens.

Team Bush was also helping him. As the *Washington Star* reported, "His wife, Barbara, has practically lived in Iowa. Four of his five children are working full-time on the campaign with two sons, Marvin and Jeb, living in Iowa; his daughter Dorothy, and son, Neil, are living in Boston and New Hampshire, respectively."[7]

The campaign marched on. Bush thought Jimmy Carter would ultimately win renomination, but his staff thought otherwise. David Keene told reporters that with Kennedy as the nominee, he'd be seen as more liberal than Carter, thus opening up the middle to Bush, who was more moderate than Reagan. "It makes the perception of Bush as a centrist versus Reagan on the right more important. It also emphasizes the age issue and the question of experience in Washington and the government, which Reagan, like Carter, doesn't have."[8]

Ronald Reagan moved ahead, albeit again lackadaisically. In Arkansas, when Iran took Americans hostage, he labeled it a "rabble," and said that any "further harm or danger to our people there will be followed by immediate punishment." He continued, taking a threatening but necessary tone: Iran would "suffer," he said, if "there is such a thing as an execution of any of those people—even one."[9] Reagan was getting extensive coverage from the media, as twenty reporters and two network crews accompanied him to Little Rock.[10] He was making noise but not getting ahead.

Carter abruptly decided to forestall any "campaigning" because of the ongoing hostage crisis in Tehran. A problem that started off as looking like something that could be negotiated with a few firm threats to Iran was shaping up as an obsessive subject for Carter, Washington, and America. More than fifty Americans had now been held for nearly two months and Walter Cronkite, whose real passion had been covering the countdowns of the American space program, now began a "count up" of the days Americans were being held hostage in Iran.

The actual political contest would begin shortly. So far, the candidates had gone through the preliminaries, including the nonsensical Florida straw poll, but Iowa's caucuses were just over the horizon. Then came Puerto Rico, but most were ignoring this to concentrate on the New Hampshire primary on February 26, when the winnowing-out process would begin to accelerate. Only Bush was staking a claim in Puerto Rico, in part because his son Jeb spoke Spanish fluently.

March was loaded with nine Republican primaries, beginning in Massachusetts, wending through Florida, and finishing in New York. April was fairly light, with only four primaries, but it did include the big Pennsylvania contest near the end of the month. May? It was staggering, with no fewer than twelve primaries in many important states, including Texas; and June was no better, with eight primaries, including California. All in all, thirty-eight primaries would be held in 1980, eight more than four years earlier and more than twice as many as in 1968.

These contests in the larger states that followed Iowa and New Hampshire were becoming increasingly scientific and media-centric. Nearly all efforts focused on media, media, and more media, either through paid advertising, "free media" (that is, television and newspaper coverage), and the mail. Iowa, with its caucuses, and New Hampshire, with its small number of primary voters, required the candidate to engage in "person-to-person" retail politicking. Increasingly, the states that followed had very little "retail" politicking and a great deal of "mass marketing."

The diminished personal connection between the candidate and the voter was not the only aspect of campaigning for president that was changing. Convention delegates, which had in the past been personally wooed by top campaign staff and, in some cases, even by the candidate himself, were becoming even more like ceremonial functionaries as opposed to the quasi-independent decision makers they had often fancied themselves to be.

Primaries were far easier and more fun than conventions to cover, as they were dominated by advertising, especially television, and the consultants controlled all this and the candidate's schedule. It was far easier to work from a cubicle at the National Press Club or from a hotel room and call a chatty campaign pollster or consultant or manager than actually go to the hinterlands and talk to strangers. Better to be able to call your dining and drinking buddies, who often were just a few blocks away in Washington. Iowa and New Hampshire were the obvious exceptions. Reporters for years had loved covering the Granite State's primary and the colorful local characters, while dining and drinking at established spots in Manchester, Nashua, and Concord. By 1980, Iowa also would become a favorite quadrennial haunt for reporters.

Primaries, too, were more suspenseful than most state conventions, which were often dreadfully boring events held in cavernous halls where speaker after speaker droned on and on about obscure and meaningless rules and policies, and where most of the action took place behind closed doors where reporters could not go, or were rarely invited to go.

The lengthening of the primary season in 1980 had another (in hindsight) predictable (but somewhat unforeseen in advance) consequence. The long preprimary season, followed by the long primary season, put a premium on men who were essentially unemployed but financially secure, men such as Reagan, Bush, and John Connally. It wreaked havoc with those who worked for a living, such as Bob Dole and Howard Baker. Every four years, academics came out of their ivory towers wringing their hands on the general theme, "This is no way to select a president."

The chorus sang again as 1980 approached.

———————

The young, hyperkinetic twenty-nine-year-old New Yorker, Rich Bond, brought in to run George Bush's Iowa campaign, was running circles around the Reagan folks. Bond worked long days and nights, flogged his staff, and was determined to make the most of his chance in the big leagues in American politics, a level at which Bond desperately wanted to play.

Bush was attracting bigger and bigger crowds, and while the initial strategy had been to replicate Carter in 1976, they had modified this somewhat and would not run in every primary as Carter had done. The Bush operation was in good shape in Iowa, Puerto Rico, and New Hampshire, and was also beginning to turn its attention to selected states in the South, except South Carolina. Bush thought he had a good chance to be competitive in Florida and Alabama, where organizations were being assembled. Arthur Finkelstein, late of the Phil Crane campaign, came back aboard the Reagan effort, where he had been in 1976.

Sears had him do a poll in South Carolina about Connally. GOP voters were asked why they disliked Connally, and they had many reasons, including his switching parties, the milk fund scandal in 1974 (when he was charged and acquitted of taking $10,000 in bribes over fixing the price of milk), and that he reminded them of Lyndon Johnson.[11] Still, Sears never used Finkelstein to his full talents, and this was chalked up as one more example of Sears's paranoid style of management.

President Carter continued to benefit from the natural tendency of the American people to rally behind their leader at times of international crisis. His political team was buoyed when polling showed that the Iranian hostage crisis was pushing his numbers up. Those who thought this support would continue until the crisis was resolved had forgotten their history. This did not mean Carter had a bottomless well of patience for strife and "entangling" predicaments.

Carter had come to Washington with a pledge to reform the bureaucracy, but had ended up joining it. A staffer who was marinated in the bureaucratic culture of the White House by the name of Michael Cardozo wrote a long, three-paragraph memo to personal secretary Susan Clough on why it was "not appropriate" to use stationery for the campaign that read "President Jimmy Carter" and instead, the stationery must say only "Jimmy Carter."[12] It reminded some of the Washington bureaucrat who once wrote a memo discussing input, "throughput," and output.

The anti-American violence so evident in Iran also swept across the rest of the Muslim world, as the U.S. embassy in Pakistan was sacked. Renewed threats were made against the American hostages in Iran, and when the Ayatollah hinted at show trials for the Americans, Carter hinted at military action, but he soon ruled it out, taking the bat out of Uncle Sam's hands.

Four days after the hostages were taken, ABC launched a special to compete with Johnny Carson, king of late-night television on NBC. Originally entitled *The Iranian Crisis, America Held Hostage*, it was hosted by the Reagans' old friend Frank Reynolds.[13] ABC executives were surprised to see that ratings held firm, and by March of 1980, the show had a new name and a new host: *Nightline*, with Ted Koppel. *Nightline* would follow the drama and help keep the public's attention zeroed in on the events halfway around the world. The tragedy laid the foundation for the long-running program's success, and watching *Nightline* became mandatory for politicos.

Carter had the spotlight for the time being, and used it to maximum advantage. Also, many voters disapproved of Ted Kennedy's intemperate comments about the Shah of Iran. Kennedy was continuing to stumble. The Bay State senator's margin over Reagan had also diminished, although he still led the Gipper 49–44. Meanwhile, Carter was smashing Reagan 60–36.[14]

By the end of December, Carter had opened up a twenty-point lead over Kennedy, an astounding 53 percent swing in just four months. Daniel Yankelovich had conducted the poll for *Time* magazine.[15] Carter led in all categories and all regions. Seventy-four percent disapproved of Kennedy's comments about the Shah. Just days before the hostage crisis erupted, the conventional wisdom around Washington was that Kennedy was a "walking dead man." John Sears believed this, but David Gergen, defending Teddy, wrote in the *Wall Street Journal*, "It strains credulity to believe that Mr. Kennedy's candidacy will collapse so easily."[16] But Gergen also wondered how well Kennedy would hold up to the pressure of a national campaign, as Kennedy himself had once said, "I'd love to campaign against my own record."[17]

Kennedy's rationale against Carter was never clear, except that he wasn't Carter and he was a Kennedy. One underlying principle that did resonate with nervous Democrats was that Teddy stood a better chance against the Republicans in the fall than did the president. That argument went out with the bathwater, however, once Kennedy stumbled with Roger Mudd, and his comments on the Shah only confirmed its demise. Despite the

Kennedy family's only partly deserved reputation for tightly run and well-oiled national campaigns, this campaign by the last Kennedy was making the cruise of the *Titanic* look like a successful voyage.

Planes were late or never showed up. Reporters' bags kept getting lost, bookings on the media planes were mishandled, and scribes were often stranded. The schedule was haphazard, and no one seemed to know from hour to hour where Kennedy was headed next. Hotel arrangements were a disaster, and the campaign went through three chairmen in a matter of weeks. Strife and turmoil dominated the headquarters and the senior staff on the plane with Kennedy. Kennedy was frustrated. He couldn't get beyond his own mishaps and the hostage crisis to get at Carter.

At a rare appearance at a party fund-raiser in Washington in early December, Carter spoke and played it straight, thus reminding the audience of the hostage crisis weighing on him. Kennedy spoke and got off some funny lines, as when he joked that, in the Des Moines debate, he thought the candidates should stand; Carter wanted everybody to sit, and Jerry Brown wanted everybody in the "lotus position." He also said that he and the president were close to an agreement on a single six-year term for the office, and that they were only "two years apart."[18]

Both handled the situation well, but it must have been difficult for Kennedy, with his sagging poll numbers. He was now losing to Carter not only among Democrats, but also by twenty-five points among independents, according to Gallup. Carter also had zoomed to a 61 percent approval rating, his highest in almost three years.[19]

Beneath the public courtly attitude of candidates toward each other, the Kennedy and the Carter camps roiled at each other. When Chicago mayor Jane Byrne endorsed Kennedy, Carter's new secretary of transportation, Neil Goldschmidt, suggested that transportation funds headed for the Windy City might be withheld. It went downhill from there. Byrne said that when she'd gone to see Carter at Camp David the previous July, he looked "unstable." She also accused Jody Powell of bragging about rigging the straw poll results in Florida. She said jobs had been offered in the government in exchange for endorsements of Carter. She said that Carter had threatened to instigate a federal investigation of her.[20]

Nancy Reagan went to Concord, New Hampshire, to file "Ronnie's" papers making his candidacy official, and still, the "age issue" would not go away.[21] Comments from the crowds and queries by local reporters kept at Reagan, asking if he was "too old." Reagan dutifully answered the question, each and every time, patiently, but it was becoming a burr under his saddle.

The Iranian hostage situation dragged on. Reagan had been the only one on the national stage to take a principled position to give sanctuary to the Shah, but then he, too, fell silent, joining the rest of his fellow Republicans. Columnist William Safire spoke for conservatives when he wrote, "The best way to free the hostages . . . would be to stop glorifying our impotence and to start putting real pressure on Iran" along with those countries still engaged in commerce with the rogue state.[22] This was exactly in line with the rising GOP and exactly contrary to pessimistic Carterism.

An effort by Reagan foes had been undertaken in California to alter the winner-take-all primary system by making it proportional. In 1976, Reagan had swept the 167 delegates when he beat President Ford there. He was expected to do the same in 1980, provided he made it to the June primary. Behind the effort to award delegates more proportionally was again none other than John Connally, whose supporters were attempting to collect 380,000 signatures on petitions to try to change the rules. In 1980, 168 delegates would be at stake.[23] Reagan didn't call California the "Big Casino" for nothing. Helping to raise money for the thinly veiled anti-Reagan move in the Golden State was Gerald Ford.

Struggling at the back of the pack in the GOP race was Congressman John Anderson, who wasn't getting the support of what was left of "moderate" Republicans. One unremitting Reagan basher, the tiresome congressman Pete McCloskey of California, was supporting Bush. He was picking up right where he'd left off in 1976, saying odious things about the Gipper. He said he was supporting Bush because he wanted to move the GOP away from conservatism. Other GOP moderates, such as Pennsylvania governor Richard Thornburgh and Delaware's chief executive Pierre S. du Pont, were staying on the sidelines out of fear of grassroots conservatives, although both had qualms about Reagan. Evans and Novak said they "viewed a Reagan nomination as political chicken pox."[24] Thornburgh had tried to induce Ford to get into the race, but failing that, he thought the better part of virtue was to become a political virgin—again.

On the Democratic side, Kennedy had not righted his campaign, but he was not going to lose the race for lack of trying. He'd been up and down, and he might just go up again, if he got his act together or if Carter stumbled. He slogged through New Hampshire, sometimes to rave reviews but sometimes leaving his supporters and potential supporters scratching their heads with his rambling, incoherent speeches. Then, just at a time when he should have been working the hardest in Iowa and New Hampshire to stabilize his efforts, he inexplicably took two weeks off in December and into January. Carter, meanwhile, was on the network news each day, looking "presidential." He had not looked so "presidential" back in 1975, when he first started traipsing across New Hampshire, an obscure former governor of Georgia who would make his case to anyone who would listen.

One of the few New Hampshire media outlets that did listen to Carter was WMUR, the ABC television affiliate based in Manchester, which was located in an old house on Elm Street with the walls on the first floor torn out to create a primitive set. In the center was the desk for the news anchor, off to one side was the weather map, and to the other side was the set for the local children's show, hosted by an aging clown. In 1978, WMUR was still using soundless black-and-white film, when most of Western civilization was using videotape, color, and sound. Progress was viewed by the curmudgeons of New Hampshire as an ailment to be avoided. The small two-man news staff made up for lack of resources with hustle and imagination. It was that hustle that had brought Carter to the viewers' attention in 1975, as the tiny news team documented his interactions with voters and even brought the candidate into the rickety old house for interviews. After Carter won New Hampshire in 1976 and went on to become president, WMUR suddenly became a "must visit" among presidential aspirants on both sides of the aisle in 1980.

The filing deadline in New Hampshire came on December 27, and Bob Dole made it with just twenty minutes to spare. To qualify for the ballot, candidates had to submit petitions signed by one thousand voters and a check for five hundred dollars. Dole's daughter, Robin, handled the task for her father. All in all, seven Republicans would compete in the primary and five Democrats.[25] A survey of the editors of nearly one hundred major newspapers revealed that the vast majority now believed that not only would Carter be renominated over Kennedy, but that he was the stronger candidate. Alarmingly for Reagan, these editors also believed the GOP would now choose Bush over the Gipper.[26]

George Bush qualified for matching funds and received an initial check of $859,091.[27] Reagan applied for a much-needed infusion of cash. Others who qualified were Carter, Howard Baker, and the clinically paranoid cultist Lyndon LaRouche,[28] a perennial presidential candidate who argued that the Queen of England was behind the drug trade in America and that the famous British intelligence traitor Kim Philby had in fact been a triple agent.

Back in the real world, the growing consensus was that "Reagan is a political hemophiliac. Once cut . . . his essence will bleed away because it will be interpreted as a judgment on his age. We would discover the emperor has no clothes."[29] Thus sayeth Jack Germond and Jules Witcover in the *Washington Star*. A highly placed, but unnamed, GOP official echoed this sentiment, saying, "I keep having this feeling that Reagan is vulnerable."[30]

So, too, did almost everybody else in American politics.

Reagan's Dunkirk

"We proved it. The emperor has no clothing."

George Bush had broken out of the pack far earlier than his campaign expected, gaining much media attention in November 1979, two months before the Iowa caucuses and three months before the New Hampshire primary. Bush had won a series of important straw polls, and plaudits from reporters and columnists, all of which culminated in his winning the Iowa caucuses.

But Ronald Reagan continued to evolve. "The Ronald Reagan who ran for President in past years was angry—angry at student militants, anti-war demonstrators, draft dodgers, urban rioters, welfare cheats, school-bussing supporters, Soviet dictators and Washington bureaucrats."[1] In his landmark speech for Barry Goldwater in 1964, Reagan said that conservatives' opponents charged that they "were always against things, were never for anything." That was true enough sixteen years earlier, but Reagan and his conservatives were now for many things, including privacy for the individual and the expansion of freedom.

This was a sign of Reagan's growing political maturity. He had always been about "maximum freedom consistent with law and order," as he'd said in that speech for Barry Goldwater, but he came to realize that, as in the case of tax cuts, fiscal policies supporting "maximum freedom" would liberate Americans to be less dependent on government and freer to spend their money as they pleased, to save or invest. By supporting limited government, he was advocating the freeing of the American people from the yoke of regulation and control exerted over them by their government and government bureaucrats. Reagan's new focus on the future was particularly appealing to young Americans. And he still had his sense of humor.

Unfortunately, outside of John Sears, the people behind the scenes in Reagan's operation were not laughing. They were worrying. Reagan had in fact been to all fifty states over the previous two years, but the quantity of events had been vastly limited in the most recent year. He'd spent almost no time in Iowa, where he had a ready base of willing supporters, including the increasingly active religious right. These Christian activists were becoming more and more engaged in politics for a variety of reasons, but most especially because of the actions of the Carter Administration over the previous three years. Carter had made much of bringing gays into the White House and opposing Anita Bryant's "anti-gay ordinance" in Dade County, Florida. He had "done nothing to advance the pro-life cause, fought the tuition tax credit bill and had his Internal Revenue Service commissioner harassing Christian schools."[2]

Reagan was more to their liking. He had come around on abortion, and now favored an amendment to overturn *Roe v. Wade* and send it back to the states. He opposed the Equal Rights Amendment to the Constitution and spoke often of "family values." Reagan favored legislation banning "sex discrimination" but was concerned that the ERA would allow deadbeat dads to avoid paying child support, and feared the amendment could open up "a Pandora's Box." He also said he was not opposed to a female running mate.[3] These family groups were also poised to take over a White House Conference on Families, by routing liberals in the states to elect conservative delegates for the confab. Liberals had scored enormous and permanent gains in the 1960s and '70s; their big mistake was in not anticipating a massive backlash from Middle America. Reagan would directly benefit from that backlash.

However, while many assumed he was opposed to gay rights, Reagan in fact did not want to attack these Americans. His years among the creative circles of Hollywood had made him more tolerant, and his two terms as governor had taken him far and wide in the multicultural Golden State. Indeed, many in the gay community liked Reagan's strong message on economics, anticommunism, law and order, and privacy. The Kremlin had a long history of harsh treatment of homosexuals in the Soviet Union, and many gays were strong anti-Communists as a result.

―――――

Just days before the caucuses, Reagan was on the verge of defeat in Iowa. Harsh prose already filled the columns of some writers, and the balloting hadn't even taken place yet. "He might lose in the caucuses, or gain only a narrow victory. In each case the unraveling of the whole Reagan effort would proceed apace."[4]

John Sears had been guided by a dated poll conducted by Arthur Finkelstein that showed Reagan winning Iowa going away,[5] even as a new poll by ABC of Iowa's 99 GOP county chairmen showed Reagan's support dwindling down to 24 percent, as compared to 21 percent for the hard-charging Bush.[6]

John Connally later set off on a forty-hour bus tour of the state to showcase his vigor and draw a contrast with Reagan's age. Connally challenged Reagan to join him, but Reagan demurred testily, telling reporters, "Why should I help him publicize his circus trick?"[7] Connally held up well on the tour, but the reporters less so after not bathing or brushing their teeth for two days. Gallows humor took over on the bus. One reporter wanted to know if "the governor believed in the forced busing of journalists to achieve election."[8]

―――――

In the meantime, the Iranian hostage crisis had lingered for several months, and the bipartisanship behind Carter was beginning to work loose. RNC chairman Bill Brock got things started, when he went on *Good Morning America* and charged that Carter's "policy of patience" was in fact a "policy of weakness."[9] Most of the other GOP candidates followed Brock's lead and pounced on Carter, except Reagan. Indeed, his staff went out of their way to tell reporters that they "feel he can wait until the others have laid out their complaints before offering his views."[10]

Polls showed that voters around the country, and in Iowa, still supported Carter's dealing with the Iranian hostage crisis. They also, by a wide margin, supported Carter's grain embargo of the Soviets over Afghanistan, even though it hurt midwestern farmers more than any other region. The grain embargo was announced less than three weeks before the Iowa caucuses, in which Carter "cancelled contracts for the sale of 17 million metric tons (mmt) of US corn, wheat and soybeans to the Soviet Union," Paige Bryan, associate director of foreign trade policy at the U.S. Chamber of Commerce, wrote to the Heritage Foundation.[11] Reagan opposed the grain embargo, as did Ted Kennedy and most of the other Republicans. Reagan instead suggested a "quarantine" of trade with Moscow, and when queried, he said, "Why not? It sure beats a war."[12]

In Iowa, speaking before an audience in Davenport, Reagan outlined his opposition to the grain embargo. "I just don't believe the farmers should be made to pay a special price for our diplomacy."[13] But he did call for a complete cutoff in trade with the Soviets "until they decide to behave as a civilized nation."[14] Reagan was breaking precedent and making an important point: Why should grain farmers bear the brunt of an economic embargo? It was a good question.

═══════

Reagan finally put in a full day of campaigning in Iowa on the Saturday before the Monday caucuses. But rather than talking like a candidate, he sounded more like a strategist. He talked about pacing himself for all thirty-five of the upcoming primaries, instead of hustling in Iowa like the other candidates. Though he never mentioned Bush by name, it was clear he was looking over his shoulder at the oncoming Houstonian. Reagan's lackluster day and comments only served to fuel the lingering issues.

The bloom was off the rose of détente now. The dance of diplomats beginning with Richard Nixon through Gerald Ford and up to Carter was over, consigned to the ashbin of history. The agreements on medicine, science, trade, and especially armaments were a thing of the past. The original agreements included selling American wheat to the Soviets, and the world price shot up, which delighted America's farmers. But the price of bread also went up, angering America's housewives. The derided striped-pants set of the State Department were in high dudgeon.

———————

All told, Reagan had made eight stops in Iowa, and George H. W. Bush might have made around eight hundred. When the results were tallied on the evening of January 21, Bush had won a narrow victory in the Iowa caucuses over the Gipper 32–30 percent.[15] Bush knew what he'd done, telling the AP that Reagan "was hurt badly" by the results.[16] When the final results came in on Caucus Night, Bush was exuberant. Earlier, Bush shouted to the crowd, "I'm on cloud nine!"[17] They also chanted repeatedly, "George! George! George!"[18] Bush tweaked the media for simply chalking his win up to "good organization," and to make his point, the band played "I've Been Working on the Railroad."[19]

The media were apocalyptic. Tom Pettit of NBC went on national television that evening to pronounce, "We have just witnessed the political funeral of Ronald Reagan."[20] Tom Brokaw, reporting for NBC's *Today* from Des Moines the following morning, said Bush emerged with "a significant upset victory over Ronald Reagan who had been considered the front-runner coming into these Iowa caucuses."[21] Reagan had sauntered across the Hawkeye State while Bush had sprinted across it, and the results of the caucuses proved it. Reagan, local boy who made good, was defeated by the Brahmin Houston oilman, George Bush, in a stunning upset.

The *Los Angeles Times* issued a great indictment of Team Reagan: "His advisers appear to believe the Republican presidential nomination will be won on personality and organization, rather than on issues."[22]

The most damning was that "they apparently have convinced him to avoid letting his basic conservatism trigger him into saying things they feel he should not say."[23]

For many in the Republican Party and some in the national media, they hoped that Bush had dealt a fatal blow to the sixty-eight-year-old Reagan in his third and final attempt to win the Republican presidential nomination. They just plain didn't like "that actor" and wished he would go away, so a serious leader, such as Bush or Howard Baker, could lead the GOP. Bush's pollster, Robert Teeter, produced a confidential report on the caucusgoers the day of the caucus that showed overwhelmingly that they had heard from Bush's campaign far more often than from Reagan's and that, by a wide margin, Bush's experience, especially in foreign policy, was uppermost in their minds.[24] Ideology was not a factor in their support for Bush, as nothing was ever cited by the two hundred Iowans questioned by Teeter.[25] Rich Bond, who ran Bush's successful Iowa operation, told Bob Shogan of the *Los Angeles Times*, "There are no issues here. This is a test of a candidate's ability to run."[26]

The consultants and many in the media had already turned their eyes to the East, where the first-in-the-nation primary would soon be held in New Hampshire. The dark specter of "expectations" had cut down many politicians in the Granite State in the past, and some Reagan men were acutely aware of this. Sears bizarrely told the Associated Press, "George Bush has done something right. He's spent a lot of time in Iowa," an attempt, perhaps, at expectations jujitsu that simply fell flat.[27] But Bush's Bond could not contain his glee after the Iowa victory: "We proved it, the emperor has no clothing."[28]

Several months before, David Broder of the *Washington Post* incisively reviewed the merits of the Reagan challenge. "Reagan is a card-carrying conservative, but he has never been a hater or a screamer, and at his age, is not about to become one." Broder's piece was by and large complimentary of Reagan, but it ended with a foreboding note: "He ought at least to be ready to meet the test of debate."[29] Broder was right, but because Reagan had ducked all the debates and "cattle shows" of the past year, he was now reeling.

He'd been pounded mercilessly over the past two years over his age, his stamina, his hair, and his conservatism; and then, toward the end of 1979, his enemies ran a new charge up the flagpole: Reagan "does not know enough to be president."[30]

A massive poll of Republicans in America had been conducted in late 1979. The survey showed that half of Republicans called themselves "moderate" and the other half called themselves "conservative." GOP pollsters would have disputed this breakdown, arguing that their party was more right of center than the *Post* poll, but this was neither here nor there. Among moderates, Reagan had a surprising 50 percent support, which was music to John Sears's ears. But among conservatives, Reagan was the choice of only 48 percent, which should have set off clanging alarm bells inside Reagan for President. Reagan had spent the better part of thirty years wooing these conservatives and should have had a far higher percentage. Just as important, it was the conservatives, not the moderates, who voted in far greater numbers in GOP primaries.[31]

The Reagan campaign seemed to lose its footing in the immediate aftermath of the Iowa loss. Jim Wooten of ABC reported that "Iowa . . . confirmed just that Ronald Reagan is mortal."[32]

Reagan and John Sears both oddly said that devoting extra attention to a single state such as Iowa was not a good idea, instead needing to focus on the entire country. In fact, Reagan had gone to other states, but a good portion of his time had been devoted to staying at home in California, per Sears's directive.[33] Iowa was an unexpected loss for Reagan, but a two-point defeat at the hands of a hardworking underdog who visited the state one hundred more times than the Gipper was not exactly a Waterloo.

Campaigns were notoriously marathons, not sprints. You had to conserve your energy through the entire race, and not just throw it all away in the beginning to get ahead. What mattered now wasn't that Reagan had lost the first sprint; what mattered now was how he would adjust for the next event. The question now was how quickly he could regain his footing—or could he? Reagan's Iowa loss had more in common with the British loss at Dunkirk in 1940, when more than 330,000 Allied troops were evacuated from the French port and lived to fight another day.

Unlike the "Dunkirk spirit" that emerged with the moral victory that came when a makeshift fleet of vessels helped these troops cross the English Channel to avoid capture or death, the Reagan campaign's fighting spirit was not visible in the first few days after Iowa.

It would be the Gipper himself who would hopefully find that fighting spirit.

Sunshine Reaganites

"I want to keep beating Reagan as much as I can."

R onald Reagan tried to downplay the results of Iowa, although he conceded he'd been "out-organized."[1] He also pointed out that the state party was behind George Bush and not him, which was true, but he was making excuses that his conservative supporters didn't want to hear. Another member of Team Reagan did not downplay Iowa. Reagan's pollster Dick Wirthlin bluntly told reporters that "the impact of the Iowa election shouldn't be underestimated."[2] But he added defensively, "[T]here was 'definitely hope for' Reagan" to mount a comeback.[3] Back in the summer of 1979, Reagan had expressed his concerns about being the front-runner to *Washington Post* reporter Lou Cannon: "There is something to be said for coming from behind." Reagan got his wish. He was now running behind.[4]

The Bush direct mail, on the other hand, was going out in a blizzard of paper. The day after Iowa, Tim Roper, in charge of the Bush mail, "dropped" an astonishing 1.7 million pieces of fund-raising mail that proclaimed Bush to be the "frontrunner" in the GOP contest. Bush's campaign payroll had swollen to 270 staffers, but they were overtaxed with all the new demands.[5] Reagan was apparently just going through the motions, even after the huge loss.

By now, conservative columnist George Will's contempt for Jimmy Carter was complete; he wrote that the Georgian was "the most dangerous President since James Buchanan."[6] Buchanan's indecisiveness on states' rights had encouraged Southern states to begin secession, and in so doing contributed to the conditions that brought about the Civil War. Buchanan was one of the most experienced men ever to be elected president—and one of the institution's biggest and most ignoble failures.

Students at Berkeley and Stanford violently protested Carter's proposal to reinstitute the draft and burned faux draft cards. And Reagan was a day late and a dollar short in criticizing Carter's call for reinstating a military draft. All the other Republicans except Bush had already inveighed angrily against the president the day before, so Reagan ended up in a "me, too" position. Again, two issues he'd owned for years, national defense and foreign policy, had been ceded to the others. Haynes Johnson wrote in the *Washington Post*, "If the country is moving to the right, if flag waving . . . is back in vogue, if Communism once again is everyone's enemy, who bears better red-blooded credentials to lead the patriotic revival than Reagan?"[7] Who indeed? Still, his issues were being appropriated by the other candidates.

One week after his Iowa loss, he put in a rare appearance in New Hampshire, and the 750 people who attended the event were warmed by Reagan's fiery rhetoric aimed at Carter's foreign policy. One hundred more Reaganites had to be turned away from the hall because it was jam-packed and because of fire regulations, but they were able to watch their hero on a closed-circuit television set up in the school's cafeteria.[8]

Reagan's stumbling start continued unabated. Campaigning in Florida ten days after the Iowa loss, he bungled yet another foreign policy issue—this time on whether the United States should allow Pakistan to develop a nuclear bomb. At first, he told reporters yes; then, via Jim Lake, he denied he'd said that, but when confronted with tapes of his clearly saying so, he visited with the media and attempted to quell the situation.[9] He tried to tell the howling media mob that he was simply musing that the Soviets needed to be reminded that the United States had a defense treaty with Pakistan, and that the bomb or maybe an American military base in Pakistan would be considered. It didn't work, and Reagan ended up embarrassed again.

A new controversy erupted because of Reagan's bungled attempt to stanch the previous one. "Did Reagan want to put a U.S. base in Pakistan?" He then said that U.S. planes could "fly in as a good will gesture and as a neighborly thing."[10] He then got crosswise on whether Fidel Castro in Cuba had Soviet-trained troops in South Yemen and whether they were preparing to attack Oman.

In New York, liberal Republican senator Jacob Javits rudely left the dais at a Republican dinner before Reagan spoke. He was that repulsed by Reagan's conservatism. At another event in New York, a supporter gave Reagan an antique coin for luck. The coin read, "Good Luck. Four Years of Prosperity. Herbert Hoover."[11] Everything Reagan had achieved in his career came because of smart and hard work. Everything he had accomplished had come due to the sweat of his brow. Reagan was a winner because he fought to win. But his campaign was not letting him fight for the one thing he wanted the most in his professional life: the nomination of his party.

Reagan had been left in the lurch after Iowa, with no plan, no theme, no way back.

———————

An early poll of Republicans in New Hampshire came out that had Reagan and Bush statistically tied, a worse run for the Gipper than in the previous election. Even worse for Reagan, he and Bush were tied nationally in an ABC News/Harris poll among Republicans at 27 percent apiece. Reagan's lead of 26 percent nationally over Bush among Republicans just weeks earlier had melted away, and now they were locked in a dead heat.[12]

The other Republicans were falling by the wayside, and all were in the low single digits in New Hampshire. Their paths to the White House had all run into roadblocks, and the media were focused, in addition to Kennedy and Carter, on Bush and Reagan.

A *Boston Globe* poll confirmed all the others showing Reagan fading in New Hampshire in relation to Bush. "In addition to nagging questions about his age and energy, Reagan has never convinced large enough numbers of Americans that he is smart enough and deep enough to be president," wrote one chronic Reagan critic, Richard Reeves.[13]

Bush haughtily told a reporter, "Reagan hasn't yet to realize I'm in the race." Later in the story, Bush thumped his chest, Tarzan-style, and said, "I want to keep beating Reagan as much as I can."[14] Former president Gerald Ford piled on, telling the media that Reagan could not win because he was "too conservative."[15] A straw poll, sponsored by Cox Newspapers, of leading GOP officeholders meeting at their annual Tidewater Conference in Maryland "showed that 58 percent of the 65 persons who responded . . . felt George Bush would be the Republican nominee. Only 12 percent picked Ronald Reagan."[16] In Florida, where Reagan once had a commanding lead, Bush had surged ahead according to the Florida Newspaper Poll of Sunshine State Republicans.[17] The Bush brigades were successfully invading "Reagan Country." GOP leaders in the South were now openly speculating that Reagan could lose there to Bush, an unthinkable proposition just a few months earlier.

Bush was featured on the cover of *Newsweek*, a major coup. He was photographed jogging with a group of young supporters (two of whom were his sons Marvin and Neil) under the banner headline "Bush Breaks Out of the Pack."[18] How active, how fit, how successful! *60 Minutes* also did a favorable profile. In his New Hampshire appearances, a local band would play, "Hey, Look Me Over" and upon Bush's landing in the biting cold, a local mayor was there to greet him. "Hey, my gosh, look at this cold weather. Mayor, aren't you nice to be out here."[19] Bush would then shake any and every hand he could find.

He was also running stronger nationally against Carter than Reagan by a wide margin. Bush was losing to the all-powerful Carter in the trial heat, but he was running the best of all the GOP candidates against the president. In the same poll, Bush also moved ahead of Reagan among Republicans and independents.[20] He had also, amazingly, pulled out to a lead in Reagan's native Illinois.[21] The state's primary was only weeks away. A later poll had Bush extending his lead there. He was receiving the bulk of his support from the upper-income Republicans in the Chicago suburbs, while Reagan's support was coming from the lower-income Republicans in rural southern Illinois, some of the best farmland in the country.[22]

Reagan's fund-raising had dried up while Bush raised almost a million dollars in January alone, three hundred thousand more than his campaign had projected. Haynes Johnson, in a long piece for the *Washington Post*, observed of the Gipper, "[T]he old fire seems lacking."[23] Reports were coming in that Reagan's support in the South was crumbling, and sunshine Reaganites and summer conservatives were taking a second look at Bush and Connally.

The calendar that year, however, would be Ronald Reagan's best friend. Bush would be his worst enemy. Reagan didn't like Bush, even though he barely knew him. They were different in every fashion, including culturally, ideologically, and temperamentally.

Bush had turned the tables on him in Iowa on January 21. In later presidential campaign elections, New Hampshire would follow Iowa more quickly, but not this election cycle. Reagan's good fortune was that the New Hampshire primary in 1980 didn't come until five weeks later, on February 26. By the first week of February, Reagan's loss in Iowa still looked more like Waterloo than Dunkirk, and there were still few signs he could regain his former front-runner status.

Ronald Reagan had just three more weeks to make his luck change.

The Politics of Politics

*"Other candidates come and go but when George Bush
comes to New Hampshire, it's like he never left."*

Four weeks before the New Hampshire primary, liberal writer Richard Reeves, in the venerable and venerated *Washington Star*, wrote, "Now that his 'above politics' pose is collapsing, he has to come out and mix it up with the likes of Baker and Bush."[1] Ronald Reagan wasn't willing to throw John Sears to the wolves yet, but he was telling his conservative supporters that there would be no place in a Reagan White House for him, if that happenstance ever came to pass.[2]

Reagan was forecasting a victory in New Hampshire, but almost nobody was listening. He later downgraded his chances to "cautiously optimistic."[3] Bush was also predicting victory in the Granite State, and more people were paying attention to him.[4] Bush also said, somewhat presumptuously, that if he won New Hampshire, it "would give him a lock on the nomination."[5]

Not all eyes were on the Granite State. The Soviets were dominating, but few in either party or the mainstream media or the universities seemed particularly bothered by this. Some of these sophisticates, deep down, had a soft spot for the Soviet collectivist system and were naïve about the horrors of the Gulag and the KGB. If only the Kremlin could be just a little bit . . . nicer. George McGovern had once lamented that John Foster Dulles, Eisenhower's secretary of state and a confirmed anti-Communist, had attempted "to bring Communism down by encouraging dissent and revolt in Eastern Europe."[6] As if that were a bad thing.

As far as Americans were concerned, Reagan was the first political leader since John Kennedy with the guts to stand up to the Commies. Reagan also taunted the Kremlin, saying, "[T]he Soviet Union is not ready for that all-out confrontation."[7] Reagan knew it was essential to condemn the Soviet Union, and understood that it was also essential for his campaign.

―――――

"In an effort to deflate the age issue Reagan's campaign strategists are confronting it head on," ABC News reported.[8] Gerald Carmen, the former New Hampshire GOP state chairman who owned a tire dealership in Manchester, had come on to serve as Reagan's New Hampshire campaign manager, and he applied his get-the-job-done-whatever-it-takes small-business skills to the effort. He was taking control of Reagan's future by having him campaign nonstop in the Granite State.

A young conservative activist, Jim Roberts—who had just completed a book, *The Conservative Decade*, for which Reagan had agreed to pen the foreword—was aiding Carmen. He flew up from Washington and simply showed up, volunteering full-time, writing press releases, and keeping his head low as Carmen was "barking out orders" and throwing one hand grenade after another at anyone who stood in his way.[9] Carmen confided in Roberts that he thought Sears believed Reagan did not have the "intellectual ability" to hold his own in a debate. To save money, Roberts stayed in a fleabag motel outside Manchester and hitchhiked back and forth to the Reagan offices each day.[10]

Gerald Carmen, the self-made businessman and hometown boy made good, took pleasure in tweaking and attacking the preppy George Bush, scion of the storied New England family who had used his financial head start to bankroll his successful business ventures in his adopted Texas. Though Carmen stood only five foot and change, there was no getting around the fact that he intimidated all.

At the end of January, the feisty Carmen told the national press that Reagan was challenging Bush to a two-man debate. It was exactly what the New England–born Texan said he had wanted for so long. The next day the unrelenting Carmen increased the pressure on Bush, telling reporters that Reagan would accept all debate invitations, including the one from the *Nashua Telegraph*, which had proposed a *High Noon*–type showdown between Governor Reagan and Ambassador Bush, exactly the two-man debate Reagan had just said he wanted. The other candidates were not invited.[11] Predictably, they went ballistic. Bob Dole filed a complaint with the FEC, claiming that the *Nashua Telegraph* could not sponsor a debate that excluded several major candidates. It was a violation of FEC law. Bush and Reagan, for their part, both ultimately accepted the *Nashua Telegraph* debate offer, until the FEC swiftly ruled it illegal.[12]

Bush, the front-runner, was getting bigger crowds than Reagan across the state. He was also running ads that claimed "Other candidates come and go but when George Bush comes to New Hampshire it's like he never left." To the combative Carmen, this sounded more like a threat than a promise.[13] Carmen wasn't the only combative conservative fighting for Reagan. Senator Orrin Hatch of Utah despised Jimmy Carter and made no bones about. He lambasted the president as a "liar" and an "incompetent, whose presidency is 'dead in the water.'"[14] Hatch was normally an amiable man, a faithful Mormon whose religion restricted his liquid consumption to water and milk. However, Hatch was lactose intolerant, which could be a problem for those of the Mormon faith. But Carter's bad track record brought out the inner fire and brimstone in Hatch.

Unlike Iowa, where he barely showed up, Reagan soon determined that if he was going to win this New Hampshire fight for his political life, he was going to have to do it on the ground in the Granite State. Reagan finally told campaign manager John Sears that his cherished "above the fray" strategy was over and done with. *Finito.* Reagan told Sears he was going to "campaign the way I like to campaign," which meant as much time face-to-face with the voters as possible.[15] Reading between the lines, many thought it wasn't just Sears's strategy that was done. Sears himself, they suspected, might soon be done as well. The former Nixon aide had, after all, taken the most surefire path to creating a small "political fortune" for Reagan. He had started with a large fortune, as the old axiom went, and blown most of it.

On February 4, two days shy of his sixty-ninth birthday, Reagan flew from Los Angeles to Boston. On the flight, he explained to *Washington Post* reporter Lou Cannon that his hearing loss was not related to the age issue. "I was doing a shoot-'em-up movie back when I was the Errol Flynn of the Bs. An actor put a .38 up to my ear and fired it."[16] Cannon may not have been entirely persuaded, but he nonetheless reported Reagan's account. For the next three weeks, up to Election Day on the twenty-sixth, "Reagan campaigned nearly uninterrupted for twenty-one days in New Hampshire, a display of stamina that quieted concerns about his age."[17]

Before the polls opened for the critical primary vote, Reagan would be able to demonstrate to voters in the Granite State that he could match, and even surpass Bush, fourteen years his junior, when it came to physical vigor.

But that all lay ahead.

Reagan's sixty-ninth birthday was February 6, 1980, two weeks after the ruinous Iowa loss, but still three weeks before New Hampshire. An early party was planned in Los Angeles with friends and celebrities (Dean Martin and Frank Sinatra among them), a fund-raiser where a much-needed $90,000 was raised. The idea of a very public celebration came from Lorelei Kinder, a devoted Reagan aide. She wisely suggested they hide in plain sight and not make an issue of Reagan's age by making an issue of it. The strategy worked.[18] Reagan's horoscope for his sixty-ninth birthday was both disquieting and instructive. "Be more concerned with money matters," and "You find some delays or obstacles attend your efforts to forge ahead."[19] Another was perceptive: "Though you have advanced ideas, you like to work within traditional frameworks."[20]

At two high school stops in New Hampshire, students asked Reagan about his age. Either they were put up to it or the issue had sunk in that deep. Either way, it was bothersome. The entire day, Senator Humphrey accompanied him; the tour included a factory and concluded at China Dragon Restaurant in Hooksett.[21] The candidate showed off his wit and nerve. At one stop, he was told that a volunteer in Bush's office in Exeter wanted one of Reagan's campaign buttons. Reagan walked across the street and into his opponent's headquarters, button in hand, smilingly gave it to the woman, and told her that after she'd worn it for a time, maybe she'd come around and see things in a different light.[22]

All told, Reagan made nine campaign stops in New Hampshire on his sixty-ninth birthday.[23]

George Bush, meanwhile, was enjoying his front-runner status. The new scrutiny also brought with it criticism, including of his charisma. He told a crowd in neighboring Maine, "I always thought I was a charismatic devil."[24] Bush "clings to some of his Andover preppy phrases: 'Fantastic,' 'super,' 'gee whiz,'" and was considered charisma-free by the press and many politicos.[25] Clearly, Bush was not ready for prime time. He jokingly told reporters that he was trying to wear fewer button-down collared shirts "so you fellows won't think I'm an elitist."[26] Truth be told, the "preppy" charge got under his skin.

In his autobiography *Looking Forward*, he complained about the whole issue of shirts, striped wristwatch bands, and the fact that he played tennis. He also complained that though JFK, FDR, and other political figures had gone to Ivy League schools, they were not raked over the coals for being "elitists."[27] However, he also complained that he did not wear Brooks Brothers suits—he wore Arthur Adler![28]

The differences between Reagan and Bush were as much economic and cultural as they were ideological. Bush had grown up in a home with a live-in maid and a cook, and enjoyed summer vacations at the family home in Kennebunkport, Maine. Reagan had grown up dirt poor. Bush, some Republican insiders believed, did not "have the stomach for confrontation or tough decisions" despite being a highly decorated war hero.[29]

Bush would not move on from the "age issue," or the "Big Momentum," for that matter, telling reporters he was able to "serve two full terms." The implication of who and what he was referring to was not lost on anyone: "If Bush's digs at Reagan's age . . . are lost on the audiences, Bush's supporters help [them] out," James Dickenson said.[30]

———————

Bush and Reagan were going all out in New Hampshire. After a grueling fourteen-hour day, Bush finished up at Plymouth State, when he got in a good shot at Carter. Telling the big crowd he would make no promises, he said, "Jimmy Carter ruined the campaign promise business for everybody."[31] Everybody believed the nonspecific Bush.

His campaign also seemed hung up on the minutiae of politics rather than the message. It let everybody know that in 1979 alone, Bush had spent 329 days on the road, attending 853 events and traveling 250,000 miles.[32] His many events often opened to the theme song of the movie *Rocky*.

When he did swerve into policy areas, he made only shallow passes, as when he said he believed the hostage crisis would be resolved shortly, that the Kremlin would engage in a "peace offensive" soon, and that American relations with Saudi Arabia "must [be] shore[d] up," but he opposed a permanent American military presence in the Middle East.[33] There wasn't much meat on those bones.

Peter Teeley, Bush's press secretary, tried to downplay the attacks on Bush as a preppy elitist as simply the machinations of the power-hungry William Loeb of the *Manchester Union-Leader*. Almost every day, it seemed, Loeb had an editorial blasting Bush on the front pages of the most widely read and influential daily newspaper in the Granite State. "If we lose New Hampshire, it will be because of Loeb," Teeley told the *Washington Post*.[34] But the thirty-nine-year-old Teeley, born in England, had made a career as a wordsmith for establishment Republican politicians, first with Michigan's senator Robert Griffin, then as communications director for the Republican National Committee as well as for the über-liberal Republican senator from New York, Jacob K. Javits. It was clear that he knew how to read what his bosses wanted. It was less clear, however, if he knew how to read what Republican voters wanted.

He had just recently come on as press secretary for the Bush campaign, and was still two months away from coining the phrase *voodoo economics* that his boss would make famous in his public criticisms of Reagan's supply-side economics policies.[35] Teeley, a serial liberal, never really came to grips with the power Reagan's message of individual liberty and lower taxes had within the rank and file of the Republican Party. But it wasn't only Teeley in the Bush campaign who misread the GOP. Bush did, too. A softball question was lobbed asking Bush if he stood to Reagan's left on the political spectrum. Astonishingly, Bush said, "I'm perceived that way. If you had to choose between right and left, I suppose I'd be left."[36]

In New Hampshire, a revitalized Reagan was hitting the campaign trail hard, and the demanding pace did not appear to be slowing him down. On the contrary, he seemed to be gaining energy from the crowds. But Reagan still needed more than stamina to win there. Longtime friend and adviser Senator Paul Laxalt of Nevada, Reagan's 1976 national campaign chairman, thought he needed much better staff work as well. So did other close friends of Reagan's.

In mid-February, William Loeb tore into John Sears for his political mismanagement of Reagan's campaign.[37] Making Loeb's point was the poor advance work done for Reagan a couple of days earlier. Reagan's plane had been delayed getting to Berlin, New Hampshire, due to engine problems. Because he was late, the crowd gathered for his headquarters opening had melted away. When he arrived to speak before the local Chamber of Commerce, men sporting "Bush" buttons, including the local congressman, Jim Cleveland, greeted him. In the introduction, it was made clear that Reagan's invitation was an afterthought, as they really wanted President Carter for their speaker. Then, after speaking, the first question Reagan got from the floor was "Would you tell us how old you feel?"[38]

———

Bush had to face his first major controversy. One of Reagan's staffers, Don Devine, took it upon himself to charge George Bush publicly with dirty tricks, accusing him of taking illegal money during his failed Senate campaign in 1970, from a slush fund controlled by Richard Nixon called Townhouse Operation. Devine was head of Reagan's Maryland operations but was in New Hampshire to do interviews and talk to conservatives, championing Reagan's cause.

No one at the Reagan operation refuted Devine's attempt to link Bush and Watergate, and it was clear that they were attempting to slow down Bush's surge in the polls. The man who ran the funds, Jack A. Gleason, came forward the next day and said that, indeed, not only did Bush receive $106,000 (which was reported and which was legal at the time), but Gleason had personally given Bush another $6,000 in cash, which was not reported. This was, if true, illegal. The extra cash was part of a nefarious plot by Nixon henchman H. R. Haldeman.[39] The stench of the Nixon epoch and crime spree had not entirely been vented from the party. The issue was never resolved, as it was Gleason's word against Bush's, and though nobody thought Bush personally capable of doing something dishonest, the controversy did slow his momentum.

———

By February 1980, the American people had swung into a very hawkish and conservative mood. Kennedy was definitely seen as too liberal, and a large minority wanted Carter to be tougher on the Soviets.[40] Carter was dropping hints, almost on a daily basis, that a "breakthrough" was just around the corner for the despairing hostages in Tehran. Nothing developed, but this did not stop the White House from holding a press conference in which Carter once again dangled the possibility of a resolution. All three networks carried the president's press conference live. The executive branch was bringing down poor Kennedy, and he must have wondered what had made him get into this thing in the first place. He was being mauled by the bulldog from Georgia. The *Washington Star* observed, "Politics is a cruel business and if Kennedy is beaten here, no one is going to pay too much attention to the fact that the deck was stacked for President Carter."[41]

———

George Bush won the Puerto Rico primary, just nine days before New Hampshire, giving added confidence to his efforts.[42] Barbara Bush was campaigning aggressively for her husband, as were the Bush children. After one campaign swing, Dave Keene called the prematurely graying but feisty lady. Mrs. Bush occasionally teased Keene, who was fighting a losing battle with his weight, about his expanding girth. Keene told Mrs. Bush, "I have good news and bad news for you . . . the good news is that you're doing a great job out there." Suspiciously, Mrs. Bush asked, "What's the bad news?" Keene said teasingly, "They think you are George's mother." After a pregnant pause, Mrs. Bush told him, "David, I think maybe we should call a truce." Keene agreed.[43] The Bush

staff was fanatically loyal to her. An aide of Keene's, Tish Leonard, remembered fondly her many kindnesses.[44]

Reagan fine-tuned his new conservative message in New Hampshire
as well. He spelled out his support for the Kemp-Roth tax-cutting plan,
for abolishing all inheritance taxes, for eliminating taxes on the interest
earned on savings accounts, and for redistribution of federal taxes back
to the states. He laid into "welfare queens" and bureaucrats in Washington. The national media thought they'd heard it all before. Indeed, the
Community Players of Concord were performing the stage version of
Play It Again, Sam, based on the Woody Allen movie.[45] These reporters
thought it was *Play It Again, Ron*.

Up north in a very cold New Hampshire, Bush was stumping hard,
showing good discipline in keeping expectations low. Reagan was stumping hard as well, but his team knew he needed something more to make
up for lost time since their candidate was still "above the fray." Bush
still appeared to have the edge. "Even veteran Republican politicians
shrugged off any prospect of a major Reagan victory," *Time* reported.[46]

Desperate for a way to get back into the two-man debate with Bush,
the cash-strapped Reagan campaign came up with a creative way to
get the debate back on schedule. The Reagan campaign would pay for
the entire event, about $3,500. The FEC, the Bush campaign, and the
Nashua Telegraph offered no objections, and the debate was back on for
Saturday, February 23. The primary was only three days after. The other
contenders were not happy, and they let Reagan know about it in the way
they knew would hit him the hardest. Dole, Baker, Crane, Anderson—
all had one word to describe the two-man debate: "unfair."[47]

What happened next was the stuff of political legend.

———————————

Even though he could read the tea leaves and probably suspected he
was on his way out as Reagan's campaign manager, the wily John Sears
showed that he still had at least one big trick up his sleeve. On the day
of the debate, he announced his desire that the other candidates be in-

cluded in it, which was too much for Bush. Almost like an insufferably perfect schoolboy, he complained about this suggestion, saying that he "played by the rules."[48]

The *Telegraph*'s editor, Jon Breen, was none too pleased, either, when he was informed of the Reagan campaign's plan. So, after consulting colleagues, he declared the matter finished. It would be their debate and their rules. It would be one on one, he declared: Bush v. Reagan.

After some back-and-forth between the campaigns' staff, nearly twenty-five hundred people came to the Nashua High School gymnasium. You could feel the tension, and the temperature, in the air rise as more than forty-five minutes passed without any peep—until, finally, Bush appeared onstage with Jon Breen. His supporters went wild.[49]

Now it was the Gipper's turn. Arriving onstage, he waved the other candidates to come up. What a showstopper. Now there were six candidates onstage, when Bush wanted just two. He looked visibly uncomfortable.

The most compelling visual optics of the entire 1980 campaign were about to play out in front of the attentive audience and the news media: Reagan and Breen, candidate and host, interrupting each other and not budging.[50] Breen had had enough of the Gipper of California and demanded that his microphone be shut off. The stage was perfectly set for the veteran movie star to deliver the most memorable line of his career thus far. Reagan looked to Breen, furious, his voice rising, and exclaimed, "I am paying for this microphone, Mr. Green!"[51] The misnaming, though noticed by all, did not matter to the audience.

And in that moment, Ronald Reagan regained his front-runner status with thundering applause.

It was over before it began.

———

Though the other four candidates, Bob Dole, Howard Baker, Phil Crane, and John Anderson, all agreed to leave the stage before the debate started, it didn't matter. Reagan had already won the night in a knockout. Bush would answer the questions posed that evening, but it didn't matter. He was, politically, already a walking dead man in New Hampshire; it just hadn't completely registered with him.

Emerging from the debate that night, it was clear that the momentum of the campaign had swung dramatically from Bush to Reagan. The rising tide for Reagan did not abate for the next three days, as the Gipper stayed in the Granite State and kept campaigning. Bush still didn't realize the political danger he was in.

On February 26, the day of the New Hampshire vote, Reagan could feel the electoral momentum shifting his way. Ronald Reagan, who had decided three weeks earlier that his friends who'd advised him that it was time to "let Reagan be Reagan" were right, had now made it official.

For the rest of the 1980 election, Ronald Reagan would campaign the way he wanted.

And then, after 1980, he would govern the way he wanted to govern.

Island of Freedom

"We begin our crusade joined together in a moment of silent prayer."

The Ronald Reagan camp was rejuvenated after the Nashua debate. Truly, the Gipper's voice was finally heard.

The next five months until the Republican National Convention were almost win after win for the Gipper. The New Hampshire primary resulted in Reagan receiving overwhelming support over George H. W. Bush, a 27 percent difference.[1] Even more surprisingly, Reagan won each demographic, by a large margin.[2]

Reagan won as many as 16 of New Hampshire's 22 delegates, and it felt good. "This is the first and it sure is the best," Reagan exclaimed to supporters. In perhaps a glimpse into the future, the prophetic Gerald Carmen, the Granite State's GOP leader, told reporters that "I think George Bush is mortally wounded and not just in New Hampshire. We always knew that if we could puncture George Bush's balloon, there would be nothing there but hot air."[3] Bob Dole, who received just over 600 votes in total in the state, agreed: "Reagan is going to be tougher and tougher. This race may end soon."[4]

Both Carmen and Dole were right. Of the remaining states (and Washington, DC), George Bush won only five. The rest went to Reagan. Funnily enough, the next victory for Bush was only a week after the Granite State upset. But it didn't do much for morale.

March Madness was under way.

The Massachusetts primary was held on Tuesday, March 4, result-ing in an extremely close race: Bush won 14 delegates with 31 percent of the vote, and Reagan and Anderson won 13 each, with around 29 percent.[5] However, mid-February polling showed that Bush was to win overwhelmingly.[6] In the end, there was only a 1,236 difference in votes between Bush and Reagan.[7] It was barely a win. It was his home state; it'd have been more newsworthy if he hadn't won. Further, Reagan spent only one day in Massachusetts; that state wasn't a priority.[8] Even when Bush won, he . . . well, just couldn't win.

In a three-day period, Reagan won four state primaries. On March 8, in South Carolina, Reagan won 54 percent of the vote; John Con-nally came in second, and Bush in third. The Gipper couldn't have been happier, and it certainly picked up his spirits. That evening, when it was clear that he had won the first big southern state, he chimed to reporters: "I have been telling some of you that I'm cautiously optimistic. Now I'm cautiously ecstatic."[9] The three primaries on March 11, Alabama, Florida, and Georgia, all went to Reagan. The highest difference, in Georgia, had Reagan win 73 percent of the vote versus Bush's 13 percent. Adding insult to injury, Florida was a closed state. Bush's beloved Republicans kicked him out of the country club and accepted Reagan's application.

The Texan John Connally's withdrawal from the race three days earlier helped Reagan win these southern states, and it gave Reagan "a boost" over Bush's more moderate policies. Even then, before the results, Bush was hopeful for his chances: "I think we're going to do respectably [enough that] people in the press would write 'he did well.'"[10] Quite the opposite happened.

At this point, it seemed that Bush was sometimes in denial. At the very least, he was going to go down swinging. "My theory always was that someone had to beat Reagan before Illinois. I've beaten him twice," he said before the Illinois primary.[11] Yes, anything could happen, but the trend wasn't looking good; Reagan had 167 committed delegates against Bush's 45.[12]

The Illinois primary came and went on March 18, and Reagan won

49 percent against Bush's 11 percent.[13] Among independents, Reagan received nearly half the votes. This reflected similar percentages in independent voters, save for Massachusetts.[14] It was theorized that Reagan's offense against inflation and higher taxes under the Carter Administration contributed to the support of middle-class Illinoisians.[15]

That Tuesday night, when asked in an interview with NBC about who was a formidable opponent, Reagan smiled and said, "They'll find someone."[16]

They never found anyone. From that day forth, in every vote until the GOP convention, Bush won four: Connecticut (where he was raised, by only 5 percent against Reagan), Pennsylvania, Washington, DC (unsurprising, given the establishment mentality), and Michigan. The rest went to Reagan.[17]

By late May, it was Reagan who adopted Bush's Big Mo, and as the 1980 primaries went on, the Gipper was the clear victor. Bush had 255 delegates and Reagan had 919, a mere 89 short of the party nomination. Bush couldn't win, and both he and the people saw that. As the expression goes, if you can't beat them, join them. On May 26, George Bush withdrew from the nomination, giving over his pledged delegates and securing Reagan's nomination for the GOP candidacy. "I've never quit a fight in my life," he asserted, placing the blame partly on lack of funds and partly on lack of support.[18] Bush sent the following telegram to Reagan after his withdrawal:

> Congratulations on your superb campaign for our party's 1980 presidential nomination. I pledge my wholehearted support in a united party effort this fall to defeat Jimmy Carter and elect not only a Republican president but Republican senators, congressmen and state and local officials who will work toward our common goal of restoring the American people's confidence in their government and our nation's future.[19]

With Bush gone, the next day in the Idaho primaries saw Reagan win 83 percent of the vote and Bush only 4 percent.[20] Kentucky, the same day, had similar results.[21] This was indicative of the final phase of Reagan's big momentum. Nine states voted on June 3, the last day of the primaries, with Reagan receiving no less than 64 percent of the vote.[22] Though Bush's dropping out effectively gave Reagan the nomination, this only sealed the deal.

With the nomination in hand, Reagan looked to the Republican convention.

———

July 14–17, 1980. Detroit, Michigan. These four days of the Republican National Convention defined a presidency. Toward the end of the convention, a roll call vote was taken to make Reagan the nominee de jure. A total of 1,939 delegates went to Reagan; John Anderson received 37, Bush received 13, and 4 were uncommitted. The entire convention booed when the rebel delegates chose Anderson.[23]

George Bush, previously reluctant to be Reagan's running mate, was "honored" to be on the ticket with Reagan when Reagan, after toying with the idea of a Reagan-Ford or a Reagan-Kemp presidency, offered him the spot. Bush, who did have ideological differences with the more conservative Reagan, promised to support him "across the board."[24] According to polls reported by the *Washington Post*, Bush was the right guy to pick. He was supported by "high-church Protestants and did relatively well with Catholic voters," two important voting blocks that Reagan wanted. Pollster Richard B. Wirthlin also supported the move, saying that Bush checked all the boxes that Reagan wanted in a veep, including "an ability to help the ticket and to take over as president if necessary."[25]

———

July 17, 1980: On the last night of the convention, Ronald Wilson Reagan, former governor of California, former actor, former president of the Screen Actors Guild, former host of CE Theater, and failed presidential nominee in 1968 and 1976, accepted the Republican presidential nomination. As he delivered his acceptance speech—it clocked in at more than 4,500 words and took approximately 45 minutes—one could see joy and humility beaming from him. He used the first-person plural pronouns *we, us,* and *our* more than two hundred times; first-person singular pronouns *I, me,* and *my* were fewer than fifty in number. Echoing his announcement to run for president less than a year earlier, he declared that the United States had a "rendezvous with destiny" as it looked to America's and the world's troubles: "Can we doubt that only a Divine Providence placed this land, this island of freedom, here as a refuge for all those people in the world who yearn to breathe freely."[26]

He closed his statement saying, "I'll confess that I've been a little afraid to suggest what I'm going to suggest—I'm more afraid not to—that we begin our crusade joined together in a moment of silent prayer."[27] Seventeen thousand stood silently.

"God bless America," he ended. A roar erupted.[28]

And Reagan charged into his destiny. Though few understood it or realized it, save the Gipper himself, the age of Reagan was just beginning to dawn over America and the world.

Author's Note

"Force of nature . . ."

Running for president is never easy, and it was especially hard for Ronald Reagan, as he had not just the usual obstacles to overcome, but also those of the skeptics in his own party and a very hostile and often malicious national media. He made a halfhearted attempt in 1968, ran full out in 1976, and even more so in 1980. By then, he was a fully formed American conservative. Many times, however, he heard from critics in the GOP establishment that he was "just an actor." But as he wisely said years later, in the waning days of his presidency, after being asked if he'd learned anything in Hollywood that had helped him be a good president, "I've wondered how you could do this job and not be an actor."[1]

Like a force of nature, Reagan rose inexorably over a period of time. From failure to receive his party's nomination in 1976 to winning forty-four states in the 1980 general election and forty-nine states in his re-election in 1984, he rose to prominence as an unforgettable world leader. It infuriates the Soviet apologists left at the Soviet sycophant the *Washington Post* and liberal academia to say that Reagan put his cowboy boot on the neck of corrupt and evil Soviet communism, destroying it, but that is exactly what he did do, infuriating the American left even after freeing millions. Across Eastern Europe are many statues of Reagan.

There are none for American liberal journalism or for an American liberal professor.

Reagan's rise was unexpected, potent, and difficult, and these were eventful and often fearsome years. A 1977 speech helped redefine a party's platform. In 1978 the Panama Canal Treaties passed and were ratified, a major source of contention and a national embarrassment in the eyes of the California governor. Within the next year, the world erupted as the Soviet Union invaded Afghanistan, crippling any sort of progress in relations between the Soviets and the West, leading to a petty boycott of the 1980 Moscow Olympics. The last year of Jimmy Carter's presidency was overshadowed, diplomatically, by the Iranian hostage crisis. An additional crisis of a sharp economic recession hit closer to home. Not a high point to sign off one's first term. Clearly, if Reagan were to win the 1980 election, this was to be the world he'd inherit.

Attacks came from all sides: left, right, and center. Everything from policy to age to experience was fair game. Reagan was a relative outsider. In the '80 election, he ran against Republicans, such as George H. W. Bush, whose credentials were certainly not unimpressive. Once Reagan secured the nomination, he'd be challenging an incumbent president. Only three incumbent presidents in the twentieth century had lost reelection—William Howard Taft against Woodrow Wilson and Teddy Roosevelt in 1912, Herbert Hoover against FDR in 1932, and most recently, the unelected Gerald Ford against Jimmy Carter in 1976.[2] Excepting 1932, all had been anomalous. It was an uphill battle that seemed not to cease.

Reagan remains one of the most fascinating figures of history and the American presidency, in part because he was a constantly evolving individual. His worldview in 1964 was not his worldview in 1980. His conservatism had changed, from being simply against the intrusions of big government to the more positive advance of individual freedom.

The Gipper stepped into a world of crises in 1980: war, inflation, and recession. Yet he looked forward, unblinking, to his and his country's future with confidence and sureness. He entered the White House as the fortieth president. He left as one of the greatest and most consequential presidents in history, having won both guns and butter. Reagan is and will always be worthy of serious academic study.

Even today, years after his election, politicians and politicos still study him, still channel him, and still have a difficult time understanding him.

Acknowledgments

"A statesman cannot create anything himself. He must wait and listen until he hears the steps of God sounding through events; then leap up and grasp the hem of his garment."

As always, I am indebted to my beloved wife and best friend, Zorine, to whom I owe so much. All I am, all I've accomplished, I owe to Zorine. She is the best editor, the best critical thinker, the best friend any husband and writer ever had. Zorine is my alpha and omega. Other writers, from Mark Twain to James Thurber, relied heavily upon the loving judgment and advice of their wives, and I am no exception. Thank you, Zorine, darling.

Thank you to my mother, Barbara Shirley Eckert, for always being there for me; my children, Matt, Taylor, and Mitch, for your moral support; and especially Andrew, who came through in a pinch to help me with the final editing for *Reagan Rising*.

The same goes for our dear ally and business partner Diana Banister, who has been a steadfast friend and confidante for many years and who, God willing, will be for many years to come.

Coming to the rescue was Scott Mauer, a splendid editor and fact checker, and now a friend and confidant. Scott worked tirelessly to perfect this manuscript. Thank you, Scott.

Michael Patrick Leahy is a superb writer and editor and friend, and

I am deeply in his debt as well. Like the cavalry coming over the hill in the nick of time, Mike showed up to help me put the finishing touches on *Reagan Rising*. Thank you, Mike.

Thank you to Bridget Matzie, my agent and my friend, to whom I owe so much. Thank you, Adam Bellow, my publisher, and Eric Meyers, my editor, at Harper, for having the confidence in this book.

Thank you to Jon Meacham for writing the foreword to this book, and for being one of our greatest historians. And thanks to my fellow Reagan biographers Lou Cannon, Kiron Skinner, Steven Hayward, Lee Edwards, and Paul Kengor. And thanks to some of Reagan's favorite writers, including our dear departed friend Peter Hannaford. And to Reagan's other favorite writers, including Ken Khachigian, Mari Maseng Will, Landon Parvin, Clark Judge, Ben Elliot, Peggy Noonan, and Aram Bakshian.

Thank you to Mrs. Ronald Reagan, who is now with her loving husband and her loving God. She was kind and helpful and supportive to me, and I will be eternally grateful to this kind, generous, and thoughtful woman. RIP Nancy Davis Reagan.

Thank you also to Duke Blackwood, Fred Ryan, John Heubusch, Joanne Drake, Jennifer Mandel, and Melissa Giller, at the Reagan Library; John Morris, David Arnold, Mike Murtagh, and Jamel Bell at Eureka College; the Honorable Jim Baker, honorary chairman of the James A. Baker III Institute for Public Policy at Rice University; Lou Cordia, head of the Reagan Alumni Association; the Honorable Ed Meese, former attorney general of the United States and close friend and adviser to Ronald Reagan for so many years; and Frank Donatelli, chairman of the Reagan Ranch. Thank you to my close and dear friends Laura Ingraham and Mark Levin, lifetime and always true-blue Reaganites. Thank you to Joe Scarborough and Gay Gaines as well, for being such good and true friends.

Thank you all.

Notes

PREFACE

1. "What Would He Make of Today's Right?" *Orange County Register*, March 21, 2005.

INTRODUCTION

1. Craig Shirley, *Reagan's Revolution: The Untold Story of the Campaign That Started It All* (Nashville, TN: Thomas Nelson, 2005), 91.
2. Ibid., 211.
3. *Face the Nation*, CBS, June 1, 1975.
4. Shirley, *Reagan's Revolution*, 91.
5. R. W. Apple Jr., "Ford-Reagan Battle Recalls 1952 Convention Struggle," *New York Times*, July 4, 1976, 28.
6. Peter Goldman, Thomas M. DeFrank, Hal Bruno, et al., "How Tight Can It Get?," *Newsweek*, July 12, 1976, 16.
7. John A. Conway, "The GOP Early Bird," *Newsweek*, July 12, 1976, 11.
8. "Reagan Plays a Wild Card," *Economist*, July 31, 1976, 33.
9. David S. Broder, "Ford Passes Critical Test," *Washington Post*, August 18, 1976.
10. "Transcript of Reagan's Remarks to the Convention," *New York Times*, August 20, 1976.
11. Al Cardenas, in discussion with the author, July 13, 2004.
12. Jack Germond, in discussion with the author, July 2, 2004.
13. "Reagan Tells His Supporters, 'Cause Goes On,'" *Morning Record* (Meriden, CT), August 20, 1976, 12.
14. Ibid.

CHAPTER 1: ALL BY HIMSELF

1.   Lou Cannon and Margot Hornblower, "A Graceful Departure," *Washington Post*, August 20, 1976, A1.
2.   Alan L. Otten, "The Party's Future After November Vote, the GOP May Take a Different Form," *Wall Street Journal*, August 16, 1976, 1.
3.   Ronald Reagan, *An American Life: The Autobiography* (New York: Simon and Schuster, 1990), 202.
4.   Lou Cannon and Margot Hornblower, "A Graceful Departure," *Washington Post*, August 20, 1976, A1.
5.   "Conclave in Chicago," *Time*, September 6, 1976, 11.
6.   "The Trouble with Conservatives," *Wall Street Journal*, August 17, 1976, 18.
7.   "Coming Out Swinging," *Time*, August 30, 1976, 8.
8.   Nancy Reynolds, in discussion with the author, October 3, 2006.
9.   John Nordheimer, "Aides Say Reagan Will Campaign for Ford If He Gets Assignments," *New York Times*, September 9, 1976, 32.
10. Ibid.
11. Dennis Farney, "Political Fumbling Match," *Wall Street Journal*, September 9, 1976, 12.
12. Frank van der Linden, *The Real Reagan: What He Believes, What He Has Accomplished, What We Can Expect from Him* (New York: William Morrow, 1981), 144.
13. Douglas E. Kneeland, "Dole Suggests That Ford Will Err If He Fails to Campaign Actively," *New York Times*, September 19, 1976, 38.
14. Christopher Lyndon, "Dole Makes Joking a Campaign Device," *New York Times*, September 19, 1976, 38.
15. "Campaign Kickoff," *Time*, September 13, 1976.
16. R. W. Apple Jr., "Contrasting Campaign Symbols," *New York Times*, September 7, 1976, 1.
17. Vermont Royster, "Thinking Things Over," *Wall Street Journal*, September 22, 1976, 20.
18. "Text of Prosecutor's Statement on Probe," circa October 1976, Harold Weisberg Archive, Hood College, Frederick, MD.
19. "Prosecutors Clear Ford in Campaign Fund Probe," *Wisconsin State Journal*, October 15, 1976, 3.
20. Ibid.
21. Jerry Landauer and Christopher A. Evans, "An IRS Analysis Tax Audit Indicates Ford's Pocket Money in '75 Was $5 a Week," *Wall Street Journal*, October 6, 1976, 1, 26.
22. Fred L. Zimmerman, "Topic A in Capital: Special Prosecutor and President Ford," *Wall Street Journal*, October 1, 1976, 1.

23. Nicholas M. Horrock, "FBI Is Said to Find No Evidence of Wrongdoing in Ford '72 Drive," *New York Times*, October 2, 1976, 38.

24. "The Plight of the G.O.P.," *Time*, August 23, 1976.

25. Ibid.

26. Ibid.

27. Karen Elliot House, "GOP Outlook in South for 1976 and Beyond Is Darkened by Carter," *Wall Street Journal*, October 11, 1976, 1.

28. Douglas E. Kneeland, "Reagan Joins Dole in New Haven and Praises the Republican Platform," *New York Times*, October 3, 1976, 33.

29. Brendan J. Lyons, "Court Gives Thumbs-Up to Middle-Finger Lawsuit," *Times Union*, January 4, 2013, http://www.timesunion.com/local/article/Court-gives -thumbs-up-to-middle-finger-lawsuit-4165770.php.

30. "Slugfest in Houston Alley," *Time*, October 25, 1976, 24.

31. Ibid.

32. Douglas E. Kneeland, "Dole Succeeds in Meeting Reagan to Display G.O.P. Unity on Coast," *New York Times*, September 27, 1976, 37.

33. Ibid.

34. Charles Mohr, "The Glow of Carter's Primary Drive Has Faded in His Contest with Ford," *New York Times*, October 29, 1976, A1.

35. Ibid.

36. Charles Mohr, "Carter, on Morals, Talks with Candor," *New York Times*, September 21, 1976, 1, 26.

37. Joseph Lelyveld, "His Campaign Cheated Voters," *New York Times*, November 1, 1976, 44.

38. Anthony Lewis, "Presidential Mystique," *New York Times*, October 11, 1976, 27.

39. "Reagan Sees Carter Lagging," *New York Times*, October 2, 1976, 10.

40. Norman Miller, "The Search for Carter," *New York Times*, September 26, 1976, 15.

41. James T. Wooten, "Carter Accepts Help of Stennis, Eastland," *New York Times*, September 18, 1976, 10.

42. Jules Witcover, *Marathon: The Pursuit of the Presidency, 1972–1976* (New York: Viking, 1977), 270.

43. "A Letter from the Publisher," *Time*, October 18, 1976.

44. "Exit Earl, Not Laughing," *Time*, October 18, 1976, 25.

45. William Safire, "Jugular Jimmy," *New York Times*, October 11, 1976, 27.

46. "How Populist Is Carter?," *Time*, August 2, 1976.

47. Confidential interview with the author.

48. Jules Witcover, *Marathon: The Pursuit of the Presidency, 1972–1976* (New York: Viking, 1977), 523–24.

49. John P. Sears, "Give 'em Hell, Jerry!," *New York Times*, September 19, 1976.

50. Ibid.

51. "Jostling for the Edge," *Time*, September 27, 1976, 12.

52. "Ford vs. Carter II," *Wall Street Journal*, October 7, 1976, 2.

53. Dennis Farney, "New Converts?," *Wall Street Journal*, October 18, 1976, 1.

54. Linda Charlton, "Mondale Marches in Two Parades, Sharing Spotlight with Republicans," *New York Times*, October 11, 1976, 1.

55. Michael McShane, in discussion with the author, 2006.

CHAPTER 2: WHILE WE WERE MARCHING THROUGH GEORGETOWN

1. John Herbers, "Washington an Insider's Game," *New York Times*, April 22, 1979, SM9.

2. Wallace Turner, "Reagan Shuns Role in Ford's Campaign," *New York Times*, October 30, 1976, 10.

3. Ibid.

4. "Reagan to Campaign for Ford," *New York Times*, October 26, 1976, 28.

5. "Reagan Bars Aiding Ford in 3 Key States," *New York Times*, October 28, 1976, 47.

6. "Reagan Will Not Bar Another Try in 1980," *New York Times*, November 3, 1976, 17.

7. Ronald Reagan, *An American Life: The Autobiography* (New York: Simon and Schuster, 1990), 203.

8. John Nordheimer, "Reagan Hints at Active Role in Shaping G.O.P. Future," *New York Times*, November 5, 1976, 16.

9. Ibid.

10. "Congress Declares Carter the Winner," *Facts on File World News Digest*, January 15, 1977.

11. Warren Weaver Jr., "Carter Asks Leaders of Congress to Help in a Reorganization," *New York Times*, November 18, 1976, 1.

12. Drummond Ayres, "Carter Gets Briefing by Kissinger and Sees a Smooth Transition," *New York Times*, November 21, 1976, 1.

13. Christopher Andrew and Vasili Mitrokhin, *The Sword and the Shield: The Mitrokhin Archive and the Secret History of the KGB* (New York: Basic Books, 1999), 180, 209.

14. "Kennedy-KGB collaboration," *Washington Times*, October 27, 2006.

15. Warren Weaver Jr., "Politicians Find G.O.P. Fighting for Its Survival," *New York Times*, November 24, 1976, 1.

16. Ibid.

17. Ibid., 15.

18. "The Shape of the Next Four Years," *Time*, November 8, 1976, 24.

19. Weaver, "Politicians Find G.O.P. Fighting for Its Survival," 15.

20. "Presidential Race Cost Government $72 Million," *New York Times*, January 1, 1977, 4; "Election Panel Suggests Rise in Campaign Funds," *New York Times*, February 9, 1977, 14.

21. "Reagan Outspending Ford," *New York Times*, August 14, 1976, 13.

22. Richard D. Lyons, "Howard Henry Baker Jr.," *New York Times*, January 5, 1977, 14.

23. Dick Cheney, in discussion with the author, March 19, 2007.

24. James M. Naughton, "Most of Ford Staff Uncertain About Future," *New York Times*, January 21, 1977, 31.

25. Walter Mondale, in discussion with the author, February 28, 2007.

26. Charles Mohr, "Carter Assails Role of Lobbyists in 'Bloated Mess' of Government," *New York Times*, September 28, 1976, 1.

27. Peter Baker and Kent Jenkins Jr., "Fairfax GOP Dinner Featured Black, Gay Jokes," *Washington Post*, March 18, 1993, A1.

28. Peter Goldman, Eleanor Clift, Thomas M. DeFrank, et al., "The Drive to Dump Carter," *Newsweek*, August 11, 1980, 18.

29. Mary Russell, "Jordan Meets with Rep. O'Neill, Gets Taken Down a Peg or Two," *Washington Post*, July 19, 1979, A13.

CHAPTER 3: INTO THE WILDERNESS

1. Jack Bass, "Southern Republicans: Their Plight Is Getting Worse," *Washington Post*, July 12, 1977, A4.

2. "A Dying Party? 5 Republican Leaders Speak Out," *U.S. News & World Report*, August 29, 1977, 22.

3. Ibid.

4. Steven F. Hayward, *The Age of Reagan: The Fall of the Old Liberal Order, 1964–1980* (Roseville, CA: Forum, 2001), 505.

5. Ibid.

6. Carter Mondale Presidential Committee, "Preliminary Debate Discussions," August 21, 1980, Jimmy Carter Presidential Library and Museum, Atlanta, GA.

7. Theodore H. White, *America in Search of Itself: The Making of the President, 1956–1980* (New York: Harper & Row, 1982), 233.

8. Lou Cannon, "Ford, in Farewell, Urges Arms Balance," *Washington Post*, January 13, 1977, A25.

9. Robert Hullihan, "Transition Tales," *Washington Post*, January 2, 1977.

10. Lou Cannon, "Ford, in Farewell, Urges Arms Balance," *Washington Post*, January 13, 1977, A25.

11. Lawrence L. Knutson, "What's a Former President to Do?," Associated Press, January 22, 2001.

12. Rodney Smith, in discussion with the author, September 7, 2006.

13. Ibid.

14. Marvin Stone, "Save the G.O.P.?," *U.S. News & World Report*, May 23, 1977, 100.

15. Mark Feeney, "Richard Nixon: He's Tan, He's Rested—and About to Turn 80," *Boston Globe*, January 3, 1993.

16. "A Look Ahead from the Nation's Capital," *U.S. News & World Report*, February 7, 1977, 5.

17. Peter Hannaford, in discussion with the author, March 24, 2006.

18. Peter Hannaford, in discussion with Stephen F. Knott and Russel L. Riley, 2005, *Miller Center*, http://www.millercenter.org/president/reagan/oralhistory/peter-hannaford.

19. Walter R. Mears, Associated Press, June 13, 1977.

20. Nofziger, *Nofziger*, 189.

21. Ed Meese, in discussion with the author, May 2006.

22. Peter Hannaford, *The Reagans: A Political Portrait* (New York: Coward-McCann, 1983), 142.

23. Hannaford, in discussion with Knott and Riley.

24. Associated Press, February 3, 1977.

25. Ibid.

26. "The Rejection of Sorensen: A Drama of Human Failing," *New York Times*, February 2, 1977, 1.

27. Robert H. Williams, "PostScript," *Washington Post*, February 14, 1977, A3.

28. David M. Alpern, Lloyd H. Norman, and Nicholas Horrock, "An Admiral for the CIA," *Newsweek*, February 21, 1977, 17.

29. Richard Bergholz, "Reagan Advises Carter to Keep Bush as CIA Chief," *Los Angeles Times*, February 7, 1977, B4.

30. Tip O'Neill and William Novak, *Man of the House: The Life and Political Memoirs of Speaker Tip O'Neill* (New York: Random House, 1987), 308.

31. Murrey Marder, "Carter to Inherit Intense Dispute on Soviet Intentions," *Washington Post*, January 2, 1977, A1.

32. Ibid.

33. George C. Wilson, "US Constituency Favoring Arms Buildup Seen Growing," *Washington Post*, January 2, 1977, 6.

34. Don Oberdorfer, "Study Says US Showed Force 215 Times Since '45," *Washington Post*, January 3, 1977, A1.

35. Robert Evans, "Record Soviet Grain Harvest Seen—but Imports Continue," *Washington Post*, January 6, 1977, C1.

36. Ibid.

37. Craig Shirley, *Reagan's Revolution: The Untold Story of the Campaign That Started It All* (Nashville, TN: Thomas Nelson, 2005), 36.

38. Ronald Reagan, "The New Republican Party," February 6, 1977, *Reagan 2020.US*, http://reagan2020.us/speeches/The_New_Republican_Party.asp.

39. Ibid.

40. Ibid.

41. Ibid.

42. Ibid.

43. Don McLeod, Associated Press, February 4, 1977.

44. "Chicago Relief 'Queen' Guilty," *New York Times*, March 19, 1977, 8.

45. Dan Miller, "The Chutzpa Queen," *Washington Post*, March 13, 1977, A3.

46. William Chapman, "Welfare Program Enigma: The Poor Always with Us," *Washington Post*, May 8, 1977.

47. Ibid.

48. Ibid.

49. Warren Weaver Jr., "Carter Proposes End of Electoral College in Presidential Votes," *New York Times*, March 23, 1977, A1.

50. Kenneth Reich, "Carter Proposals Could Kill GOP, Reagan Says," *Los Angeles Times*, April 16, 1977, A18.

51. James C. Roberts, "A Conservative's Advice for the Future," *Washington Post*, January 11, 1977, A17.

52. Lou Cannon, "Bittersweet Farewell for a Man Who Has 'Best of Both Worlds,'" *Washington Post*, January 21, 1977, A10.

53. Rowland Evans and Robert Novak, "Nominee in Trouble: Mary King Action," *Washington Post*, January 22, 1977, A15.

54. Donald M. Rothberg, Associated Press, January 14, 1977.

55. Lee Lescaze, "Warnke to Head Arms Control Agency," *Washington Post*, February 1, 1977, A2.

56. Lou Cannon, "Reagan Criticizes Carter for Proposing Defense Budget Cuts," *Washington Post*, February 6, 1977, 20.

57. Tom Raum, Associated Press, September 8, 1977.

58. Joseph Kraft, "Renewing the Chinese Connection," *Washington Post*, February 6, 1977, C7.

59. Peter Hannaford, *The Reagans: A Political Portrait* (New York: Coward-McCann, 1983), 144.

60. Arthur Schlesinger Jr., "The Politics of Confusion," *Wall Street Journal*, June 6, 1977, 16.

CHAPTER 4: THE BEAR IN THE ROOM

1. Donnie Radcliffe, "Reagan Feeling 39 at 66," *Washington Post*, February 7, 1977, B1.

2. Ibid., B3.

3. Peter Milius, "OSHA: A 4-Letter Word," *Washington Post*, February 12, 1977, A1.

4. Ibid.

5. "Overhauling OSHA," *Washington Post*, May 26, 1977, A14.

6. Ronald Reagan, *Reagan's Path to Victory: The Shaping of Ronald Reagan's Vision: Selected Writings*, ed. Kiron K. Skinner, Annelise Anderson, and Martin Anderson (New York: Simon and Schuster, 2004), 109–10.

7. Rowland Evans and Robert Novak, "Another Round for Reagan?," *Washington Post*, March 2, 1977, A19.

8. Ibid.

9. Lou Cannon, "Conservatives Form a 'Shadow Cabinet,'" *Washington Post*, February 25, 1977, A3.

10. Warren Weaver Jr., "G.O.P. Liberal Group Opens a Fund Drive," *New York Times*, April 1, 1977, 13.

11. James P. Gannon, "Coalition Politics on the Right," *Wall Street Journal*, January 3, 1978, 12; Lou Cannon, "Tapping the Little Guy," *Washington Post*, March 6, 1977, A1.

12. Lou Cannon, "Tapping the Little Guy," *Washington Post*, March 6, 1977, A22.

13. Lou Cannon, "Ford, Reagan Demonstrate Their Party's Persistent Split," *Washington Post*, April 18, 1977, A5.

14. Ibid.

15. Neal Peden, in discussion with the author, July 6, 2006.

16. "A Spy for the GOP Says His Job Was Easy," *San Francisco Chronicle*, October 11, 1973, Harold Weisberg Archive, Hood College, Frederick, Maryland.

17. Rowland Evans and Robert Novak, "Young Republicans, Old Tricks," *Washington Post*, April 27, 1977, A19.

18. Carl Bernstein and Bob Woodward, "FBI Finds Nixon Aides Sabotaged Democrats," *Washington Post*, October 10, 1972, A1.

19. Les Seago, Associated Press, June 10, 1977.

20. Rowland Evans and Robert Novak, "Young Republicans, Old Tricks," *Washington Post*, April 27, 1977, A19.

21. John T. Dolan, in discussion with the author.

22. Roger Stone, in discussion with the author, April 10, 2006.

23. Dennis Farney, "GOP's Old Pros Take Stock of Jimmy Carter," *Wall Street Journal*, June 27, 1977, 10.

24. Lou Cannon, "Carter Energy Plan Assailed by Brock," *Washington Post*, April 30, 1977, A4.

25. "Gas Tax Goal Is to 'Make' Believers of Americans," *Washington Post*, April 18, 1977, A9.

26. J. P. Smith and David S. Broder, "The High-Stakes Fight over Energy," *Washington Post*, April 18, 1977, A8.

27. Edward Walsh, "Carter: Energy Outlook Grim," *Washington Post*, April 19, 1977, A1.

28. "Carter: 'Oil and Natural Gas . . . Are Running Out,'" *Washington Post*, April 19, 1977, A14.

29. Paul Laxalt, in discussion with the author, July 31, 2006.

30. George F. Will, "Seeking Sacrifice in Good Times," *Washington Post*, April 21, 1977, A23.

31. Albert R. Hunt, "House Unit Approves Most of Carter's Energy Plan, but Votes to End Price Control of New Natural Gas," *Wall Street Journal*, June 28, 1977, 3.

32. William Claiborne, "A School for Losers," *Washington Post*, May 1, 1977, A6.

33. Paul L. Martin and W. Robert Shoup, "Republican Strategy Back to Grass Roots," *U.S. News & World Report*, May 16, 1977, 34.

34. George Gallup, "GOP Affiliation Falls to 40-Year Low Point," *Washington Post*, August 21, 1977, A12.

35. Lou Cannon, "Rhodes Withdraws Support to Register Voters Election Day," *Washington Post*, May 17, 1977, A15.

36. Ibid.

37. Betty Anne Williams, Associated Press, April 14, 1977.

38. Myra McPherson, "Florida Rejects ERA; Ratification Chances Dim," *Washington Post*, May 17, 1977, A1.

39. Peter G. Bourne, *Jimmy Carter: A Comprehensive Biography from Plains to Postpresidency* (New York: Scribner, 1977), 398.

40. Bart Barnes, "All-Boy Choir Approved," *Washington Post*, April 14, 1977, A1; "A Bravo for Boys' Choirs," *Washington Post*, April 18, 1977, A18.

41. Hobart Rowen, "Carter Backs Off on Bid to Spur Western Growth," *Washington Post*, May 8, 1977, A1; James L. Rowe Jr., "Food, Fuel Pace 1.1 April Rise in Wholesale Prices," *Washington Post*, May 6, 1977, A1; Mike Doan, Associated Press, May 10, 1977.

42. Walter R. Mears, Associated Press, May 13, 1977.

43. David S. Broder, "Ford Voices Concerns on Successor's Budget," *Washington Post*, May 21, 1977, A2.

44. Warren Weaver Jr., "Ford Sidesteps Criticism of Carter," *New York Times*, May 20, 1977, 14.

45. Craig Shirley, *Reagan's Revolution: The Untold Story of the Campaign That Started It All* (Nashville, TN: Nelson Current, 2005), xxiv.

46. Associated Press, "People in the News," August 3, 1977.

47. Doug Willis, Associated Press, May 29, 1977.

48. Rowland Evans and Robert Novak, "Clamping Down Anew on Russian Dissidents," *Washington Post*, June 3, 1977, A27.

49. Peter Osnos, "New Charter Stresses Party Role, Limits Expression," *Washington Post*, June 5, 1977, A1; Peter Osnos, "Drive on Dissidents Creates 'Ominous Atmosphere,'" *Washington Post*, June 5, 1977, A1.

50. Edward Walsh, "Carter Stresses Social Justice in Foreign Policy," *Washington Post*, May 23, 1977, A1.

51. Ibid.

52. Murrey Marder, "Kissinger Warns Carter of West European Communism," *Washington Post*, June 10, 1977, A23.

53. Joseph B. Treaster, "Reagan Is Critical of Carter on Rights," *New York Times*, June 10, 1977, A5.

54. Max Lerner, "Singlaub Affair Raises Questions," *Eugene Register-Guard*, May 26, 1977, 18A.

55. Warren Brown, "Carter Had 'No Choice' on Singlaub, Byrd Says," *Washington Post*, May 23, 1977, A3.

56. Helen Dewar, "B-1 Plane Too Costly, Byrd Tells Carter," *Washington Post*, June 26, 1977, A5.

57. Peter Osnos, "Brezhnev Attacks 'Interference,'" *Washington Post*, March 22, 1977, A1.

58. Austin Scott, "Carter Invites Soviets to Debate Principles," *Washington Post*, March 23, 1977, A1.

59. "The Retreat from Moscow," *Washington Post*, April 5, 1977, A18.

60. Lawrence Feinberg, "Kissinger Backs Carter Arms Stand," *Washington Post*, April 6, 1977, A4.

61. "President Hopeful on Brezhnev Trip," *New York Times*, July 1, 1977, A1.

62. David K. Shipler, "Moscow Protests Viewing of 'Zhivago,'" *New York Times*, April 16, 1977, 8.

63. "FTC Cites Ford on Auto Engine Defect," *Facts on File World News Digest*, February 4, 1978.

64. Joseph F. Sullivan, "Reagan Agrees with McCarthy on Deficit in '81," *New York Times*, June 11, 1977, 9.

65. Roper Center for Public Opinion Research, "Carter Presidential Approval," Cornell University, https://ropercenter.cornell.edu/polls/presidential-approval/.

66. Stan Benjamin, Associated Press, June 1, 1977.

67. Edward Cowan, "World Oil Shortage Is Called Inevitable," *New York Times*, May 17, 1977, 1; Joanne Omang, "Dire Oil Forecast," *Washington Post*, May 17, 1977, A1.

68. Thomas O'Toole, "CIA Foresees Global Oil Shortage," *Washington Post*, April 16, 1977, A1.

69. Stan Benjamin, Associated Press, June 1, 1977.

70. "GOP Leaders Assail Carter's Energy Policy," *Washington Post*, May 7, 1977, A6.

71. Warren Weaver Jr., "G.O.P. Energy Reply Features New Faces," *New York Times*, June 3, 1977, D14.

72. David S. Broder, "GOP Film to Rebut Carter on Energy," *Washington Post*, June 2, 1977, A4.

73.   Weaver, "G.O.P. Energy Reply Features New Faces," D14.

74.   Haynes Johnson, "ADA Warns Against Promises Unkept," *Washington Post*, May 8, 1977, A1.

75.   Ibid.

76.   Stan Benjamin, Associated Press, June 1, 1977.

77.   Warren Brown, "Anderson Rules Out Reagan, Ford in '80," *Washington Post*, June 6, 1977, A7.

78.   *U.S. News & World Report*, June 13, 1977, 16.

79.   Associated Press, August 25, 1977.

80.   Shirley, *Reagan's Revolution*, 197.

81.   Joseph Lelyveld, "The Path to 1980," *New York Times*, October 2, 1977, SM28.

82.   "The Panama Compromise," *Wall Street Journal*, February 1, 1978, 6.

CHAPTER 5: CANAL ZONE DEFENSE

1.   Ronald Reagan, *Reagan, in His Own Hand*, ed. Kiron K. Skinner, Annelise Anderson, and Martin Anderson (New York: Free Press, 2001), 385–86.

2.   Robert D. Novak, *The Prince of Darkness: 50 Years Reporting in Washington* (New York: Crown Forum, 2007), 270.

3.   Associated Press, August 25, 1977.

4.   Stefan H. Leader, "The Canal as Republican Rallying Point," *Washington Post*, September 6, 1977, A21.

5.   Reagan, *Reagan, in His Own Hand*, 198–99.

6.   Hedrick Smith, "White House Opens Drive to Win Senate Approval of Canal Accord," *New York Times*, August 12, 1977, A1.

7.   "Panama Canal Treaties; Carter's High-Risk Move," *U.S. News & World Report*, September 12, 1977, 18.

8.   "Summaries for the Department of Energy Organization Act," GovTrack, https://govtrack.us/congress/bills/95/s826/summary.

9.   Warren Brown, "Antipoverty Agency Reported Riddled with Waste and Fraud," *Washington Post*, August 12, 1977, A14.

10.   Norman C. Miller, "Administration's Honeymoon Is Now Finally Over," *Wall Street Journal*, August 30, 1977, 12.

11.   William Safire, "Carter's Broken Lance," *New York Times*, July 21, 1977, 23.

12.   "National Express Cards," *Saturday Night Live*, NBC, September 24, 1977.

13.   "Washington Wire," *Wall Street Journal*, September 23, 1977, 1.

14.   Phillip Shabecoff, "Black Aide Charges Carter with Racism," *New York Times*, July 2, 1977, 13.

15.   "White Driver Rams Klan Rally in Plains, Injuring 30," *New York Times*, July 3, 1977, 1.

16. Mary Russell, "Administration, Hill Leaders Gut Voting Bill," *Washington Post*, July 16, 1977, A1.

17. "Panama: A Doomed Treaty?," *U.S. News & World Report*, September 19, 1977, 18.

18. Peter G. Bourne, *Jimmy Carter: A Comprehensive Biography from Plains to Postpresidency* (New York: Scribner, 1997), 392.

19. Mike Shanahan, Associated Press, September 8, 1977.

20. Bob Beckel, *I Should Be Dead: My Life Surviving Politics, TV, and Addiction* (New York: Hachette Books, 2015), 100.

21. Richard Pipes, "Team B: The Reality Behind the Myth," *Commentary Magazine*, October 1, 1986, https://www.commentarymagazine.com/articles/team-b-the-reality-behind-the-myth/.

22. Helen Thomas, "President Reagan, Who Strongly Opposed the Panama Canal Treaty . . . ," United Press International, October 1, 1982.

23. "Conservatives Begin Campaign," *New York Times*, August 22, 1977, 5.

24. Tom Raum, Associated Press, September 8, 1977.

25. Ibid.

26. Lou Cannon, "Supporters See Reagan Opposing Panama Treaties," *Washington Post*, August 24, 1977, A2.

27. Bourne, *Jimmy Carter*, 393.

28. "Ford Gives Support to New Agreement on Panama Canal," *New York Times*, August 17, 1977, A1.

29. James Reston, "Reagan on Panama," *New York Times*, August 28, 1977, 17.

30. "Warning from Panama," *New York Times*, August 20, 1977, 5.

31. Bob Livingston, in discussion with the author, April 19, 2007.

32. Albert R. Hunt, "Reagan: 'A Man, a Plan, a Canal . . .'" *Wall Street Journal*, August 26, 1977, 8.

33. George F. Will, "Those Little Waves of Détente," *Washington Post*, July 14, 1977, A27.

34. Rowland Evans and Robert Novak, "Conceding Defeat in Europe," *Washington Post*, August 3, 1977, A19.

35. "Conceding a Third of Germany," *Washington Post*, August 5, 1977, A20.

36. Michael Getler, "Bonn Is Expected to Allow Stationing of Neutron Arms," *Washington Post*, September 27, 1977, A15.

37. Alan L. Otten, "Politics and People," *Wall Street Journal*, September 1, 1977, 8.

38. Michael G. Gartner, "Vigorous but Erratic President," *Wall Street Journal*, October 24, 1977, 20.

39. Merwin Sigale, "Helms Says Baker Is 'Squirming' on Canal Treaty Issue," *Washington Post*, September 12, 1977, A2.

40. Terence Smith, "G.O.P. Anti-Treaty Vote Expected," *New York Times*, September 30, 1977, A19.

41. Lou Cannon, "GOP Convention in Calif. Haunted by Lingering Differences," *Washington Post*, October 3, 1977, A2.

42. Ibid.

43. Ibid.

44. Richard Viguerie, in discussion with the author, December 6, 2006.

45. Adam Clymer, "TV Campaign Begun Against Panama Canal Treaty," *New York Times*, October 31, 1977, 20.

46. Douglas E. Kneeland, "Seeds of Bush Candidacy Are Recalled," *New York Times*, January 24, 1980, B9.

47. James Baker, in discussion with the author, September 6, 2006.

48. Associated Press, "Names in the News," October 27, 1977.

49. Ibid.

50. Terence Smith, "Reagan, Connally and Baker Open a Broad Attack on Administration," *New York Times*, December 4, 1977, 1.

51. Donald M. Rothberg, Associated Press, November 28, 1977.

52. Sally Quinn, "'The Pedestal Has Crashed': Pride and Paranoia in Houston," *Washington Post*, November 23, 1977, B1.

53. George Lardner Jr., "Intra-Party Fight on Panama Endangers GOP, Baker Warns," *Washington Post*, November 14, 1977, A3.

54. Terence Smith, "Opinion in US Swinging to Right, Pollsters and Politicians Believe," *New York Times*, December 4, 1977, 1.

55. Ibid.

56. Associated Press, December 20, 1977.

57. Craig Shirley, *Reagan's Revolution: The Untold Story of the Campaign That Started It All* (Nashville, TN: Nelson Current, 2005), 336.

58. Don McLeod, Associated Press, January 4, 1978.

59. James P. Gannon, "Sen. Baker Uses Post of GOP Leader to Aid Party—and Himself," *Wall Street Journal*, November 23, 1977, 1.

60. Ibid.

61. Don McLeod, Associated Press, January 12, 1978.

62. Jim Roberts, in discussion with the author, July 6, 2006.

63. Rowland Evans and Robert Novak, "Reagan in '80?," *Washington Post*, January 16, 1978, A21.

64. "Verdict from Experts on Jimmy Carter's First Year," *U.S. News & World Report*, January 9, 1978, 16.

65. Peter Hannaford, *The Reagans: A Political Portrait* (New York: Coward-McCann, 1983), 151–53.

66. James Gerstenzang, Associated Press, February 20, 1978.

67. "Presidential Aide Jordan Denies Story on Spitting Incident in Bar," *Washington Post*, February 19, 1978.

68. Edward Walsh and Lou Cannon, "Key Constituencies Alienated as President Begins Journey," *Washington Post*, October 21, 1977, A1.

69. Albert R. Hunt, "Carter and the '78 Elections," *Wall Street Journal*, December 27, 1977, 8.

70. Ibid.

71. Jody Powell, in discussion with the author, April 10, 2006.

72. Dennis Farney, "Carter's Troubled First Year," *Wall Street Journal*, December 23, 1977, 3.

CHAPTER 6: DRINKING THE KOOL-AID

1. Adam Clymer, "Jesse Jackson Tells Receptive G.O.P. It Can Pick Up Votes of Blacks," *New York Times*, January 21, 1978, 4.

2. Ibid.

3. James M. Perry, "The GOP and Black Voters," *Wall Street Journal*, December 7, 1977, 24.

4. Ibid.

5. Howell Raines, "Republicans Courting Black Voters in South After Years of Inactivity," *New York Times*, August 1, 1978, A13.

6. Perry, "The GOP and Black Voters," 24.

7. William Claiborne, "Bid to Stiffen GOP Stand on Canal Pacts Dropped," *Washington Post*, January 22, 1978, A13.

8. "Wooing the Black Vote," *Time*, January 30, 1978.

9. Ibid.

10. Bill Peterson, "GOP Targets Deep South Governorship," *Washington Post*, August 7, 1979, A2.

11. James W. Singer, "Taking Aim at Youth Unemployment," *National Journal*, January 28, 1978, 143.

12. Edward Walsh, "NAACP Opposition to Carter Policies Cited by Reagan," *Washington Post*, January 22, 1978, A12.

13. Peter Hannaford, *The Reagans: A Political Portrait* (New York: Coward-McCann, 1983), 170.

14. James P. Gannon, "Heat on Key Senators over Canal Treaties Is Becoming Intense," *Wall Street Journal*, January 19, 1978, 1.

15. Terence Smith, "Carter, in TV Talk, Asks Canal Backing," *New York Times*, February 2, 1978, A1.

16. Hannaford, *The Reagans*, 150.

17. Donald M. Rothberg, Associated Press, February 8, 1978.

18. Peter G. Bourne, *Jimmy Carter: A Comprehensive Biography from Plains to Postpresidency* (New York: Scribner, 1997), 382.

19. George de Lama, "Baker on the Left, Buchanan on the Right Lead a Fight for Reagan's Soul," *Chicago Tribune*, March 29, 1987, C6.

20. Charles B. Seib, "Training Journalists, Conservative Style," *Washington Post*, February 17, 1978, A17.

21. David S. Broder, "End of the Washington-Can-Do-It Era," *Washington Post*, February 15, 1978, A23.

22. Graham Hovey, "Canon 'Truth Squad' Plans a 5-Day," *New York Times*, January 10, 1978, 11.

23. Ward Sinclair, "Nobody Budges in Buckley-Reagan Canal Treaty Debate," *Washington Post*, January 15, 1978, A11.

24. Ibid.

25. James P. Gannon, "Man on the Run? Senator Baker Uses Post of GOP Leader to Aid Party—and Himself," *Wall Street Journal*, November 23, 1977, 1.

26. Steven R. Roberts, "Panama Treaties at Stake in Bitter Propaganda War," *New York Times*, January 20, 1978, A3.

27. Walter Mondale, in discussion with the author, February 28, 2007.

28. Lee Edwards, *The Essential Ronald Reagan: A Profile in Courage, Justice, and Wisdom* (Lanham, MD: Rowman and Littlefield Publishers, 2005), 77.

29. Lawrence L. Knutson, Associated Press, February 25, 1978.

30. Ibid.

31. William Robbins, "Dole Softens Wit in Quest for Top Spot on 1980," *New York Times*, January 29, 1978, 24.

32. Thomas J. Foley, "Bob Dole Shoots for the Top with a New Image," *U.S. News & World Report*, September 17, 1979, 40.

33. Associated Press, March 28, 1978.

34. Don McLeod, Associated Press, April 6, 1978.

35. Ibid.

36. Walter Pincus, "Democrats Fall Way Behind Money-Making GOP," *Washington Post*, December 18, 1977, A1.

37. William Claiborne, "Little in 'New Right' War Chest Finding Its Way to Candidates," *Washington Post*, March 20, 1978, A1.

38. Richard Pyle, Associated Press, March 16, 1978.

39. "Discussion on Second Canal Treaty Vote," *PBS NewsHour*, PBS, April 18, 1978.

40. Peter Hannaford, in discussion with the author, April 24, 2007.

41. Ronald Reagan, *Reagan, in His Own Hand*, ed. Kiron K. Skinner, Annelise Anderson, and Martin Anderson (New York: Free Press, 2001), 208–9.

42. Donald M. Rothberg, Associated Press, June 9, 1978.

43. Tom Wicker, "The Republicans Try to Get Their Act Together," *New York Times Magazine*, February 12, 1978, SM2.

44. Ibid.

45. Ronald Reagan, "America's Purpose in the World," March 17, 1978, Reagan2020, http://reagan2020.us/speeches/Americas_World_Purpose.asp.

46. Hannaford, *The Reagans*, 173.

47. Albert R. Hunt, "Carter and the '78 Elections," *Wall Street Journal*, December 27, 1977, 8.

48. Associated Press, May 25, 1978.

49. Ibid.; Adam Clymer, "Kennedy Leads Carter 53% to 40% as Democrats' 1980 Choice in Poll," *New York Times*, May 7, 1978, 26.

50. Marquis Childs, "Reagan: More Serious Than Ever," *Washington Post*, May 16, 1978, A13.

51. Anthony Marro, "Fraud in Federal Aid May Exceed $12 Billion Annually, Experts Say," *New York Times*, April 16, 1978, 1.

52. Ibid.

53. Ibid.

54. Ibid.

55. Thomas G. Donlan, Associated Press, June 6, 1978.

56. Rowland Evans and Robert Novak, "Reagan the Party Man," *Washington Post*, May 10, 1978, A27.

57. Arthur Laffer, in discussion with the author, January 18, 2007.

58. Don Holt, Susan Agrest, and Hal Bruno, "A Vote for New Faces," *Newsweek*, June 19, 1978, 33.

59. *Mad as Hell: The Taxpayers Revolt*, NBC, June 16, 1978.

60. *The Angry Taxpayer*, CBS, June 15, 1978.

61. Tom Wicker, "Just a Kid in China," *New York Times*, June 16, 1978, A27.

62. Albin Krebs, "Notes on People," *New York Times*, July 19, 1978, C2.

63. Hannaford, *The Reagans*, 177.

64. David S. Broder, "1980: They're Off and Running Already," *Washington Post*, March 5, 1978, C7.

65. Louis Harris, "Democrats Lead, 49–29%, in Race for House Seats," *Washington Post*, May 8, 1978, A4.

66. Donald M. Rothberg, Associated Press, June 23, 1978.

67. Lyn Nofziger, in discussion with the author.

68. James Willwerth, "In Los Angeles: Prisoners of War—and Peace," *Time*, July 3, 1978.

69. Ibid.

70. Tom Wicker, "The Republicans Try to Get Their Act Together," *New York Times Magazine*, February 12, 1978, SM2.

71. "Washington Whispers," *U.S. News & World Report*, October 9, 1978, 18.

72. Adam Clymer, "Ford Hires an Aide to Coordinate Campaign Activities," *New York Times*, June 2, 1978, A12.

73. Adam Clymer, "G.O.P. Officeholders Confer on Policies," *New York Times*, April 10, 1978, 19.

74. Walter R. Mears, Associated Press, July 19, 1978.

75. Ibid.

76. Adam Clymer, "Reagan Urges Party to Support Tax Cuts," *New York Times*, June 25, 1978, 27.

77. Adam Clymer, "Ford Hails California Tax Vote as a Signal to Curb Government," *New York Times*, July 9, 1978, 12.

78. R. W. Apple Jr., *New York Times*, April 23, 1978, 1.

79. David Margolick, "Legal Issues Hinge on Whether Korean Airliner Was Warned," *New York Times*, September 2, 1983, A4.

80. Peter Osnos, "Moscow Replies: Carter's Policies Blocking Détente," *Washington Post*, June 8, 1978, A1.

81. "Carter Under Pressure to Respond to Soviets," *Washington Post*, July 12, 1978, A1.

82. Majorie Hyer, "Soviets 'Cold but Courteous' to US Church Group," *Washington Post*, July 21, 1978, C12.

83. Adam Clymer, "Rep. Crane of Illinois Is the First in Republican Race for President," *New York Times*, August 3, 1978, B17.

84. "$200 Million Price Tag on November Elections," *U.S. News & World Report*, October 16, 1978, 26.

85. Walter R. Mears, Associated Press, July 10, 1978.

86. Jack W. Germond and Jules Witcover, "Reagan Planning Early Blitz to Snare GOP Nomination for 1980," *Washington Star*, July 3, 1979, A1.

87. Thomas Wells, Associated Press, 1978.

88. Associated Press, July 30, 1978.

89. Associated Press, August 21, 1978.

90. Don McLeod, Associated Press, August 2, 1978.

91. Ibid.

92. Associated Press, August 18, 1978.

93. Donald M. Rothberg, Associated Press, September 7, 1978.

94. Ibid.

95. Edward Walsh and Warren Brown, "Pennsylvania GOP Heading for Upset in Governor's Race," *Washington Post*, November 8, 1978, A19.

96. "Reagan Sees a Nixon Role After History's Review," *New York Times*, May 15, 1978, A17.

97. Jonathan D. Salant, "Tapes: Nixon Called Reagan 'Strange,'" Associated Press, December 10, 2003.

98. David S. Broder, "New Hampshire Already Churning for '80 Primary," *Washington Post*, August 27, 1978, A1.

99. Walter R. Mears, Associated Press, September 27, 1978.

100. Associated Press, October 6, 1978.

101. Rowland Evans and Robert Novak, "The Ford and Reagan Show," *Washington Post*, September 15, 1978, A15.

102. Associated Press, October 26, 1978.

103. Ibid.

104. Bill Peterson, "GOP Unit Disbursing Extra $900,000 for Tight Senate Races," *Washington Post*, November 5, 1978, A2.

105. "Unofficial Senate Election Results," *National Journal*, November 11, 1978, 1813.

106. Martin Waldron, "Bradley Is Favored over Bell in Jersey Race for Senate," *New York Times*, November 6, 1978, 41.

107. Ward Sinclair, "Anti-Abortion Activists Help Scuttle Clark in Iowa," *Washington Post*, November 9, 1978, A3.

108. Jim Adams, Associated Press, November 8, 1978.

109. "Making Democracy More Interesting," *New York Times*, November 27, 1978, A18.

110. Nicholas D. Kristof, "Learning How to Run: A West Texas Stumble," *New York Times*, July 27, 2000, A1.

111. David S. Broder, "Texas Runoff Strains GOP Relations; Some Sour Feelings in Bush Camp over Reagan's Role," *Washington Post*, June 3, 1978, A2.

112. Edward Cody, "The Shah of Iran Gives Assurance of US Support," *Washington Post*, November 1, 1978, A1.

113. Jim Hoagland, "He Also Criticizes Record on Rights," *Washington Post*, December 8, 1978, A1.

114. Jonathan C. Randal, "Tensions in Revolutionary Afghanistan," *Washington Post*, November 7, 1978, A17.

115. Ibid.

116. Edward Walsh, "Carter Asserts Human Rights Is 'Soul of Our Foreign Policy,'" *Washington Post*, December 7, 1978, A2.

117. George F. Will, "SALT Echoes," *Washington Post*, December 14, 1978, A23.

118. Associated Press, May 15, 1978.

119. David S. Broder and Bill Peterson, "Credibility of US Hurt, Critics Say," *Washington Post*, December 16, 1978, A1.

120. Edward Walsh, "US to Normalize Ties with Peking, End Its Defense Treaty with Taiwan," *Washington Post*, December 16, 1978, A1.

121. George F. Will, "A Price for Taiwan," *Washington Post*, December 21, 1978, A15.

122. Hannaford, *The Reagans*, 192.

123. Jonathan Wolman, Associated Press, November 9, 1978.

124. "The Delay on SALT," *Washington Post*, December 24, 1978, D6.

125. David S. Broder, "GOP Strategist, Labor Chief Agree: Elections Hurt Carter," *Washington Post*, November 10, 1978, A2.

126. Tom Braden, "Carter's Midterm Crisis," *Washington Post*, December 30, 1978, A11.

127. "The GOP's Grass Roots," *Washington Post*, November 13, 1978, A22.

CHAPTER 7: REAGAN ON ICE

1. Adam Clymer, "Reagan Backers Decide to Meet Issue of Age Early and Head On," *New York Times*, January 23, 1979, A14.

2. Jack W. Germond and Jules Witcover, "Reagan Targets Issues as Key Strategy for '80," *Washington Star*, May 11, 1978

3. Peter Goldman and Tony Fuller, "The Early Bird," *Newsweek*, March 19, 1979, 37.

4. Peter Hannaford, *The Reagans: A Political Portrait* (New York: Coward-McCann, 1983), 195.

5. Berkeley University, "Life Expectancy in the USA, 1900–98," http://demog.berkeley.edu/~andrew/1918/figure2.html.

6. Ibid.

7. *U.S. News & World Report*, February 5, 1979, 4.

8. Hannaford, *The Reagans*, 196–98, 210.

9. Adam Clymer, "Poll Finds Reagan Keeping Lead; More Democrats Back Kennedy," *New York Times*, November 6, 1979, A1.

10. "The Most Influential Americans: How They Rank," *U.S. News & World Report*, April 16, 1979, 33.

11. Tom Wicker, "Just a Kid in China," *New York Times*, June 16, 1978, A27.

12. Michael Deaver, in discussion with the author, October 18, 2006.

13. Donald Lambro, in discussion with the author, January 24, 2007.

14. Norman C. Miller, "Can a 69-Year-Old Win Presidency? Reagan Eyes 1980," *Wall Street Journal*, July 20, 1978, 1.

15. Adam Clymer, "New Hampshire G.O.P. Aligning for Presidential Primary," *New York Times*, December 23, 1978, 7.

16. John W. Masheck, Thomas J. Foley, and Jack McWethy, "'80 Sweepstakes; At Starting Gate," *U.S. News & World Report*, June 4, 1979, 28.

17. John W. Mashek, "Preview '80 Reagan: Leading Contender, but Age Looms as Issue," *U.S. News & World Report*, May 7, 1979, 54.

18. Lou Cannon, "Reagan Hustling to Defuse Age Issue," *Washington Post*, May 27, 1977, A2.

19. "The Reagan Smear," *Los Angeles Herald-Examiner*, April 6, 1979, A18.

20. George F. Will, "Visions of 1980," *Newsweek*, March 5, 1979, 108.

21. David S. Broder, "Republican Governors: A Wholesome Species," *Washington Post*, December 3, 1978, C7.

22. David S. Broder, "1980 GOP Presidential Field Is Already Crowded," *Washington Post*, December 21, 1979, A1.

23. Robert D. Novak, *The Prince of Darkness: 50 Years of Reporting in Washington* (New York: Crown Forum, 2007), 334.

24. Clymer, "Reagan Backers Decide to Meet Issue of Age Early and Head On," A1.

25. Broder, "1980 GOP Presidential Field Is Already Crowded," A1.

26. Ronald Reagan, *Reagan: A Life in Letters*, ed. Kiron K. Skinner, Annelise Anderson, and Martin Anderson (New York: Simon and Schuster, 2004), 586.

27. Barry Schweid, Associated Press, January 29, 1979; Kenneth Bredemeier, "Goldwater, Other Law Makers Filed Suit over Repeal at Taiwan Defense Pact," *Washington Post*, December 23, 1978, A2.

28. Author's recollection.

29. "White House Solar Heater No Bargain," *Washington Post*, April 6, 1979, A12.

30. Neal R. Pearce and Jerry Hagstrom, "Solar Advocates Damn Carter, Brown with Faint Praise," *National Journal*, August 4, 1979, 1289; Mike Feinsilber, Associated Press, May 7, 1979.

31. Mashek, "Preview '80 Reagan," 54.

32. "Poll Shows Ford Trailing Reagan in a 1980 Race," *New York Times*, January 11, 1979, B7.

33. Germond and Witcover, "Reagan Targets Issues as Key Strategy for '80," A8.

34. Peter Hannaford, in discussion with the author, March 24, 2006.

35. David S. Broder, "Connally Takes the Big Plunge in '80 Campaign," *Washington Post*, January 25, 1979, A1.

36. Edward Mahe, in discussion with the author, May 24, 2006.

37. "George H. W. Bush: Life Before the Presidency," Miller Center, http://millercenter.org/president/biography/bush-life-before-the-presidency.

38. Tom Wicker, "Just a Kid in China," *New York Times*, June 16, 1978, A27.

39. Robert Reinhold, "Poll Indicates Congress Candidates Were More Extreme than Voters," *New York Times*, November 9, 1978, A21.

40. "Carter Leads Ford, Reagan in Poll," *New York Times*, January 21, 1979, 44.

41. Martin Schram, "Reagan Campaign Pays Rent to a Bush Adviser," *Washington Post*, May 12, 1979, A8.

42. Ibid.

43. John Whitcomb and Claire Whitcomb, *Real Life at the White House* (London: Routledge, 2002), 424.

44. David Keene, in discussion with the author, March 17, 2006.

45. Ibid.

46. "Pope Beams Mass to Homeland," *Washington Post*, January 8, 1979, A9.

47. George F. Will, "Carter's Tougher Soviet Standard," *Washington Post*, March 3, 1977, A23.

48. George F. Will, "Promises, Promises," *Washington Post*, January 14, 1979, G7.

49. National Archives, "Executive Order 11967—Relating to Violations of the Selective Service Act, August 4, 1964 to March 28, 1973," http://www.archives.gov/federal-register/codification/executive-order/11967.html.

50. Jack Anderson, "President's Brothers," *Washington Post*, January 14, 1979, G7.

51. Phil Gailey, "Billy Carter's Libyan Affair Causes a Furor in Atlanta," *Washington Star*, January 11, 1979, A1.

52. Fred Barnes, "Carter 'Embarrassed' by Billy's Libyan Ties," *Washington Star*, January 12, 1979, A3.

53. David S. Broder, "GOP to Target SALT for Major Debate on Policy," *Washington Post*, February 4, 1979, A1.

54. Andrew Schneider, Associated Press, December 2, 1978.

55. Adam Clymer, "Five G.O.P. Hopefuls to Aid New Senator," *New York Times*, November 15, 1978, A17.

56. Gerald Carmen, in discussion with the author, July 25, 2006.

57. Andrew Schneider, "Today's Topic: Early Days in the Presidential Circus," Associated Press, November 23, 1978.

58. Michael Knight, "1980 Primary Off to Early Start for G.O.P. in New Hampshire," *New York Times*, December 2, 1978, 12.

59. Gerald Carmen, in discussion with the author, July 25, 2006.

60. Charles Black, in discussion with the author, March 20, 2006.

61. Seth S. King, "Reagan Urges an Alliance to Defy Big Government," *New York Times*, January 16, 1979, A7.

62. Tom Morganthau, James Doyle, Tony Fuller, et al., "A Very Hopeful GOP," *Newsweek*, February 5, 1979, 31.

63. Adam Clymer, "Republicans Reach Out to a Wider Audience," *New York Times*, January 28, 1979, E2.

64. Don McLeod, Associated Press, December 14, 1978.

65. Ibid.

66. Richard Bergholz, "Reagan Edges Closer to Making White House Bid," *Los Angeles Times*, January 14, 1979, A3.

67. David Smick, in discussion with the author, October 18, 2006.

68. David S. Broder, "GOP Chiefs Shun Balancing Budget by Amendments," *Washington Post*, February 5, 1979, A1.

69. Evans Witt, Associated Press, February 11, 1979.

70. Don Oberdorfer, "SALT Would Lock US into Inferiority, Military Men Warn," *Washington Post*, April 12, 1979, A24.

71. Warren Weaver Jr., "Conservatives Find Reason for Elation," *New York Times*, February 10, 1979, 7.

72. "Carter's Vow on Taiwan Is Demanded by Reagan," *New York Times*, February 11, 1979, 9.

73. Richard Bergholz, "Reagan Demands Taiwan Guarantees," *Los Angeles Times*, January 13, 1979, A21.

74. Richard Bergholz, "Reagan to Form Campaign Unit," *Los Angeles Times*, February 6, 1977, B3.

75. "People in the News," Associated Press, August 16, 1977.

76. Adam Clymer, "G.O.P. Presidential Aspirants Tour Nation to Denounce Carter's Foreign Policy," *New York Times*, February 20, 1979, A12.

77. Hedrick Smith, "More Than a Feeling You Can't Push America Around," *New York Times*, February 25, 1979, E4.

78. Wayne King, "President Gets Low Ratings in Survey of Georgia Voters," *New York Times*, February 20, 1979, A12.

79. Smith, "More Than a Feeling You Can't Push America Around," E4.

80. Douglas E. Kneeland, "Republican Hopefuls Tread the Carter Path to Iowa," *New York Times*, March 4, 1979, E4.

81. Craig Shirley, "Does Jeremiah Wright Still Matter?," *Politico*, May 17, 2012, http://www.politico.com/arena/archive/jeremiah-wright-chickens-coming home-to-roost.html.

82. Bill Paterson, "Reagan for President Committee Is Formed, but He Hasn't Announced Candidacy—Yet," *Washington Post*, March 8, 1979, A2.

83. Robert Shogan, "Reagan Takes Big Step Toward 1980 Race," *Los Angeles Times*, March 8, 1979.

84. Cindy Canevaro, in discussion with the author, March 11, 2006; Helene von Damm, in discussion with the author, May 17, 2006.

85. Richard Bergholz, "Reagan Campaign Fund Solicitation Begins," *Los Angeles Times*, March 19, 1979, A3.

CHAPTER 8: BREAD AND CIRCUSES

1. Bruce Reed, "Panic Button," *Slate*, January 20, 2006, http://www.slate.com/ articles/news_and_politics/the_hasbeen/2006/01/panic_button.html.

2. David Frum, *How We Got Here: The 70's, the Decade That Brought You Modern Life—for Better or Worse* (New York: Basic Books, 2000), 155.

3. William H. Jones, "Food Here Up 7% in 3 Months," *Washington Post*, March 24, 1979, A1.

4. David S. Broder, "A President Confronts an Old Coalition," *Washington Post*, January 15, 1979, A1.

5. Barry Schweld, Associated Press, February 17, 1979.

6. Jim Hoagland, "Soviet Role Alleged in Dubs' Death," *Washington Post*, February 22, 1979, A1.

7. Rowland Evans and Robert Novak, "Moscow: Dirty Tricks and Disinformation," *Washington Post*, February 16, 1979, A15.

8. George F. Will, "The Administration's Unpredictable Voice," *Washington Post*, February 22, 1979, A17.

9. David S. Broder, "Democratic Party in Transition, but the Question Is, to What?," *Washington Post*, January 14, 1979, A1.

10. Martin Schram, "Birth of a Notion," *Washington Post*, January 24, 1979, A1.

11. Ibid.

12. T. R. Reid, "GOP Leaders Move to Parry Carter Speech," *Washington Post*, January 25, 1979, A4.

13. Theodore H. White, *America in Search of Itself: The Making of the President, 1956–1980* (New York: Harper & Row, 1982), 157.

14. John Herbers, "Washington an Insider's Game," *New York Times Magazine*, April 22, 1979, SM9.

15. Ibid.

16. Ibid.

17. Isabelle Shelton, "Percy to Ask Parking Levy for Capitol Hill Aides Again," *Washington Star*, April 7, 1979, A4.

18. Jack W. Germond, "Carter Has a Tough Act to Follow—Himself," *Washington Star*, May 5, 1979, A6.

19. James R. Dickenson, "Crane's Fledgling Presidential Campaign Beset by Full Share of Woes," *Washington Star*, May 13, 1979, A4.

20. Tom Morganthau, James Doyle, Tony Fuller, et al., "Jostling in the GOP," *Newsweek*, May 14, 1979, 38.

21. Peter Goldman and Tony Fuller, "The Early Bird," *Newsweek*, March 19, 1979, 37.

22. Adam Clymer, "Reagan Camp Scoffs at Reports Support for His Nomination Is Dwindling," *New York Times*, April 30, 1979, A15.

23. Associated Press, March 12, 1979.

24. Rowland Evans and Robert Novak, "Open Stage for Connally at a GOP Show," *Washington Post*, March 14, 1979, A23.

25. Diane Henry, "Weicker Joins Race for the Presidency," *New York Times*, March 13, 1979, A1.

26. Ibid.

27. Andrew Schneider, Associated Press, March 12, 1979; Stacy Jolna, "Prime Ribs, Hoopla for GOP Hopefuls," *Washington Post*, April 9, 1979, A3.

28. Jolna, "Prime Ribs, Hoopla for GOP Hopefuls," A3.

29. Clyde Haberman and Albin Krebs, "Notes on People," *New York Times*, April 13, 1979, C15.

30. Lynn Darling, "From 'Butch Cassidy' to the Sun Lobby Kid," *Washington Post*, March 21, 1979, B1.

31. John Tierney, "Anti-Nuke Rally to Draw Thousands to Capitol," *Washington Star*, May 5, 1979, A6.

32. Howell Raines, "G.O.P. Contenders Begin Efforts Early in the South," *New York Times*, April 20, 1979, A14.

33. William Safire, "Dudley W. Dudley," *New York Times*, September 13, 1979, A27.

34. "US President—R Convention: State Level Results," Our Campaigns, http://www.ourcampaigns.com/RaceDetail.html?RaceID=57992.

35. Albert R. Hunt, "Reagan Campaign Staff Squabbles Seen Hurting Him; One Aide Switches to Rival," *Wall Street Journal*, February 20, 1979, 15.

36. Raines, "G.O.P. Contenders Begin Efforts Early in the South," A14.

37. Tish Leonard, in discussion with the author, September 12, 2006.

38. Martin Schram, "Reagan Adviser Quits to Direct Bush Campaign," *Washington Post*, February 20, 1979, A4.

39. Ibid.

40. Rowland Evans and Robert Novak, "Trouble for Reagan," *Washington Post*, February 26, 1979, A21.

41. James Reston, "The New Decline of the West," *New York Times*, April 22, 1979, E19.

42. Ibid.

43. Hedrick Smith, "Carter Shows a Lot of '76 Political Magic Remains," *New York Times*, April 27, 1979, A1.

44. Ibid.

45. Frank Cormier, Associated Press, April 23, 1979.

46. Terence Smith, "Jimmy Carter, Now the Insider, Dusts Off the Outsider Appeal," *New York Times*, April 29, 1979, A4.

47. Donald M. Rothberg, Associated Press, May 1, 1979.

48. Bill Curry, "Bush Declares Presidential Candidacy," *Washington Post*, May 2, 1979, A3.

49. Jack W. Germond and Jules Witcover, "Bush Family Opens '80 Drive," *Washington Star*, May 1, 1979, A1.

50. Jack W. Germond and Jules Witcover, "The Coaching of Candidate Bush," *Washington Star*, May 8, 1979, A4.

51. Bill Curry, "Bush Takes Hard Line on Foreign Policy," *Washington Post*, May 3, 1979, A2.

52. Nick Thimmesch, "Bush: 'Confidence Itself,'" *Washington Post*, May 3, 1979, A19.

53. Bill Curry, "Stumbles Aside, Bush Off on Right Foot," *Washington Post*, May 6, 1979, A16.

54. Bill Peterson, "Bickering Racks Crane's Presidential Campaign," *Washington Post*, May 5, 1979, A2; Bill Peterson, "Beleaguered Crane Campaign Takes Another Jolt," *Washington Post*, May 12, 1979, A2.

55. Daniel Yergin, "Margaret Thatcher: Convictions of a Conservative," *Washington Post*, March 26, 1979, A23.

56. Bill Curry, "Dole Opens Presidential Campaign," *Washington Post*, May 15, 1979, A2.

57. David Keene, in discussion with the author, March 17, 2006.

58. Ibid.

59. Adam Clymer, "Senator Dole Joins G.O.P. Race for '80 Presidential Elections," *New York Times*, May 15, 1979, A1.

60. David S. Broder, "GOP Is Haunted by Ghost of '76: Ford vs. Reagan," *Washington Post*, July 23, 1978, A1.

61. Ibid.

62. Ibid.

63. Diane Henry, "Weicker Pulls Out of G.O.P. Race," *New York Times*, May 17, 1979, A20.

64. Richard Bergholz, "High-Ranking Aide in Reagan Campaign Quits," *Los Angeles Times*, August 28, 1979, B3.

65. Ronald Reagan, *Reagan: A Life in Letters,* edited by Kiron K. Skinner, Annelise Anderson, and Martin Anderson (New York: Free Press, 2003), 227.

66. Michael K. Deaver and Mickey Herskowitz, *Behind the Scenes: In Which the Author Talks About Ronald and Nancy Reagan . . . and Himself* (New York: William Morrow, 1987), 85.

67. Bob Colacello, *Ronnie and Nancy: Their Path to the White House, 1911 to 1980* (New York: Warner Books, 2004), 477.

68. Martin Schram, "Reagan Would Accept 2nd Spot on GOP Ticket," *Washington Post*, June 23, 1979, A5.

69. Richard Ebeling, "The Lasting Legacy of Ronald Reagan," Foundation for Economic Education, July 1, 2005, https://fee.org/articles/the-lasting-legacy -of-the-reagan-revolution/.

70. Thomas W. Evans, *The Education of Ronald Reagan* (New York: Columbia University Press, 2006), 77.

71. "NASA Takes 'Meatball' over 'Worm,'" *Roanoke Times*, May 24, 1992, C15.

72. Peter Gwynne and Alfred Friendly Jr., "Link-Up in Space," *Newsweek*, July 21, 1975, 46.

73. Mary McGrory, "Carter's at War with His Party, Estranged from the People," *Washington Star*, May 25, 1979, A4.

CHAPTER 9: UP FROM CARTERISM

1. Kevin Klose, "Brezhnev Demonstrates His Political Grip Is Still Strong," *Washington Post*, December 3, 1978, A23.
2. George F. Will, "The Morals of the Market Place," *Washington Post*, May 3, 1979, A19.
3. Adam Clymer, "Ford Cites Economy as Major G.O.P. Issue," *New York Times*, May 20, 1979, 27.
4. Robert G. Kaiser, "Jackson Rips 'Appeasement' of Moscow," *Washington Post*, June 13, 1979, A1.
5. Vernon C. Thompson and Courtland Milloy, "Gas Lines Long, Tempers Short in Panic Buying," *Washington Post*, June 9, 1979, A1.
6. Peter Goldman, James Doyle, Thomas M. DeFrank, et al., "The Politics of Gas," *Newsweek*, July 9, 1979, 29.
7. Donald P. Baker, "Odd-Even Ration Plan Begins in Metro Area," *Washington Post*, June 21, 1979, A1.
8. Thomas Grubisich, "Labor Day Swim-In, Ski-In to Dramatize 'Clean Potomac,'" *Washington Post*, September 1, 1978, A1.
9. Bill Peterson, "Gas Pumps on Hill Cater to Wheels," *Washington Post*, June 6, 1979, A1.
10. Ibid.
11. Bradley Graham, "Gold Soars to $440, Then Falls Sharply in Frenzied Trading," *Washington Post*, October 3, 1979, A7.
12. Art Pine, "Major Recession Now Forecast by Hill Budget Office," *Washington Post*, June 10, 1979, A1.
13. Bill Peterson, "Kennedy Candidacy: A Phantom Haunts the White House," *Washington Post*, June 4, 1979, A3.
14. Goldman, Doyle, and DeFrank, et al., "The Politics of Gas."
15. Helen Dewar, "Labor Leader Eyes Switch to the GOP," *Washington Post*, July 13, 1979, A1.
16. Richard Bergholz, "Let Gasoline Prices Rise, Reagan Says," *Los Angeles Times*, June 21, 1979, A1.
17. Jack W. Germond and Jules Witcover, "GOP 'Cattle Shows' Lack Stars," *Washington Star*, June 27, 1979, A9.
18. "Iowa Republican Straw Poll," *Des Moines Register*, May 23, 1979, 3.
19. Douglas E. Kneeland, "Bush Is Buoyed by Iowa Straw Poll; Pleased by Results of Poll," *New York Times*, May 24, 1979, B12.
20. "Ford Wins His Battle: Now for the War," *Economist*, August 21, 1976.
21. Kenneth Reich, "Reagan's Problem in Mississippi," *Los Angeles Times*, October 27, 1979, B1.

22. Robert Shogan, "Connally Takes Aim at Reagan's Big Lead," *Los Angeles Times*, August 12, 1979, A1.

23. Charles Pickering, in discussion with the author.

24. "Reagan and Ford About Even in Gallup Poll," *New York Times*, June 3, 1979, 27.

25. John F. Stacks, "Carter, Reagan Equal in Voter Popularity, New Survey Indicates," *Washington Star*, April 23, 1979, A1.

26. Ibid.

27. Rowland Evans and Robert Novak, "Reagan: Losing Luster in Dixie," *Washington Post*, June 13, 1979, A17.

28. Jim Mason, *No Holding Back: The 1980 John Anderson Presidential Campaign* (Lanham, MD: University Press of America, 2011), 3.

29. Adam Clymer, "Carter's Standing Drops to New Low in Times-CBS Poll," *New York Times,* June 10, 1979, 1.

30. Ibid.

31. "Unusual Prospects Seeking Presidency," *New York Times*, June 10, 1979, 25.

32. Adam Clymer, "Role of Issues Held Minor for '80 Race," *New York Times*, June 11, 1979, A13.

33. C. Gerald Fraser, "Wayne Hailed as Symbol of American Heroism," *New York Times*, June 13, 1979, B10.

34. Edward Walsh, "Carter in California: Gasoline Lines and Mex-American Disenchantment," *Washington Post*, May 6, 1979, A12.

35. Hedrick Smith, "Candidates Running for Visibility in New Hampshire," *New York Times*, September 17, 1979, A22.

36. John Herbers, "G.O.P. Ends Its National Meeting with an Optimistic View of 1980," *New York Times*, June 27, 1979, A14.

37. Richard Willing, "Republican Leaders Rate Reagan Party's Strongest Candidates," *Washington Star*, June 24, 1979, A7.

38. Ed Blakely, in discussion with the author, April 17, 2007.

39. Elizabeth Drew, *Portrait of an Election: The 1980 Presidential Campaign* (New York: Simon and Schuster, 1981), 108–9.

40. Merrill Brown, "Reagan Wins Challenge to Radio Show," *Washington Star*, July 19, 1979, A4.

41. Ronald Koven, "John Paul: Only So Many Change," *Washington Post*, June 2, 1979, A1.

42. Ibid.

43. "The Millions in Poland," *Washington Post*, June 10, 1979, D6.

44. Adam Clymer, "G.O.P. Candidates Differ on Gas-Shortage Policy," *New York Times*, June 28, 1979, B9.

45. Hedrick Smith, "Intimates Fearful for Re-election If Carter Fails to Tame Gas Crisis," *New York Times*, June 29, 1979, A1.

46. Peter D. Hart, "Carter Has a Strength Not Shown in the Polls," *Chicago Tribune*, August 12, 1979, 23.

47. "A Look Ahead from the Nation's Capital," *U.S. News & World Report*, August 27, 1979, 13.

48. Theodore H. White, *America in Search of Itself: The Making of the President, 1956–1980* (New York: Harper & Row, 1982), 137–38.

49. Associated Press, August 2, 1979.

50. Peter Gwynne and Mary Lord, "Chicken Little Day," *Newsweek*, June 4, 1979, 57.

51. "SALT Pact Supported by 69% of Americans in Poll by ABC News," *Washington Star*, June 18, 1979, A6.

52. Zofia Smardz, "Connally Gains on Reagan in Virginia GOP Meeting," *Washington Star*, July 1, 1979, A6.

53. Karlyn Barker, "Connally Captures Interest of Virginia GOP Delegates," *Washington Post*, July 1, 1979, A14.

54. Adam Clymer, "Connally and Reagan Offer 2 Brands of Conservatism to Virginia G.O.P.," *New York Times*, July 2, 1979, A14.

55. Albin Krebs, "Notes on People," *New York Times*, July 2, 1979, B7.

56. Steven F. Hayward, *The Real Jimmy Carter* (Washington, DC: Regnery Gateway, 2004), 145.

57. Ibid.

58. Harry Kelly, "Nation's Morale Torments Carter," *Chicago Tribune*, July 15, 1979, B12.

59. Jeremiah O'Leary, "Seeking People's Opinions," *Washington Star*, July 13, 1979, A1.

60. "Many Businessmen Say Carter's Actions Strengthen Doubts of His Ability to Lead," *Wall Street Journal*, July 9, 1979, 18.

61. Fred Barnes, "Carter Considers Staff Shake-up, Sees Governors," *Washington Star*, July 7, 1979, A1.

62. John F. Stacks, *Watershed: The Campaign for the Presidency, 1980* (New York: Times Books, 1981), 19.

63. Jack W. Germond, "Falling Polls Convinced Him Confidence Lost," *Washington Star*, July 14, 1979, A1.

64. Peter Goldman, Thomas M. DeFrank, James A. Doyle, et al., "To Lift a Nation's Spirit," *Newsweek*, July 23, 1979, 20.

65. Steven F. Hayward, *The Real Jimmy Carter* (Washington, DC: Regnery Gateway, 2004), 145.

66. "Transcript of President's Address to Country on Energy Problems," *New York Times*, July 16, 1979, A10.

67. Dominic Sandbrook, *Mad as Hell: The Crisis of the 1970s and the Rise of the Populist Right* (New York: Alfred A Knopf, 2011), 303.

68. "Transcript of President's Address to Country on Energy Problems," *New York Times*, July 16, 1979, A10.
69. Adam Clymer, "Reagan: The 1980 Model," *New York Times*, July 29, 1979, 24.
70. Jeremiah O'Leary and Jack W. Germond, "Energy Plan Cost Put at $140 Billion," *Washington Star*, July 16, 1979, A1.
71. Ibid.
72. Robert G. Kaiser and Richard L. Lyons, "Speech Draws Both Applause, Partisan Gibes," *Washington Post*, July 16, 1979, A1.
73. George F. Will, "A Reluctant Broker," *Washington Post*, July 19, 1979, A19.
74. Rowland Evans and Robert Novak, "A National Revival," *Washington Post*, July 16, 1979, A19.
75. Ibid
76. Walter Mondale, in discussion with the author, February 28, 2007.
77. Stacks, *Watershed*, 20–21.
78. Peter Goldman, Thomas M. DeFrank, Rich Thomas, et al., "Jimmy Carter's Cabinet Purge," *Newsweek*, July 30, 1979, 22.
79. Joanne Omang, "Carter Urged to Broaden Inner Circle," *Washington Post*, July 22, 1979, A7.
80. David S. Broder, "President, Rosalynn and Amy Planning a Trip on the Riverboat *Delta Queen*," *Washington Post*, August 10, 1979, A2.
81. Ron Alexander, "Some of the Skylab Fallout Is on the Light Side," *New York Times*, July 7, 1979, 42.
82. Associated Press, July 16, 1979.
83. Lance Gay, "Winisinger Blames 'Crisis' on President," *Washington Star*, July 19, 1979, A4.
84. Glenn Ritt, Associated Press, August 2, 1979.
85. Lance Gay, "Consumer Prices Climb 1% Because of Energy Costs," *Washington Star*, July 26, 1979, A1.
86. White, *America in Search of Itself*, 161.
87. Leonard Curry, "Recession Is Here, Miller Says," *Washington Star*, July 28, 1979, A1.
88. Steven V. Roberts, "How Carter Can Win," *New York Times Magazine*, November 25, 1979, SM9.
89. Jeremiah O'Leary, "Carter Takes His Case to Kentuckians," *Washington Star*, July 31, 1979, A1.
90. "Poll on President Drops to New Low," *Washington Star*, July 31, 1979, A1.
91. Stanley W. Cloud, "Jordan Accusers 'Telling Lies' Carter Says," *Washington Star*, September 2, 1979, A3.
92. Jack Kemp, "'The People Aren't Fooled,'" *Washington Post*, June 5, 1979, A17.
93. George F. Will, "Thatcher Governing as Thatcher," *Washington Post*, June 28, 1979, A19.

94. Mike Shanahan, Associated Press, September 1, 1979.

95. Alan Ehrenhalt, "GOP to Try Tories' Tactic That Won in British Voting," *Washington Star*, July 31, 1979, A5.

96. Newt Gingrich, in discussion with the author, May 21, 2006.

97. Cynthia Kadonaga, Associated Press, July 31, 1979.

98. Kenneth A. Briggs, "Marcuse, Radical Philosopher, Dies," *New York Times*, July 31, 1979, A1.

99. Leonard Silk, "Carter Expects Rise in Joblessness; Believes G.O.P. Will Pick Reagan," *New York Times*, August 1, 1979, A1.

100. Ibid.

101. Dean Reynolds, "Does Reagan Owe Victory to Carter?," United Press International, November 8, 1980.

102. Bill Peterson, "Reagan Tests Political Water in a Shower of Fees," *Washington Post*, August 15, 1979, A2.

103. Jack W. Germond and Jules Witcover, "Reagan on Abortion," *Washington Star*, August 26, 1979, A9.

104. Douglas E. Kneeland, "Republican 'Presidential Forums' Get Big Turnouts, but Candidates Grumble," *New York Times*, August 5, 1979, 26.

105. Rowland Evans and Robert Novak, "Baker: Hardball in the Deep South," *Washington Post*, June 6, 1979, A21.

106. Ibid.

107. George C. Wilson, "'Counterforce' Arms Attract US, Soviets," *Washington Post*, June 1, 1979, A1.

108. "Mr. Gromyko's Gratuitous Advice," *Washington Post*, June 28, 1979, A18.

109. Doug Willis, "Millionaire Candidate Secretive About Finances," Associated Press, July 5, 1980.

110. "Reagan Got $72,840 in 18 Months for His Talks to G.O.P. Groups," *New York Times*, August 16, 1979, A20.

111. "Reagan's Candidacy Is Endorsed by Young Americans for Freedom," *New York Times*, August 19, 1979, 37.

112. Ibid.

113. Warren Weaver, "Conservative Youth Group Finds Itself in a More Centrist Position," *New York Times*, August 19, 1979, A1.

114. Jules Witcover, "It's Laughter for Reagan at YAF's Mock Convention," *Washington Star*, August 19, 1979, A1.

115. John Russell, "Books: Raymond Massey on the Stage of the Past," *New York Times*, August 20, 1979, C18.

116. Rowland Evans and Robert Novak, "Reagan's Age," *Washington Post*, June 22, 1979, A19.

117. Jules Witcover, "Reagan Enlists Key Pennsylvanian Who Got Away in 1976 Campaign," *Washington Star*, June 20, 1979, A4.

118. "A Look Ahead from the Nation's Capital," *U.S. News & World Report*, June 25, 1979, 13.

119. Bill Peterson, "Kennedy's Strength May Be Exaggerated," *Washington Post*, June 25, 1979, A4.

120. John F. Stacks, "Reagan Leads Carter in Newest Time Poll," *Washington Star*, July 11, 1979, A3.

121. Ibid.

122. Robert Pear, "Connally Leads Pack in Fund-Raising," *Washington Star*, July 11, 1979, A3.

123. Associated Press, July 12, 1979.

124. Mike Glover, "Dole Opens Intense Iowa Drive," Associated Press, December 30, 1987.

125. White, *America in Search of Itself*, 245, 249.

CHAPTER 10: "BIG JOHN" VERSUS "POPPY"

1. "Again, Connally for Veep?," *Time*, August 2, 1976, 12.

2. Albert R. Hunt, "John Connally Stirs Up Many Strong Emotions Among Friend and Foe," *Wall Street Journal*, August 6, 1979, 1.

3. Thomas M. DeFrank, *Write It When I'm Gone: Remarkable Off-the-Record Conversations with Gerald R. Ford* (New York: G. P. Putnam's Sons, 2007), 110.

4. Elizabeth Drew, *Portrait of an Election: The 1980 Presidential Campaign* (New York: Simon and Schuster, 1981), 108.

5. Hedrick Smith, "Ford, Warning on US Security, Seems to Encourage G.O.P. Draft," *New York Times*, September 28, 1979, A16.

6. Helene von Damm, in discussion with the author, May 17, 2006.

7. Ibid.

8. Adam Clymer, "Bush a Hot Property on G.O.P. Dinner Circuit," *New York Times*, February 19, 1978, 26.

9. Lou Cannon, Bill Curry, and Margot Hornblower, et al., "George Bush: Hot Property in Presidential Politics," *Washington Post*, January 27, 1980, A1.

10. Norman C. Miller, "Perspective on Politics," *Wall Street Journal*, February 14, 1980, 26.

11. Paul Hendrickson, "Into the Marathon with Earnest George Bush," *Washington Post*, May 24, 1979, D1.

12. John F. Stacks, *Watershed: The Campaign for the Presidency, 1980* (New York: Times Books, 1981), 97.

13. Allan J. Mayer, Stryker McGuire, Jerry Buckley, et al., "Bush Breaks Out of the Pack," *Newsweek*, February 4, 1980, 30.

14. Ibid.

15. Rowland Evans and Robert Novak, "George Bush, Jogging," *Washington Post*, August 8, 1979, A17.

16. Stacks, *Watershed*, 97.

17. William Safire, "A Primary Paradox," *New York Times*, June 8, 1979, A27.

18. Nicholas D. Kristof, "Learning How to Run: A West Texas Stumble," *New York Times*, July 27, 2000, A1.

19. Ibid.

20. Bill Minutaglio, *First Son: George W. Bush and the Bush Family Dynasty* (New York: Times Books, 1999), 187; Lynn Nofziger, *Nofziger* (Washington, DC: Regnery Gateway, 1992), 224.

21. "A Dying Party, 5 Republican Leaders Speak Out," *U.S. News & World Report*, August 29, 1977, 22.

22. Cannon, Curry, and Hornblower, et al., "George Bush: Hot Property in Presidential Politics," A1.

23. John W. Mashek, "Preview '80 Texas' Connally: Bucking Odds, Coming Up Fast," *US News & World Report*, July 2, 1979, 29.

24. Cannon, Curry, Hornblower, et al., "George Bush: Hot Property In Presidential Politics," A1.

25. Timothy S. Robinson, "Jury Selection Under Way: Connally Bribe Trial Opens," *Washington Post*, April 2, 1975, A4.

26. Ibid.

27. Victor Gold, in discussion with the author, April 5, 2007.

28. George F. Will, "The Joy of Politics," *Newsweek*, November 26, 1979, 128.

29. Steven Brill, "Connally: Coming on Tough," *New York Times Magazine*, November 18, 1979, 10.

30. David S. Broder, "Reagan Defeats Connally in Florida Test," *Washington Post*, November 18, 1979, A1.

31. Robert Shogan, "Florida Republicans Hold Early Lottery on Delegates to Presidential Convention," *Los Angeles Times*, August 19, 1979, A4.

32. Fred Barbash, "Florida GOP Convention Criticized," *Washington Post*, August 13, 1979, A3.

33. Shogan, "Florida Republicans Hold Early Lottery on Delegates to Presidential Convention," A4.

34. Rita Kempley, "'Blaze': Louisiana's Long Hot Story," *Washington Post*, December 13, 1989, B1; Charles E. Cook, "'Blaze' Is the Best Movie About Politics, If Not Ever," *Roll Call*, December 18, 1989.

35. Fred Barbash, "Florida GOP Convention Criticized," *Washington Post*, August 13, 1979, A3.

36. "One Slip Too Many for President Carter," *U.S. News & World Report*, August 27, 1979, 7.

37. Warren Weaver Jr., "Conservatives Plan $700,000 Drive to Oust 5 Democrats from Senate," *New York Times*, August 17, 1979, A1.

38. Mark Green, "Financing Campaigns," *New York Times*, December 14, 1980, 21.

39. Fred Barbash, "GOP Pre-Season Exhibition Matches Continue in Florida," *Washington Post*, August 26, 1979, A4.

40. Adam Clymer, "Connally Organizing Campaign Workers in Florida," *New York Times*, September 16, 1979, 35.

41. Jack W. Germond, and Jules Witcover, "Bush Trying to Sneak Up on Foes in Florida," *Washington Star*, December 19, 1979, A3.

42. Roy Reed, "George Bush on the Move," *New York Times Magazine*, February 10, 1980, SM5.

43. Adam Clymer, "Both Reagan and Rivals Are Depending on Early Victories," *New York Times*, October 8, 1979, A14.

44. "Bush, in Manhattan, Describes P.L.O. as Similar to the Klan," *New York Times*, October 19, 1979, A18.

45. Bernard Weinraub, "Connally Urges Israelis to Leave Occupied Lands," *New York Times*, October 12, 1979, A1.

46. Ronald Reagan, "Recognizing the Israeli Alert," *Washington Post*, August 15, 1979, A25.

47. Weinraub, "Connally Urges Israelis to Leave Occupied Lands," A1.

48. "Foreign Policy—Where the Candidates Stand," *U.S. News & World Report*, March 17, 1980.

49. Associated Press, October 13, 1979.

50. Adam Clymer, "Baker and Dole Castigate Connally for Talk on Israeli Role in Mideast," *New York Times*, October 13, 1979, 44.

51. James R. Dickenson, "It's Numbers at Iowa GOP Parley," *Washington Star*, October 14, 1979.

52. Douglas E. Kneeland, "Bush Easily Wins Iowa Straw Vote for Republican Presidential Choice," *New York Times*, October 16, 1979, B10.

53. Dickenson, "It's Numbers at Iowa GOP Parley."

54. Douglas E. Kneeland, "9 G.O.P. Hopefuls Vie for Attention at an Iowa Dinner," *New York Times*, October 15, 1979, A14.

55. Dickenson, "It's Numbers at Iowa GOP Parley."

56. James R. Dickenson, "Bush Completes Sweep of Five Straw Votes in Iowa," *Washington Star*, October 15, 1979.

57. Adam Clymer, "Connally Is First to Air a 1980 Campaign Commercial," *New York Times*, October 31, 1979, A16.

58. Ibid.

59. "PostScript," *Washington Post*, March 26, 1979, A2.

60. Don McLeod, Associated Press, October 31, 1979.

61. Adam Clymer, "Connally Is Seeking Victory in Florida," *New York Times*, November 17, 1979, 12.

62. David S. Broder, "No Longer 'George Who?'" *Washington Post*, November 21, 1979, A17.

63. Adam Clymer, "Reagan Wins a Poll of G.O.P. in Florida," *New York Times*, November 18, 1979, 1.

64. Broder, "No Longer 'George Who?,'" A17.

65. Richard Bergholz, "Reagan Expected to Win Straw Vote That He Considers Meaningless in Florida Today," *Los Angeles Times*, November 17, 1979, A13.

CHAPTER 11: GEORGIA VERSUS GEORGETOWN

1. Tom Shales, "Teddy's Torment: A TV Soap Opera," *Washington Post*, November 22, 1979, C1.

2. Peter Goldman, Eleanor Clift, Thomas M. DeFrank, et al., "The Politics of Austerity," *Newsweek*, January 29, 1979, 20.

3. Steven V. Roberts, "Ted Kennedy: Haunted by the Past," *New York Times*, February 3, 1980, SM114.

4. Phil Gailey, "Kennedy-Carter Rift Is Steadily Widening," *Washington Star*, August 5, 1979, A1.

5. Charles Mohr, "Carter Scores Ford on Missiles Sale," *New York Times*, October 1, 1976, 19.

6. Clayton Fritchey, "The 'Real' Carter: A Republican at Heart," *Washington Post*, September 2, 1978, A17.

7. Jack W. Germond and Jules Witcover, "Brown Goes to N.H. but Ted Is the Star," *Washington Star*, September 10, 1979, A1.

8. John Persinos, in discussion with the author.

9. Gailey, "Kennedy-Carter Rift Is Steadily Widening," A1.

10. Donald M. Rothberg, Associated Press, September 8, 1979.

11. Phil Gailey, "Kennedy-Carter Rift Is Steadily Widening," A1.

12. "Carter's Job Rating at 19%, the Lowest Since the 1950's," *Washington Post*, September 14, 1979, A2.

13. Phil Gailey and Jeremiah O'Leary, "Doctor Says President in Excellent Health," *Washington Star*, September 16, 1979, A1.

14. Ibid.

15. "Economy Is Key to Run by Kennedy," *Washington Star*, September 11, 1979, A1.

16. "Carter Sinks Deeper in Latest ABC Poll," *Washington Star*, September 11, 1979, A10.

17. "How the Race Is Run," *New York Times*, September 16, 1979, E18.

18. "Jackson Insists Soviet Withdraw Planes in Cuba," *New York Times*, September 12, 1979, A1.

19. Ibid.

20. Richard Harwood, "How to Resolve the Cuba 'Crisis': Views of the Candidates," *Washington Post*, October 18, 1979, A7.

21. Associated Press, October 14, 1979.

22. Ronald Reagan, *Reagan: A Life in Letters*, ed. Kiron K. Skinner, Annelise Anderson, and Martin Anderson (New York: Free Press, 2003), 228.

23. James Reston, "Carter and Kennedy," *New York Times*, September 14, 1979, A25.

24. "Carter Denies Kennedy Asked Him to Quit Race," *New York Times*, September 10, 1979, C20.

25. Howell Raines, "Carter Campaign Struggles to Regain Ground in South," *New York Times*, September 24, 1979, A17.

26. Jack W. Germond and Jules Witcover, "Bad Day for Carter with Southern Governors," *Washington Star*, October 3, 1979, A3.

27. "Atheist's Suit Seeks to Bar Mass on Mall," *Washington Post*, September 18, 1979, A3.

28. "News Summary," *New York Times*, October 15, 1979, B1.

29. James Reston, "Politics as Theater," *New York Times*, October 17, 1979, A27.

30. "Kennedy Near a Campaign," *Detroit Free Press*, October 19, 1979.

31. James R. Dickenson, "Carter Sets Campaign Themes on Illinois Trip," *Washington Star*, October 17, 1979, A3.

32. Caroline Rand Herron, Michael Wright, and Daniel Lewis, "Candidacies Get Somewhat Nearer the Real Thing," *New York Times*, October 21, 1979, E4.

33. Phil Gailey, "Carter Defends Record, Plays Down '80 Race," *Washington Star*, October 10, 1979, A1.

34. Jules Witcover, "Carter Wins as Expected in Florida Straw Poll," *Washington Star*, November 19, 1979, A3.

35. Phil Gailey and Jules Witcover, "Political Note Touches JFK Library Dedication," *Washington Star*, October 21, 1979.

36. John F. Stacks, "Kennedy's Huge Lead over Carter Dwindles to 10 Points, Poll Shows," *Washington Star*, November 5, 1979, A1.

37. Adam Clymer, "Many Republicans Are Fearful of Kennedy Candidacy," *New York Times*, October 24, 1979, A18.

38. Steven V. Roberts, "How Carter Can Win," *New York Times Magazine*, November 25, 1979, SM9.

39. Paul Taylor, "City Seeks to Bury the Past," *Washington Post*, November 23, 1983, A3.

40. Edward Walsh, "Kennedy Gets Secret Service Protection," *Washington Post*, September 21, 1979, A1.

41. Jo Thomas, "Secret Service Protection for 5 Candidates to Begin Jan. 11," *New York Times*, October 30, 1979, B12.

42. Guy Darst, Associated Press, November 5, 1979.

43. "Citibank Jumps Its Prime Rate to 15 1/4 Percent," *Washington Star*, October 26, 1979, A1.

44. "Poll on Primary Finds Kennedy and Reagan Lead New Hampshire," *New York Times*, October 29, 1979, A17.

45. Donald Lambro, "Mrs. Carter's Staff of 21 Is Largest for First Lady," November 4, 1979, *Washington Post*, A3.

46. Associated Press, November 3, 1979.

47. Adam Clymer, "Poll Finds Reagan Keeping Lead; More Democrats Back Kennedy," *New York Times*, November 6, 1979, A1.

48. Edward Walsh, "Carter, in '80 Campaign Preview, Defends Record," *Washington Post*, October 25, 1979, A6.

49. "Television," *Time*, December 8, 1967.

50. Jamin Soderstrom, *Qualified: Candidate Resumes and the Threshold for Presidential Success* (Bloomington, IN: iUniverse, 2011), 267.

51. Tom Shales, "Teddy's Torment: A TV Soap Opera," *Washington Post*, November 22, 1979, C1.

52. Tom Wicker, "Reprieved by 'Jaws,'" *New York Times*, November 9, 1979, A35.

53. Shales, "Teddy's Torment," C1.

54. Associated Press, November 7, 1979.

55. Jules Witcover, "Kennedy Launches His Campaign with Blast at Carter Leadership," *Washington Star*, November 7, 1979, A1.

56. Walter Mondale, in discussion with the author, February 28, 2007.

57. Mark Bowden, *Guests of the Ayatollah: The Iran Hostage Crisis, the First Battle in America's War with Militant Islam* (New York: Grove Press, 2006), 2.

58. "Anti-Iranian Protests Continue Across US," *Washington Star*, November 11, 1979, A1.

59. Zofia Smardz and Joan Lowy, "Agency Calls Action Its 'No. 1 Priority,'" *Washington Star*, November 12, 1979, A1.

60. Peter Goldman, John Walcott, John J. Lindsay, et al., "Kennedy's Blooper," *Newsweek*, December 17, 1979, 46.

61. "Black Hostage Reports Abuse," *New York Times*, January 27, 1981.

62. Reagan, *Reagan: A Life in Letters*, 234.

63. Hedrick Smith, "Price for Kennedy's Quick Entry into Campaign Was a Ragged Start," *New York Times*, November 15, 1979, B16.

64. Jules Witcover, "Chappaquiddick Interest Surprises Kennedy Backer," *Washington Star*, November 16, 1979, A9.

65. Jules Witcover, "Carter Wins as Expected in Florida Straw Poll," *Washington Star*, November 19, 1979, A9.

66. Barry Sussman, "Post Poll Shows Potentially Sharp Democratic Split," *Washington Post*, November 18, 1979, A1.

67. David Espo, "Kennedy Says He Should Have Been More Outspoken," Associated Press, February 2, 1980.

68. B. Drummond Ayres Jr., "Kennedy, After Criticizing Shah, Supports Carter's Efforts on Iran," *New York Times*, December 4, 1979, A1.

69. "Public Would Back Retaliation, Poll Shows," *Washington Star*, November 30, 1979, A1.

70. "Ted Kennedy's Amateur Hour," *Washington Post*, December 7, 1979.

71. Ibid.

72. "In New Hampshire, They're Off!," *Time*, February 25, 1980.

CHAPTER 12: ADRIFT

1. Albert R. Hunt, "Reagan Strategist Surveys Scene—and Picks a Winner," *Wall Street Journal*, August 30, 1979, 18.

2. Charles Black, in discussion with the author, March 20, 2006.

3. Ronald Reagan, *The Reagan Diaries*, ed. Douglas Brinkley (New York: Harper-Collins, 2007), 257.

4. "Primary Results: California," *Facts on File World News Digest*, June 16, 1978.

5. Adam Clymer, "Kennedy Candidacy a Big Topic at Michigan Republicans' Parley," *New York Times*, September 24, 1979, A17.

6. Richard Bergholz, "Reagan Adamant on Not Moving Toward Center," *Los Angeles Times*, September 30, 1979, 1.

7. James R. Dickenson, "Reagan Is Favorite at GOP Women's Meeting," *Washington Star*, September 30, 1979, A1.

8. S. J. Horner, "Simon on the Run and Maybe Running," *New York Times*, October 7, 1979.

9. "Reagan to Raise Funds with a Sinatra Concert," *New York Times*, October 1, 1979, A11.

10. Adam Clymer, "Both Reagan and Rivals Are Depending on Early Victories," *New York Times*, October 8, 1979, A14.

11. Douglas E. Kneeland, "9 G.O.P. Hopefuls Vie for Attention at an Iowa Dinner," *New York Times*, October 15, 1979, A14.

12. Ibid.

13. Helene von Damm, in discussion with the author, May 17, 2006.

14. Joseph E. Persico, *Casey: From the OSS to the CIA* (New York: Viking, 1990), 174.

15. James R. Dickenson, "Reagan Won't Debate Party Rivals," *Washington Star*, November 14, 1979, A1.

16. Helene von Damm, *At Reagan's Side* (New York: Doubleday, 1989), 39–42, 46, 78, 98.

17. Ibid., 101–3.

18. Maurice Carroll, "Reagan Confident in New York Area," *New York Times*, November 13, 1979, A1; Michael Long, in discussion with the author, April 19, 2007.

19. "Reagan, Already the Front Runner, Formally Enters GOP Race Tomorrow," *Washington Star*, November 11, 1979, A4.

20. Roger Stone, in discussion with the author, April 10, 2006.

21. George F. Will, "Connally's Recklessness Makes Reagan Look Better for GOP," *Los Angeles Times*, October 22, 1979, C7.

22. Dan Balz, "Connally Decides Not to Accept US Matching Funds," *Washington Post*, December 12, 1979, A5.

23. "Poll on Primary Finds Kennedy and Reagan Lead in New Hampshire," *New York Times*, October 29, 1979, A17.

24. Adam Clymer, "Maine Republicans in Informal Ballot Give Bush a Victory," *New York Times*, November 4, 1979, A1.

25. Adam Clymer, "Bumper Crop of Straw Polls," *New York Times*, November 7, 1979, A31.

26. John Sears, in discussion with the author, July 26, 2006.

27. Adam Clymer, "Poll Finds Reagan Keeping Lead; More Democrats Back Kennedy," *New York Times*, November 6, 1979, A1.

28. "Cleveland Striving to Polish Its Image," *New York Times*, October 12, 1980, A59.

29. Brent Larkin, "A Time to Recall—Almost 30 Years Ago This Week, a Surly City Nearly Sacked the Boy Mayor," *Plain Dealer* (Cleveland), August 10, 2008, G1.

30. James Reston, "'On the Other Hand,'" *New York Times*, November 7, 1979, A31.

31. Tom Mathews and Gerald C. Lubenow, "The Leading Man," *Newsweek*, October 1, 1979, 20.

32. Al Martinez, "Presidential Campaigns: Money Plus Manpower," *Los Angeles Times*, November 1, 1979, 15.

33. Paul Hendrickson, "Ronald Reagan: Rugged Runner in the Biggest Race," *Washington Post*, November 13, 1979, B1.

34. Les Brown, "Political Campaigns: Woes in Broadcasting," *New York Times*, November 10, 1979.

35. "Reagan's Old Films Off TV for a While," *Washington Star*, November 23, 1975.

36. Richard L. Madden, "The Incarnations of a Campaign Worker," *New York Times*, November 11, 1979.

37. Lou Cannon, "Reagan Announces, Urges Strength at Home, Abroad," *Washington Post*, November 14, 1979, A1.

38. Robert Lindsey, "Reagan, Entering Presidency Race, Calls for North American 'Accord,'" *New York Times*, November 14, 1979, A1.

39. Lou Cannon, "Ronald Reagan: All-American Individualist," *Washington Post*, April 22, 1991, C1.

40. Lindsey, "Reagan, Entering Presidency Race, Calls for North American 'Accord,'" A1.

41. Ibid.

42. Ibid.

43. Ibid.

44. "Reagan: Oblivious to Iran?" *New York Times*, November 15, 1979, A30.

45. William Slate, "Reagan's Growing War Chest," *Newsweek*, December 3, 1979, 43.

46. Cannon, "Reagan Announces, Urges Strength at Home, Abroad," A1.

47. Helene von Damm, *At Reagan's Side* (New York: Doubleday, 1989), 109.

48. Ibid., 112.

49. Rowland Evans and Robert Novak, "Shaky Start," *Washington Post*, November 16, 1979, A21.

50. Lindsey, "Reagan, Entering Presidency Race, Calls for North American 'Accord,'" A1.

51. David Smick, in discussion with the author, October 18, 2006.

52. "Reagan: Oblivious to Iran?" *New York Times*, November 15, 1979, A30.

53. Paul Hendrickson, "Ronald Reagan: Rugged Runner in the Biggest Race," *Washington Post*, November 13, 1979, B1.

54. Ibid.

55. Peter Goldman, Gerald C. Lubenow, Tony Fuller, et al., "A Royal Progress," *Newsweek*, November 26, 1979, 50.

56. Lou Cannon, "Reagan Refuses to Debate His GOP Opponents," *Washington Post*, November 15, 1979, A3.

57. Adam Clymer, "Reagan off to a Fast Start Ahead of the Field," *New York Times*, November 18, 1979, E4.

58. Rowland Evans and Robert Novak, "Not So Inevitable," *Washington Post*, November 12, 1979, A17.

59. James J. Kilpatrick, "Ronald Reagan's Weakness," *Washington Star*, December 4, 1979, A10.

CHAPTER 13: IOWA AGONISTES

1. Peter Goldman, Gerald C. Lubenow, Tony Fuller, et al., "A Royal Progress," *Newsweek*, November 27, 1979.

2. Douglas L. Hallett, "The Reagan Candidacy," *Wall Street Journal*, October 30, 1979, 24.

3. "Israelis Support Carter's Stand on Olympics," Associated Press, January 23, 1980; "Countries Definitely Boycotting Still Few," Associated Press, February 20, 1980.

4. George F. Will, "Games People Shouldn't Play," *Washington Post*, January 6, 1980, B7.

5. "Olympic Boycott Could Bring Claim," *Globe and Mail*, January 22, 1980.

6. Elisabeth Bumiller, "Mike Deaver, the Man Who Looks After the Man," *Washington Post*, March 8, 1981, K1.

7. "George Bush: Urbane, Unflappable, Unknown," *Washington Star*, October 28, 1979.

8. Ibid.

9. William Endicott, "Reagan Toughens Stand on Crisis in Iran," *Los Angeles Times*, November 28, 1979, B23.

10. Ibid.

11. Jack W. Germond, "Reagan Poll Pinpoints Connally's 'Negatives,'" *Washington Star*, December 9, 1979.

12. Memorandum from Michael Cardozo to Susan Clough, Jimmy Carter Library and Museum, Atlanta, GA.

13. Bill Peterson, "The Greening of Patriotism in America; Iran Crisis Revives Patriotism in US," *Washington Post*, December 8, 1979, A1.

14. Barry Sussman, "Gallup Poll Cites Big Turnaround in Carter's Strength," *Washington Post*, December 12, 1979, A1.

15. "New Polls Show Kennedy Decline," *Washington Post*, December 25, 1979.

16. David Gergen, "Kennedy: A Profile in Power," *Wall Street Journal*, October 15, 1979, 24.

17. Ibid.

18. Jeremiah O'Leary, "Carter, Kennedy Exchange Light Quips at Dinner for House Speaker O'Neill," *Washington Star*, December 10, 1979.

19. "Carter's Popularity Zooms, Poll Suggests," Associated Press, December 9, 1979.

20. "Carter-Kennedy Fight Gets Bitter in Chicago," *Washington Star*, December 17, 1979, A3.

21. Margaret Nelson, "Filings Begin for New Hampshire Presidential Primary," Associated Press, December 13, 1979.

22. William Safire, "A Call for Disunity," *New York Times*, December 20, 1979, A27.

23. "Anti-Reagan Forces Fight to End Winner-Take-All Primary," Associated Press, January 25, 1980.

24. Rowland Evans and Robert Novak, "Cooling on Connally," *Washington Post*, November 23, 1979, A23.

25. "Baker, Dole Qualify," *Washington Post*, December 28, 1979, A6.

26. "Editors Say Carter Will Get 1980 Democratic Presidential Nod," Associated Press, December 28, 1979.

27. "Brown Seals Federal Matching Funds," Associated Press, January 2, 1980.

28. "Campaign News," *Washington Post*, January 3, 1980, A3.

29. Jack W. Germond and Jules Witcover, "After Full Year of Campaigning, Reagan Remains GOP Front-Runner," *Washington Star*, December 31, 1979, A2.

30. Ibid.

CHAPTER 14: REAGAN'S DUNKIRK

1. Alan Baron, "Ronald Reagan Trumpets His Traditional Verities in a New Key," *Los Angeles Times*, November 18, 1979, F1.

2. John Dart, "50 Christians Agree to Raise Reagan Funds," *Los Angeles Times*, December 3, 1979, B3.

3. Lou Cannon, "Reagan Poses Option of Suspending All Trade to 'Quarantine' Soviets," *Washington Post*, January 17, 1980, A3.

4. Joseph Kraft, "Iowa: The Beginning of the End," *Los Angeles Times*, January 18, 1980, D7.

5. Personal archives collection of Kenny Klinge.

6. Donald M. Rothberg, Associated Press, January 21, 1980.

7. Bill Peterson, "Reagan Returns to WHO . . . What Hit Him?" *Washington Post*, January 19, 1980, A1.

8. Lawrence Walsh, "Connally Sets Grueling Pace in Iowa in Marathon 40-Hour Campaign Trip," *Washington Post*, January 21, 1980, A3.

9. David S. Broder, "Bipartisan Support Wanes for Carter's Iran Policy," *Washington Post*, January 3, 1980, A1.

10. Adam Clymer, "Brock Declares President Deceives Nation on Foreign Policy Weakness," *New York Times*, January 2, 1980, A1.

11. Paige Bryan, "The Soviet Grain Embargo," Heritage Foundation, January 2, 1981, http://www.heritage.org/research/reports/1981/01/the-soviet-grain-embargo.

12. Lou Cannon, "Reagan Poses Option of Suspending All Trade to 'Quarantine' Soviets," *Washington Post*, January 17, 1980, A3.

13. William Endicott, "Reagan's Strategy: Say What They Like to Hear," *Los Angeles Times*, January 11, 1980, B6.

14. Ibid.

15. "Caucus History: Past Years' Results," *Des Moines Register*, http://caucuses .desmoinesregister.com/caucus-history-past-years-results.

16. Walter R. Mears, Associated Press, January 22, 1980.

17. "Surprise Harvest in Iowa," *Time*, February 4, 1980.

18. Michael Holmes, "Bush Says He's Ready to Go All the Way," Associated Press, January 22, 1980.

19. Ibid.

20. Craig Shirley, *Rendezvous with Destiny: Ronald Reagan and the Campaign That Changed America* (Wilmington, DE: Intercollegiate Studies Institute, 2009), 104.

21. "George H. W. Bush Wins 1980 Iowa Caucus," *Today*, NBC, January 22, 1980.

22. Richard Bergholz, "Reagan Campaign Staff Seems to Shield Him More," *Los Angeles Times*, November 16, 1979, B11.

23. Ibid.

24. "Iowa Caucuses Executive Summary 1980—Michigan Presidential Study," George Bush Presidential Library and Museum, College Station, TX.

25. Ibid.

26. Robert Shogan, *None of the Above: Why Presidents Fail—and What Can Be Done About It* (New York: New American Library, 1982), 19.

27. "Reagan Ambushed in Iowa; Carter Whips Kennedy, 2–1," *Concord Monitor*, January 22, 1980, 1.

28. James M. Perry and Albert R. Hunt, "Carter and Bush Win 1st Round but Have No Knockdowns Yet," *Wall Street Journal*, January 23, 1980, 1.

29. David S. Broder, "Is Reagan Right for the Age?" *Washington Post*, November 11, 1979, B7.

30. Lou Cannon, "Reagan Displays Vitality but Bobbles," *Washington Post*, November 19, 1979, A8.

31. Barry Sussman, "Reagan Gaining with Moderates, Post Poll Shows," *Washington Post*, November 19, 1979, A1.

32. *World News Tonight*, ABC, January 22, 1980.

33. Walter R. Mears, Associated Press, January 22, 1980.

CHAPTER 15: SUNSHINE REAGANITES

1. Martin Schram, "Candidates' Strategies Shift with the Political Landscape," *Washington Post*, January 18, 1980, A1.

2. Bud Lembke, "Race Reduced to Bush-Reagan, Pollster Says," *Los Angeles Times*, January 24, 1980, A1.

3. Ibid.

4. Lou Cannon, "Reagan Hustling to Defuse Age Issue," *Washington Post*, May 27, 1979, A2.

5. Don McLeod, "Evidence Mounts That Iowa Caucuses Turned Around Bush Campaign," Associated Press, February 10, 1980.

6. Thomas Griffith, "The Well-Balanced Fight Card," *Time*, February 25, 1980.

7. Haynes Johnson, "Reagan Finds Himself Racing Clock," *Washington Post*, February 4, 1980, A2.

8. George Esper, "Reagan Steps Up N.H. Campaign," *Nashua Telegraph*, January 29, 1980, 3.

9. Richard Bergholz, "Reagan Shifts Stand on Pakistan A-Arms," *Los Angeles Times*, February 1, 1980, B16.

10. Ibid.

11. James Feron, "Westchester Journal," *New York Times*, January 27, 1980, WC3.

12. "Bush and Reagan Tied with 27% in Republican Independent Poll," *New York Times*, January 25, 1980, D14.

13. Richard Reeves, "The President Loses Reagan," *Washington Star*, January 29, 1980.

14. George Esper, "Suddenly, It's No Longer 'George Who?'" *Nashua Telegraph*, January 28, 1980.

15. "What's News," *Wall Street Journal*, February 1, 1980, 1.

16. Tom Raum, "Tidewater Conference Links Strong Defense to Strong Economy," Associated Press, February 2, 1980.

17. Stephen C. Smith, "Connally Unbowed by Poor Arkansas Showing," Associated Press, February 3, 1980.

18. Allan J. Mayer, Stryker McGuire, Jerry Buckley, et al., "Bush Breaks Out of the Pack," *Newsweek*, February 4, 1980, 30.

19. George Esper, "Bush Emphasizes Youth and Fitness," Associated Press, February 6, 1980.

20. Associated Press, "Poll Shows Carter Well Ahead of Kennedy in Maine," February 8, 1980.

21. "Carter, Bush Lead Ill. Poll," *Washington Post*, February 5, 1980, A3.

22. "Ill. Poll Shows Bush Leads Reagan 34–23," *Washington Star*, February 14, 1980.

23. Haynes Johnson, "Reagan Finds Himself Racing Clock," *Washington Post*, February 4, 1980, A1.

CHAPTER 16: THE POLITICS OF POLITICS

1. Richard Reeves, "The President Loses Reagan," *Washington Star*, January 29, 1980, A11.

2. Bill Roeder, "The Ups and Downs of John Sears," *Newsweek*, February 18, 1980, 29.

3. "Reagan Against Drafting Women," Associated Press, January 30, 1980.

4. Tim Ahern, "Bush Predicts New Hampshire Win, Reagan May Intensify Campaign," Associated Press, January 23, 1980.

5. Ibid.

6. Steven F. Hayward, *The Age of Reagan: The Fall of the Old Liberal Order, 1964–1980* (Roseville, CA: Forum, 2001), 536–37.

7. Donald M. Rothberg, "Reagan Spells Out Soviet Policy," Associated Press, February 1, 1980.

8. *World News Tonight*, ABC News, February 6, 1980.

9. Jim Roberts, in discussion with the author, July 6, 2006.

10. Ibid.

11. David Olinger, "Bush and Reagan Succumb to Debate Fever," *Concord Monitor*, January 31, 1980, A1.

12. Lou Cannon, "Reagan Forces to Pay Costs of Debate with Bush," *Washington Post*, February 22, 1980.

13. Guy MacMillin, "New Hampshirites Are All Wooed Out," *Los Angeles Times*, February 22, 1980, C7.

14. Robert Furlow, Associated Press, February 8, 1980.

15. Craig Shirley, "Fast Times at Nashua High," *National Review*, October 19, 2009, http://www.nationalreview.com/article/228434/fast-times-nashua-high-craig-shirley.

16. Lou Cannon, "Reagan Is Feted for His Birthday During N.H. Tour," *Washington Post*, February 6, 1980, A3,

17. "Ronald Reagan: Campaigns and Elections," Miller Center, http://millercenter.org/president/biography/reagan-campaigns-and-elections.

18. "Reagan Likes Birthday, Considering Alternative," *Los Angeles Times*, February 4, 1980, A14.

19. *Washington Post*, February 6, 1980.

20. *Chicago Tribune*, February 6, 1980, E8.

21. "Reagan Changes Gears, Enter Debate," *Concord Monitor*, February 6, 1980, 1.

22. Cannon, "Reagan Is Feted for His Birthday During N.H. Tour," A3.

23. "Ronald Reagan: Campaigns and Elections," Miller Center, http://millercenter.org/president/biography/reagan-campaigns-and-elections.

24. James M. Perry, "Bush, Former Underdog, Is Riding High on Campaign Circuit After Iowa Voting," *Wall Street Journal*, January 28, 1980, 8.

25. "To the Manner Made," *Time*, February 4, 1980.

26. Ibid.

27. George Bush and Victor Gold, *Looking Forward* (New York: Doubleday, 1987), 203.

28. Ibid., 207.

29. Allan J. Mayer, Styker McGuire, Jerry Buckley, et al., "Bush Breaks Out of the Pack," *Newsweek*, February 4, 1980, 30.

30. James R. Dickenson, "Fit for 2 Terms, Bush Says, Which Makes Reagan . . . ," *Washington Star*, February 5, 1980.

31. David Espo, "Mondale Says Voters Want Carter Working, Not Debating," Associated Press, February 5, 1980.

32. Mayer, McGuire, Buckley, et al., "Bush Breaks Out of the Pack," 30.

33. George Esper, "Bush Predicts Carter Move on Iranian Economic Sanctions," Associated Press, January 23, 1980.

34. Myra MacPherson, "Who Is William Loeb and Why Is He Saying Those Mean Things About . . . ," *Washington Post*, February 24, 1980,

35.   Jon Meacham, *Destiny and Power: The American Odyssey of George Herbert Walker Bush* (New York: Random House, 2015), 235.

36.   Warren Weaver Jr., "Conservatives Fear Bush Profits from the Reagan-Crane Rivalry," *New York Times*, February 8, 1980, A17.

37.   Craig Shirley, *Rendezvous with Destiny: Ronald Reagan and the Campaign That Changed America* (Wilmington, DE: Intercollegiate Studies Institute, 2009), 128.

38.   Donald M. Rothberg, "Campaign Trail a Rocky One for Reagan," Associated Press, February 15, 1980.

39.   Scott Armstrong and Bill Peterson, "Nixon Aide Says He Gave Cash to Bush," *Washington Post*, February 16, 1980, A2.

40.   "In a Fiercely Hawkish Mood," *Time*, February 11, 1980.

41.   Jack W. Germond, "Democrats' Campaign in New England Lacks Dialogue and Debate," *Washington Star*, February 17, 1980, A3.

42.   "Bush Is the Front-Runner," *Washington Post*, February 18, 1980, A3.

43.   David Keene, in discussion with the author, March 17, 2006.

44.   Tish Leonard, in discussion with the author, September 12, 2006.

45.   Advertisement in *Concord Monitor*, February 11, 1980.

46.   "Reagan's Rousing Return," *Time*, March 10, 1980.

47.   Don McLeod, Associated Press, February 21, 1980.

48.   Shirley, *Rendezvous with Destiny*, 148.

49.   Ibid., 149–51.

50.   Ibid, 152.

51.   Myra MacPherson, "The Granite State Follies," *Washington Post*, February 26, 1980.

CHAPTER 17: ISLAND OF FREEDOM

1.   Lou Cannon, "Carter, Reagan Dominate in Heavy N.H. Vote," *Washington Post*, February 27, 1980, A1.

2.   Craig Shirley, *Rendezvous with Destiny: Ronald Reagan and the Campaign That Changed America* (Wilmington, DE: Intercollegiate Studies Institute, 2009), 160.

3.   Cannon, "Carter, Reagan Dominate in Heavy N.H. Vote," A1.

4.   Bob Fick, "It's All but Over: Dole," Associated Press, February 27, 1980.

5.   *World News Tonight*, ABC News, March 5, 1980.

6.   Donald M. Rothberg, "Reagan Looking Forward to Massachusetts and the South," Associated Press, February 27, 1980.

7.   Walter R. Mears, "Baker Quits; Ford Says Even Chance He Will Start," Associated Press, March 6, 1980.

8.   Shirley, *Rendezvous with Destiny*, 188–89; F. Richard Ciccone, "Forget Moral Victories, Massachusetts Counts," *Chicago Tribune*, March 2, 1980, 12.

9. Bill Peterson and Jack Bass, "Reagan Crushes Connally, Bush in S.C.," *Washington Post*, March 9, 1980, A1.

10. Mike Feinsilber, "Bush Claims Connally Votes but Says Reagan Will Win in South," Associated Press, March 10, 1980.

11. Ibid.

12. David S. Broder, "Reagan Captures Illinois GOP Vote," *Washington Post*, March 19, 1980, A1.

13. Ibid.

14. Evans Witt, Associated Press, March 19, 1980.

15. Ibid.

16. Donald M. Rothberg, "Reagan Beginning to Act Like He's the GOP Nominee," Associated Press, March 19, 1980.

17. "A Victory in Pennsylvania Is Crucial, Bush Concedes," *Washington Post*, April 11, 1980, A3.

18. Lawrence L. Knutson, "Bush Bows Out of GOP Race, Leaving Nomination to Reagan," Associated Press, May 26, 1980.

19. Ibid.

20. "The Vote Tally," *New York Times*, May 29, 1980, B6.

21. James R. Dickenson, "Carter, Reagan Win Primaries in Landslide," *Washington Star*, May 28, 1980, A1; Shirley, *Rendezvous with Destiny*, 300.

22. "Results of Primary Voting," *New York Times*, June 5, 1980, B9.

23. Walter R. Mears, Associated Press, July 17, 1980.

24. Lou Cannon, "Reagan Nominated, Picks Bush," *Washington Post*, July 17, 1980, A1; Shirley, *Rendezvous with Destiny*, 365.

25. Lou Cannon, "Bush Fills in Weak Spots for Reagan, Poll Shows," *Washington Post*, July 19, 1980, A11.

26. Ronald Reagan, "Address Accepting the Presidential Nomination at the Republican National Convention in Detroit," July 17, 1980, American Presidency Project, http://www.presidency.ucsb.edu/ws/?pid=25970.

27. Ibid.

28. Ibid.

AUTHOR'S NOTE

1. Lou Cannon, *President Reagan: The Role of a Lifetime* (New York: PublicAffairs, 2000), 32.

2. Hayley Peterson, "The Five Incumbent Presidents Who Lost," *Washington Examiner*, April 4, 2011, http://www.washingtonexaminer.com/the-five-incumbent-presidents-who-lost/article/143176.

Bibliography

BOOKS

Anderson, Martin. *Revolution*. New York: Harcourt Brace Jovanovich, 1988.

Anderson, Martin, and Annelise Anderson. *Ronald Reagan: Decisions of Greatness*. Stanford, CA: Hoover Institution Press Publication, 2015.

Andrew, Christopher, and Vasili Mitrokhin. *The Sword and the Shield: The Mitrokhin Archive and the Secret History of the KGB*. New York: Basic Books, 1999.

Babbin, Jed. *In the Words of Our Enemies*. Washington, DC: Regnery, 2007.

Baker, James A. III, and Steve Fiffer. *"Work Hard, Study, and Keep Out of Politics!": Adventures and Experiences from an Unexpected Public Life*. New York: G.P. Putnam's Sons, 2006.

Barrett, Laurence I. *Gambling with History: Ronald Reagan in the White House*. New York: Penguin Books, 1984.

Beckel, Bob. *I Should Be Dead: My Life Surviving Politics, TV, and Addiction*. New York: Hachette Books, 2015.

Bourne, Peter G. *Jimmy Carter: A Comprehensive Biography from Plains to Postpresidency*. New York: Scribner, 1997.

Bowden, Mark. *Guests of the Ayatollah: The Iran Hostage Crisis, the First Battle in America's War with Militant Islam*. New York: Grove Press, 2006.

Boyarsky, Bill. *The Rise of Ronald Reagan*. New York: Random House, 1968.

———. *Ronald Reagan: His Life and Rise to the Presidency*. New York: Random House, 1981.

Buckley Jr., William F. *On the Firing Line: The Public Life of Our Public Figures*. New York: Random House, 1989.

———. *The Reagan I Knew*. New York: Basic Books, 2008.

Bush, George, and Victor Gold. *Looking Forward*. New York: Doubleday, 1987.

Cannon, Lou. *Governor Reagan: His Rise to Power*. New York: PublicAffairs, 2003.

———. *President Reagan: The Role of a Lifetime*. New York: PublicAffairs, 2000.

———. *Reagan*. New York: G.P. Putnam's Sons, 1982.

———. *The Reagan Paradox: The Conservative Icon and Today's GOP.* Des Moines, IA: Time Books, 2014

———. *Ronnie and Jesse: A Political Odyssey.* New York: Doubleday, 1969.

Carter, Jimmy. *Keeping Faith: Memoirs of a President.* New York: Bantam Books, 1982.

Clymer, Adam. *Edward M. Kennedy: A Biography.* New York: Perennial, 1999.

Colacello, Bob. *Ronnie and Nancy: Their Path to the White House, 1911 to 1980.* New York: Warner Books, 2004.

Dallek, Matthew. *The Right Moment: Ronald Reagan's First Victory and the Decisive Turning Point in American Politics.* New York: Oxford University Press, 2004.

Davis, Patti. *The Long Goodbye: Memories of My Father.* New York: Alfred A. Knopf, 2011.

Deaver, Michael K. *A Different Drummer: My Thirty Years with Ronald Reagan.* New York: HarperCollins, 2001.

———. *Nancy: A Portrait of My Years with Nancy Reagan.* New York: William Morrow, 2004.

Deaver, Michael K., and Mickey Herskowitz. *Behind the Scenes: In Which the Author Talks About Ronald and Nancy Reagan . . . and Himself.* New York: William Morrow, 1987.

DeFrank, Thomas M. *Write It When I'm Gone: Remarkable Off-the-Record Conversations with Gerald R. Ford.* New York: G.P. Putnam's, 2007.

Diggins, John Patrick. *Ronald Reagan: Fate, Freedom, and the Making of History.* New York: W.W. Norton and Company, 2007.

Drew, Elizabeth. *Portrait of an Election: The 1980 Presidential Campaign.* New York: Simon and Schuster, 1981.

D'Souza, Denish. *Ronald Reagan: How an Ordinary Man Became an Extraordinary Leader.* New York: Simon and Schuster, 1997.

Edwards, Anne. *Early Reagan: The Rise to Power.* New York: William Morrow, 1987.

Edwards, Lee. *The Conservative Revolution: The Movement That Remade America.* New York: Free Press, 1999.

———. *The Essential Ronald Reagan: A Profile in Courage, Justice, and Wisdom.* Lanham, Maryland: Rowman and Littlefield Publishers, 2005.

———. *The Reagans: Portrait of a Marriage.* New York: St. Martin's Press, 2003.

———. *Ronald Reagan: A Political Biography.* Houston: Nordland Publishing, 1981.

Evans, Thomas W. *The Education of Ronald Reagan.* New York: Columbia University Press, 2006.

Ford, Gerald R. *A Time to Heal: The Autobiography of Gerald R. Ford.* New York: Harper & Row, 1979.

Frum, David. *How We Got Here: The 70s, the Decade That Brought You Modern Life—for Better or Worse.* New York: Basic Books, 2000.

Greenfield, Jeff. *The Real Campaign: How the Media Missed the Story of the 1980 Campaign.* New York: Summit Books, 1982.

Groom, Winston. *Ronald Reagan: Our 40th President.* Washington, DC: Regnery, 2012.

Halberstam, David. *The Powers That Be.* Chicago: University of Illinois Press, 2000.

Hannaford, Peter. *The Reagans: A Political Portrait.* New York: Coward-McCann, 1983.

———. *Reagan's Roots: The People and Places That Shaped His Character.* Bennington, VT: Images from the Past, 2012.

———, ed. *Recollections of Reagan: A Portrait of Ronald Reagan.* New York: William Morrow, 1997.

Hatfield, Mark O. *Vice Presidents of the United States, 1789–1993.* Washington, DC: U.S. Government Printing Office, 1997.

Hayward, Steven F. *The Age of Reagan: The Fall of the Old Liberal Order, 1964–1980.* Roseville, CA: Forum, 2001.

———. *The Real Jimmy Carter.* Washington, DC: Regnery Gateway, 2004.

Holden, Kenneth. *The Making of the Great Communicator: Ronald Reagan's Transformation from Actor to Governor.* Guilford, CT: Globe Pequot Press, 2013.

Jensen, Richard J. *Reagan at Bergen-Belsen and Bitburg.* College Station, TX: Texas A&M University Press, 2007.

Jordan, Hamilton. *Crisis: The Last Year of the Carter Presidency.* New York: G.P. Putnam's Sons, 1982.

Kengor, Paul. *God and Ronald Reagan: A Spiritual Life.* New York: ReaganBooks, 2004.

Kengor, Paul, and Patricia Clarke Doerner. *The Judge: William P. Clark, Ronald Reagan's Top Hand.* San Francisco: Ignatius, 2007.

Kopelson, Gene. *Reagan's 1968 Dress Rehearsal: Ike, RFK, and Reagan's Emergence as a World Statesman.* Los Angeles: Figueroa Press, 2016.

Leighton, Frances Spatz. *The Search for the Real Nancy Reagan.* New York: Macmillan, 1987.

Marton, Kati. *Hidden Power: Presidential Marriages That Shaped Our Recent History.* New York: Pantheon Books, 2001.

Mason, Jim. *No Holding Back: The 1980 John Anderson Presidential Campaign.* Lanham, MD: University Press of America, 2011.

Matthews, Chris. *Tip and the Gipper: When Politics Worked.* New York: Simon and Schuster, 2013.

Meacham, Jon. *Destiny and Power: The American Odyssey of George Herbert Walker Bush.* New York: Random House, 2015.

Minutaglio, Bill. *First Son: George W. Bush and the Bush Family Dynasty.* New York: Times Books, 1999.

Morell, Margot. *Reagan's Journey: Lessons from a Remarkable Career.* New York: Simon and Schuster, 2011.

Nofziger, Lyn. *Nofziger.* Washington, DC: Regnery Gateway, 1992.

Noonan, Peggy. *What I Saw at the Revolution: A Political Life in the Reagan Era.* New York: Random House, 2003.

———. *When Character Was King: A Story of Ronald Reagan*. New York: Penguin, 2002.

Novak, Robert D. *The Prince of Darkness: 50 Years of Reporting in Washington*. New York: Crown Forum, 2007.

O'Neill, Tip, and William Novak. *Man of the House: The Life and Political Memoirs of Speaker Tip O'Neill*. New York: Random House, 1987.

Perry, Roland. *Hidden Power: The Programming of the President*. New York: Beaufort Books, 1984.

Persico, Joseph E. *Casey: From the OSS to the CIA*. New York: Viking, 1990.

Ranney, Austin, ed. *The American Elections of 1980*. Washington, DC: AEI Press, 1981.

Reagan, Michael, and Joe Hyamas. *On the Outside Looking In*. New York: Zebra Books, 1988.

Reagan, Nancy. *I Love You, Ronnie: The Letters of Ronald Reagan to Nancy Reagan*. New York: Random House, 2000.

Reagan, Nancy, and William Novak. *My Turn: Memoirs of Nancy Reagan*. New York: Random House, 1989.

Reagan, Ronald. *An American Life: The Autobiography*. New York: Simon and Schuster, 1990.

———. *Reagan: A Life in Letters*. Ed. Skinner, Kiron K., Annelise Anderson, and Martin Anderson. New York: Free Press, 2003.

———. *The Notes: Ronald Reagan's Private Collection of Stories and Wisdom*. New York: HarperCollins, 2011.

———. *The Reagan Diaries*. Edited by Douglas Brinkley. New York: HarperCollins, 2007.

———. *Reagan: In His Own Hand*. Edited by Kiron K. Skinner, Annelise Anderson, and Martin Anderson. New York: Free Press, 2001.

———. *Reagan's Path to Victory: The Shaping of Ronald Reagan's Vision: Selected Writings*. Ed. Kiron K. Skinner, Annelise Anderson, and Martin Anderson. New York: Simon and Schuster, 2004.

Reagan, Ronald, and Richard G. Hubler. *Where's the Rest of Me?: The Autobiography of Ronald Reagan*. New York: Karz Publishers, 1981.

Reed, Thomas C. *The Reagan Enigma: 1964–1980*. Los Angeles: Figueroa Press, 2014.

Regan, Donald T. *For the Record: From Wall Street to Washington*. New York: Harcourt Brace Jovanovich, 1988.

Roberts, James C. *A City Upon a Hill: Speeches by Ronald Reagan Before the Conservative Political Action Conference, 1974–1988*. Washington, DC: American Studies Center, 1989.

Sandbrook, Dominic. *Mad as Hell: The Crisis of the 1970s and the Rise of the Populist Right*. New York: Alfred A. Knopf, 2011.

Schlesinger, Arthur M. *Journals: 1952–2000*. New York: Penguin Books, 2007.

———. *Robert Kennedy and His Times*. Boston: Mariner Books, 2002.

Shesol, Jeff. *Mutual Contempt: Lyndon Johnson, Robert Kennedy, and the Feud That Defined a Decade*. New York: W.W. Norton and Company, 1997.

Shirley, Craig. *Reagan's Revolution: The Untold Story of the Campaign That Started It All*. Nashville, TN: Nelson Current, 2005.

———. *Rendezvous with Destiny: Ronald Reagan and the Campaign That Changed America*. Wilmington, DE: Intercollegiate Studies Institute, 2009.

Shogan, Robert. *None of the Above: Why Presidents Fail—and What Can Be Done About It*. New York: New American Library, 1982.

Shrum, Robert. *No Excuses: Concessions of a Serial Campaigner*. New York: Simon and Schuster, 2007.

Slosser, Bob. *Reagan Inside Out*. Nashville, TN: W Publishing Group, 1984.

Soderstrom, Jamin. *Qualified: Candidate Resumes and the Threshold for Presidential Success*. Bloomington, IN: iUniverse, 2011.

Stacks, John F. *Watershed: The Campaign for the Presidency, 1980*. New York: Times Books, 1981.

Stockman, David A. *The Triumph of Politics: How the Reagan Revolution Failed*. New York: Harper & Row, 1986.

Strock, James M. *Reagan on Leadership: Executive Lessons from the Great Communicator*. Scottsdale, AZ: Serve to Lead Press, 2014.

A Time for Choosing: The Speeches of Ronald Reagan, 1961–1982. Chicago: Regnery Gateway, 1983.

Van Der Linden, Frank. *The Real Reagan: What He Believes, What He Has Accomplished, What We Can Expect from Him*. New York: William Morrow, 1981.

von Damm, Helene. *At Reagan's Side*. New York: Doubleday, 1989.

Whitcomb, John, and Claire Whitcomb. *Real Life at the White House*. London: Routledge, 2002.

White, Clifton F., and William J. Gill. *Why Reagan Won: A Narrative History of the Conservative Movement, 1964–1981*. Chicago: Regnery, 1981.

White, Theodore H. *America in Search of Itself: The Making of the President, 1956–1980*. New York: Harper & Row, 1982.

Wilber, Del Quentin. *Rawhide Down: The Near Assassination of Ronald Reagan*. New York: Henry Holt and Company, 2011.

Wilentz, Sean. *The Age of Reagan: A History, 1947–2008*. New York: HarperCollins, 2008.

Wilson, James Graham. *The Triumph of Improvisation: Gorbachev's Adaptability, Reagan's Engagement, and the End of the Cold War*. Ithaca, NY: Cornell University Press, 2014.

Witcover, Jules. *Crapshoot: Rolling the Dice on the Vice Presidency*. New York: Crown, 1992.

———. *Marathon: The Pursuit of the Presidency 1972–1976*. New York: Viking, 1977.

Yager, Edward M. *Ronald Reagan's Journey: Democrat to Republican*. New York: Rowman and Littlefield, 2006.

Bibliography

Zeller, F. C. Duke. *Devil's Pact: Inside the World of the Teamsters Union.* Secaucus, NJ:
    Birch Lane Press, 1996.

PERIODICALS

Baltimore Sun
Baron Report Newsletter
Boston Globe
Chicago Tribune
Christian Science Monitor
Commentary Magazine
Concord Monitor
Congressional Quarterly
Daily Standard (Sikeston, MO)
Decatur Herald
Des Moines Register
The Economist
Eugene Register-Guard
Facts on File World News Digest
Globe and Mail
Los Angeles Herald-Examiner
Los Angeles Times
Morning Record (Meriden, CT)
Nashua Telegraph
National Journal
National Review
New York Post
New York Times
New York Times Magazine
Newsweek
Orange County Register
Philadelphia Inquirer
Plain Dealer (Cleveland)
Politico
Roanoke Times
Roll Call
Slate
Time
Times Union
U.S. News & World Report

Virginia Living Magazine
Wall Street Journal
Washington Post
Washington Star
Washington Times
Wisconsin State Journal

NEWS WIRES

Associated Press
United Press International

ELECTRONIC MEDIA

ABC
CBS
C-SPAN
Fox News
NBC
PBS

AUTHOR INTERVIEWS

James Baker, September 6, 2006
Charles Black, March 20, 2006
Ed Blakely, April 17, 2007
Richard Bond, October 4, 2006
Jeb Bush, August 22, 2006
Al Cardenas, July 13, 2004
Gerald Carmen, July 25, 2006
Jimmy Carter, July 11, 2006
Dick Cheney, June 9, 2004, and March 19, 2007
Michael Deaver, October 18, 2006
John T. Dolan
John Fund, April 12, 2007
Jack Germond, July 2, 2004
Newt Gingrich, May 21, 2006
Victor Gold, April 5, 2007
Peter Hannaford, March 24, 2006, and April 24, 2007
David Keene, March 17, 2006

Arthur Laffer, January 18, 2007
Donald Lambro, January 24, 2007
Paul Laxalt, July 31, 2006
Tish Leonard, September 12, 2006
Bob Livingston, April 19, 2007
Michael Long, April 19, 2007
Edward Mahe, May 24, 2006
Michael McShane, 2006
Ed Meese, May 2006
Walter Mondale, February 28, 2007
Lyn Nofziger
Neal Peden, July 6, 2006
John Persinos
Charles Pickering
Jody Powell, April 10, 2006
Nancy Reagan, October 3, 2006
Jim Roberts, July 6, 2006
John Sears, July 26, 2006
David Smick, October 18, 2006
Rodney Smith, September 7, 2006
Roger Stone, April 10, 2006
Kathleen Kennedy Townsend, December 18, 2007
Richard Viguerie, December 6, 2006
Helene von Damm, May 17, 2006

WEBSITES

American Presidency Project
American Rhetoric Online Speech Bank
Berkeley University
Foundation for Economic Education
GovTrack
Heritage Foundation
Miller Center
Our Campaigns
Reagan2020
Roper Center for Public Opinion Research
United States Maritime Administration

OTHER MATERIAL

Eureka College, Eureka, IL
George Bush Presidential Library and Museum, College Station, TX
Harold Weisberg's Archives, Hood College, Frederick, MD
Jimmy Carter Library, Atlanta, GA
National Archives, Washington, DC
Private Archives Collections of Kenneth Klinge
Ronald Reagan Presidential Library and Museum, Simi Valley, CA

Index

Index

About the Author

CRAIG SHIRLEY is the author of *Last Act*, *Rendezvous with Destiny*, *Reagan's Revolution*, and the *New York Times* bestseller *December 1941*. Shirley is widely known as a leading presidential historian and important biographer of Ronald Reagan. He is a regular commentator throughout the media and a contributor to national publications, and was hailed by the *London Telegraph* as "the best of the Reagan biographers." A widely sought after speaker and commentator, he appears regularly on network and cable shows, including NewsmaxTV, Fox News, MSNBC, CNN, ABC, CBS, CNBC, C-SPAN, and many others. He has also written extensively for the *Wall Street Journal*, the *Washington Post*, *Newsmax*, the *Washington Examiner*, the *Washington Times*, the *Los Angeles Times*, *Town Hall*, the *Weekly Standard*, *Politico*, *ConservativeHQ*, *CNS News*, *Conservative Review*, Reuters, Lifezette, and many other publications. He is also the Visiting Reagan Scholar at Eureka College, Reagan's alma mater, and he lectures often at the Reagan Library and the Reagan Ranch. He and his wife, Zorine, divide their time between Ben Lomond, a three-hundred-year-old Georgian manor house in Essex County, Virginia, and Trickle Down Point, on the Rappahannock River in Lancaster, Virginia. Shirley is also the founder of Ft. Hunt Youth Lacrosse Program and was a coach there for many years, racking up a record of 119 wins and 26 losses. He is working on several more books about Reagan.

SCOTT MAUER is Craig Shirley's primary research assistant. Previously, he had presented and guest-lectured on Soviet and Russian history at international conferences and college campuses. He earned his Master of Arts in History and Humanities from Hood College in Frederick, Maryland, in 2014. Mauer has coauthored several articles with Mr. Shirley. He is currently assisting Mr. Shirley in his future projects, including a history of communist Hungary, a biography of Mary Ball Washington, and other works.